Texts in Philosophy
Volume 14

Corroborations and Criticisms
Forays with the Philosophy of Karl Popper

Volume 3
Monsters and Philosophy
Charles T. Wolfe, ed.

Volume 4
Computing, Philosophy and Cognition
Lorenzo Magnani and Riccardo Dossena, eds.

Volume 5
Causality and Probability in the Sciences
Federica Russo and Jon Williamson, eds.

Volume 6
A Realist Philosophy of Mathematics
Gianluigi Oliveri

Volume 7
Hugh MacColl: An Overview of his Logical Work with Anthology
Shahid Rahman and Juan Redmond

Volume 8
Bruno di Finetti: Radical Probabilist
Maria Carla Galavotti, ed.

Volume 9
Language, Knowledge, and Metaphysics. Proceedings of the First SIFA Graduate Conference
Massimiliano Carrara and Vittorio Morato eds.

Volume 10
The Socratic Tradition. Questioning as Philosophy and as Method
Matti Sintonen, ed.

Volume 11
PhiMSAMP. Philosophy of Mathematics: Sociological Aspects and Mathematical Practice
Benedikt Löwe and Thomas Müller, eds.

Volume 12
Philosophical Perspectives on Mathematical Practice
Bart Van Kerkhove, Jonas De Vuyst and Jean-Paul Van Bendegem, eds.

Volume 13
Beyond Description. Naturalism and Normativity
Marcin Milkowski and Kontrad Talmud-Kaminski, eds.

Volume 14
Corroborations and Criticisms. Forays with the Philosophy of Karl Popper
Ivor Grattan-Guinness

Texts in Philosophy Series Editors
Vincent F. Hendriks   vincent@hum.ku.dk
John Symons   jsymons@utep.edu
Dov Gabbay   dov.gabbay@kcl.ac.uk

Corroborations and Criticisms
Forays with the Philosophy of Karl Popper

Ivor Grattan-Guinness

© Individual author and College Publications 2010. All rights reserved.

ISBN 978-1-84890-004-2

College Publications
Scientific Director: Dov Gabbay
Managing Director: Jane Spurr
Department of Computer Science
King's College London, Strand, London WC2R 2LS, UK

http://www.collegepublications.co.uk

Original cover design by orchid creative   www.orchidcreative.co.uk
Printed by Lightning Source, Milton Keynes, UK

All rights reserved. No part of this publication may be reproduced, stored in a retrieval system or transmitted in any form, or by any means, electronic, mechanical, photocopying, recording or otherwise without prior permission, in writing, from the publisher.

'I loved Russell
This is Russell as I remembered him'

Sir Karl Popper with his favourite Russell photograph,
February 1992: see chapters 7 and 8

Photograph © I. Grattan-Guinness, 1992

TABLE OF CONTENTS

Preface — 1

PART 1 ELABORATIONS

1. Truths and contradictions about Karl Popper — 15
2. What do theories talk about? A critique of Popperian fallibilism, with especial reference to ontology — 25
3. The place of the notion of corroboration in Karl Popper's philosophy of science — 67
4. Karl Popper and the 'problem of induction': a fresh look at the logic of testing theories — 79
5. Levels of criticism: handling Popperian problems in a Popperian way — 91
6. On Popper's use of Tarski's theory of truth — 103

PART 2 COMPETITIONS

7. Russell and Karl Popper: their personal contacts — 111
8. Karl Popper for and against Bertrand Russell — 127
9. On Popper's philosophy and its prospects — 143

PART 3 APPLICATIONS

10. Notes on some methodological problems in the history of science — 171
11. Structure-similarity as a cornerstone of the philosophy of mathematics — 177
12. Solving Wigner's mystery: the reasonable (though perhaps limited) effectiveness of mathematics in the natural sciences — 195
13. History or heritage? An important distinction in mathematics and for mathematics education — 217
14. Not from nowhere. History and philosophy behind mathematical education — 231

PART 4   SPECULATIONS

15	Decline of the philosophical spirit	249
16	Psychical research versus the established sciences	251
17	Psychical research and parapsychology: notes on the development of two disciplines	261
18	What are coincidences, and what are they good for?	281
19	Is psi intrinsically non-linguistic? Some preliminary considerations	299
20	Experience or innateness? Sir Karl Popper on the origins and acquisition of natural languages	311

PART 5   ADDITIONS

21	Types of generality in and around mathematics and logics	323
22	On receiving the Kenneth O. May Medal and Prize	331
	Index	337

0 Preface

The articles reprinted in this book are centred upon the work of Sir Karl Popper, who has been widely (though not uncritically) acknowledged as a very important contributor to philosophy in our time. After indicating the principal pertinent features of his philosophy, I summarise the ways in which I develop, doubt or deploy it in the chapters to follow.

1. Entrées

1.1 Description. The work of Karl Popper (1902-1994) centred on the philosophy of science and on epistemology. It began to mature in the mid and late 1920s, partly as a reaction to the views of the famous 'Vienna Circle' of philosophers in his native city and partly from issues in the psychology of education. His main publication at that time was the book *Logik der Forschung* [Popper 1935]. In 1937 he took up a lectureship at Canterbury University College in New Zealand; his interests broadened to include social and political questions, which were discussed especially in the book *The open society and its enemies* [1945]. His life up to this point is described in detail in the excellent biography [Hacohen 2000], which is reviewed in the next chapter.

After the end of the War Popper was made Reader and later Professor of Logic and Scientific Method at the London School of Economics in the University of London. His major publications included his English translation of *Logik* as *The logic of scientific discovery* [Popper 1959], and the collection of essays *Conjectures and refutations* [1963]. Three years later appeared the second German edition of *Logik,* including a translation of all the additional material by L. Walentik, and a most interesting new preface by Popper on the influence of Immanuel Kant on his work and the differences between his general position and the British tradition in philosophy [Popper 1966].

After retirement in 1969 Popper continued to work on a wide range of subjects, producing another collection of essays, *Objective knowledge* [1972], and a muse with Sir John Eccles on *The self and its brain* [Popper and Eccles 1977]. Then, thanks to the editorial efforts of his former student W. W. Bartley, III, the three-volume *Postscript to the logic of scientific discovery* came out [Popper 1982a, 1982b and 1983]. During his final years and after his death further book collections of essays appeared under various editors, although unfortunately he never completed to his satisfaction one under his own editorship called 'Philosophy and physics',

---

¶2 and ¶3 are based upon parts of 'Notes on some methodological problems in the history of science', *Rete, 1* (1971-1972), 1-14; they reappear here with the kind agreement of Gerstenberg Verlag GmbH & Co. KG.

which was announced for 1974 by Clarendon Press. But a major commentary on his work appeared that year, [Schilpp 1974] in the series 'Library of living philosophers'; it comprised two volumes of essays and his replies, his bibliography, and an extensive autobiography that was reprinted as a separate book [Popper 1976].

1.2 *Acquaintance.* I came across Popper in 1964 when I took a new master's degree in the philosophy of science launched by his Department of Logic and Scientific Method at the London School of Economics. I came there as a refugee from an undergraduate course in mathematics, three years of non-stop bombardment of mathematical theories of various kinds. The theories as such were splendid, but the way of teaching them struck me as obnoxious. Why was so little attention paid to their heuristic background? Why were they not put forward as solutions to problems in some significant way? Where did these theories come from in the first place? And also, what sort of questions was I asking myself anyway, since obviously they were not restricted to mathematics?

Popper's philosophy, centred on the claim that knowledge is guesswork, provided an excellent background for me to tackle those questions about mathematics and its education. I started to study the history and philosophy of mathematics, and also the history and philosophy of logic, enterprises that are still in progress. A book collection of the more general articles appeared recently under the title *Routes of learning* [Grattan-Guinness 2009]. This current volume is a kind of complement, for it contains a collection of articles in which Popper's philosophy itself is examined from various points of view (Part 1), compared with some other philosophies (Part 2), and applied to mathematics and logics (Part 3) and to some parts of psychology (Part 4).

Popper stated his views in numerous places, so I shall not cite his voluminous writings for every detail: usually I use the books already mentioned. He was a fusspot with them, adding notes at the ends of chapters, and bringing out new editions and corrected reprints; in particular, some parts of *Logik* and *Logic* are difficult to follow, for the new appendices and footnotes complement and sometimes revise the old text and appendices. I have read many of these addenda and cite a few, but doubtless my scrutiny is incomplete. Normally I supply only the details of the first editions of his books in the bibliographies.

The rest of this chapter describes in more detail the contents and organization of this book; so I need to explain here the manner in which cross-references between chapters are made. Like this chapter, each one carries a number and is divided into numbered sections and maybe also subsections. To help bind the chapters together I have inserted cross-references between them, in the manner '§6.4', which cites section 4 of chapter 6. Cross-references within a chapter are rendered as '¶3.2', which cites a subsection that would be cited in other chapters as, say, '§0.3.2'. Chapters in *Routes of learning* are cited in the manner '*ROL*, §6'.

2. A selected summary of Popper's philosophy of science

I outline here those features of Popper's philosophy that are of most relevance to this book. I do not discuss his contributions to quantum mechanics, probability theory, time or social philosophy. The order of the sub-sections does not represent any law of diminishing or increasing importance.

2.1 Our knowledge is, and always remains, *tentative and conjectural*. It is created in the form of theories, which initially are designed to solve certain problems and then are tested by applying their consequences to new problems. The process is unending; and we assess our theories in terms of their ability to solve problems.

There is no ultimate authority for our knowledge, including this statement itself. The self-reference here is important: the statement 'everything is conjectural' is itself conjectural.

Philosophical problems are unavoidable, though their sources lie *outside* philosophy — in science, for example. Thus they are better discussed explicitly rather than left implicit or intuitive; for intuition, like everything else, is fallible. To draw on self-reference once more, we may critically discuss critical discussions.

2.2 The *tests* of our theories may be logical or empirical (or both). If a prediction of a theory is not fulfilled, then the theory is refuted and its falsity has been demonstrated. However, if the prediction is successful, then the theory is corroborated. But the word 'corroboration' is not to be taken as synonymous with 'verification', for this latter word has a connotation of 'making more certain' or 'establishing securely'.[1] In Popper's view, the idea that a theory can be made more certain is mistaken, for any theory, however well previously corroborated, may fail its next test. If this happens, then our task is to invent a new theory that, if possible, will preserve the corroborations of the old theory and convert this refutation into a new corroboration. The question of how one proceeds after the discovery of a refutation is methodologically very complicated; there is a variety of possible choices, and a new theory may be a minor modification of its predecessor, or radically different, or somewhere in between.

2.3 *Experiments* are an important type of test of a theory. They are motivated initially by predictions drawn from a theory, and embody theories of their own. In particular, a measurement is itself a theory, for when we wish to measure something (for example, the weight of a body) we actually measure something else (the deflection of a pointer) and then relate it to the quantity required by appealing to some theory of operation of the instrument.

---

[1] In *Logik der Forschung* Popper used the word 'Bewährung', for which Rudolf Carnap proposed the translation 'confirmation'. But Popper came to reject this translation because of its connotation, and translated 'Bewährung' as 'corroboration'. 'Verification' could be rendered as 'Bestätigung'.

The situation in technology is somewhat different; while it shares with science the same knowledge and use of critical discussion, it has restricted rather than unrestricted aims. That is, whereas a scientific theory will be tested in as many different problem-situations as possible, technological theories are directed to the production of one particular type of object and no other. Thus technology is a theory of *safety* in a sense in which science is a theory of *risk*.

2.3 There is *no* process of *verification* of theories in science, not in the sense of making them more certain. For it is no virtue of a theory that it can be verified, for we can always verify it simply by looking for what we want to find. Instead we aim at criticising our theories, in order to find their weak points. When we talk of criticism in science, we mean criticism of the consequences of a theory, rather than of its reasons for existence or its justification. The idea that a theory should be justified is mistaken, since any justification that is provided itself requires justification, ...; and thus an infinite regress is set up.

2.4 It may happen that a theory is (non-trivially) *true*, when it will explain a group of phenomena in the way in which they actually take place. Thus all tests (if properly conducted) will be corroborated. But we shall never know that the theory is in fact true, for the next test could be a refutation. Thus we may have the truth in our hands, but not know it; better, we may possess the truth, but we shall never be able to prove that we possess it.

True theories are nevertheless falsifiable, in the sense that there are possible situations (in other worlds) where they would be false. However, there are theories that are *unfalsifiable*, either logically or experimentally. All unrestricted existential propositions ('There exists a ...') are of this type. They often become involved in science, and then we aim at criticism of their consequences (such as contradiction with well-corroborated scientific theories), while bearing in mind that falsification is impossible.

The question of whether a theory is true or not is quite independent of whether it is falsifiable or not. In particular, an unfalsifiable theory may nevertheless be false. For example, the 'realist' view, that the world is an objective place independent of our sense-experience of it, is opposed by the 'idealist' view, that the world is only what we experience through our senses. Now both of these theories are unfalsifiable; but therefore at least one of them is also false.

2.5 As a theory becomes more and more corroborated, it increases its *verisimilitude*. But this word is not a synonym for 'probability', for, while the verisimilitude of a theory increases with successive corroboration, its probability — in the sense of the calculus of probabilities — decreases. In addition, 'verisimilitude' is a term used in a technical sense to denote the corroborative content of a theory, not in an intuitive (and non-probabilistic) sense to denote its alleged similarity to the truth; like 'corroboration', it is simply a report on the success of a theory as a problem-solver.

2.6 The knowledge that we obtain from all this activity is *objective* — that is, it consists of statements about the world, rather than about our hopes or beliefs. Knowledge is expressed in language, but it is not dominated by (any) language; solving the problems of language cannot solve the problems of science, though linguistic devices may be a useful aid to scientific endeavours.

2.7 More important than any state of our knowledge is its *growth* through critical discussion from one state to another. The future is open, undetermined; we do not know in what direction our knowledge will grow next (if at all). Analogies with adaptations in biology are strong enough to sustain talk of 'evolutionary epistemology', together with the metaphysics of three 'Worlds': of physical objects, of mental experiences, and of the products of thought.

3. Some important confrontations

Popper called his philosophy 'fallibilism' or 'falsificationism'; 'criticismism' would also be appropriate. It has many far-reaching consequences, for it contradicts some well-established philosophies.

3.1 Popper contrasted fallibilism with *'justificationist'* philosophies, which assign authority to some particular category or other: to sense data, perhaps, or to language. Fallibilism rejects authority-seeking, on the grounds of its unavoidable appeal to an infinite regress. Fallibilism also rejects *instrumentalist* or *conventionalist* philosophies, which deny any purpose to talking about truth and regard scientific theories merely as predicting machines. Once the distinction of ¶2.4 between truth, its possession, and proof of possession, of truth is made, the rejection of truth is seen to be unnecessary.

3.2 *Empiricism* is the philosophy that our knowledge is grounded in sense impressions. Popper called this stance 'common-sense epistemology' and rejected it while of course not denying the importance of sense impressions as *sources* of knowledge. He advocated 'common-sense realism', the quite different position that agrees that the world does indeed exist out there, and is more or less as it seems: at the moment that, say, there is a nice smell coming out of the kitchen, Dandelion cat has just scratched me yet again, some great jazz is on the radio, and so on.

Facts are actually theories, interpretations of reality by language. Contrary to the view of empiricists, facts are uncertain and so cannot be a solid foundation for our knowledge. They do, however, play a role of great importance in tests for theories, and even as stimuli to theories and problems.

Similarity, induction, or the classification of facts, does not provide a theory of knowledge. Quite apart from the uncertainty of facts themselves, the process of their classification involves theories of similarity and dissimilarity, which are conjectural and in any case presuppose interest in some particular problem. Apart from this, there are other criticisms of induction as a theory of knowledge.

3.3 *Psychologism*, or the theory that our knowledge is a set of states of mind, is a type of subjectivism. It may be of importance to the individual concerned, but has no significance for the progress of science; for our most fervent beliefs may turn out to be false, and the most surprising nonsense true — or, at least, corroborated.

Psychologistic theories are often circular; for example, if we explain the greatness of a man's work by postulating the quality of his mind, then in fact we can only know of the quality of his mind by the greatness of his work. Even psychological states, such as joy or despair, hinge on the objective outside world whose affairs have caused them as well as on the internal mind that is suffering them.

3.4 *Essentialism*, or the theory that knowledge is founded on fundamental concepts or words, assumes that these concepts have "solid" meanings specified by exact definitions. But if we give a definition of a concept, then the concepts in the defining expressions themselves require definition; and thus we enter into a process that either is infinitely regressive or returns circularly to the original defined term. For the same sort of reasons we should avoid seeking or offering supposed 'final analyses' or 'ultimate explanations'.

The role that definitions really play in our knowledge is that they specify new distinctions of sense within the rough and unclear 'meaning' of words. More generally, they express the creation of divisions of a body of knowledge into new subsections by means of newly discovered criteria of distinction. When a new concept is introduced into science, it is in fact the fresh set of distinctions that contains much of the new knowledge. These distinctions tell us which concepts are related to which other concepts, and how. Concepts have no intrinsic importance on their own; this is derived from the content of the statements in which they and other concepts appear. It is that, rather than the meaning of words, which we seek when trying to increase our knowledge.

4. Reactions to Popper's philosophy of science

Part 1 of this book comprises six chapters (§1-§6) on various aspects of Popper's own version of his philosophy.

4.1 My chief reservation concerns *ontology*. As a realist, Popper thought there are there was an objective world out there in which events and processes were happening, by some means or other. We humans observe and examine this world and put forward as conjectures our scientific theories about what is happening (and maybe why), which we subject to criticism and test.

Given this attitude to science, it seems to be quite natural to consider the 'ontologically correct' theory (as I shall call it), which accounts for the phenomena at hand in the manner that actually is the case, as the target of human theorising in the domain of reference involved. Indeed, on occasion the scientist might even think up the ontologically correct theory for some class of phenomena, although he will never be able to prove this success since the next test may falsify it (¶2.4).

However, Popper dismissed talk of ontology. I have never understood why he did so, and I was never able to get him to discuss it with me seriously; but I think that his philosophy is all the poorer for its absence. Thus §2 is an extensive wander around Popper's philosophy of science, with ontology as the common factor. Some emphasis is laid here, and also in other chapters, on the processes of reification and (de)simplification, where the scientist notes the great complications that attend the phenomena that he wishes to study and forms simplified versions of them about which he hopes to form theories and which may be rendered less simple in later work (§2.6, §3.5, §12.7). This approach is rather similar to Popper's evolutionary epistemology and theory of the three Worlds; however, the analogy is not pressed (§2.12). Further, possibilities for resolving known difficulties over verisimilitude are aired optimistically in §2.7.5 and §9.7.5.

4.2 One problematic category is that of *facts*, against which we test our theories. For facts are themselves are theory-laden, as Popper stressed; so a contradiction seems to be in the air. It can be resolved (§2.5), but examination brings to light some surprising consequences of Popper's demand that theories must be universal and take the form of logical implications (§2.4), whether they are descriptions or explanations of phenomena. Further, special care has to be taken when the phenomena involved are not directly experientable (§2.3).

Popper adopted Alfred Tarski's semantic theory of truth. Here truth (and falsehood) are the properties of propositions in a theory T; the criterion is that of corresponding to the facts (or not); and the associated propositions belong to the metatheory of T. Now Tarski restricted this theory to formalized languages whereas Popper extended it to natural languages and gave it a central place in his philosophy; Tarski's caution seems to be judicious (§2.5.2, and especially §6). But Popper's recognition of metaphilosophy was very fruitful; for example on a large scale, of arguing in [1982b] for indeterminism and against determinism.

4.3 Popper underestimated the place of *corroboration* of theories; the chapter on it in his first book [1935, ch. 10] is last and decidedly the least of them, and also out of place. There are lots of highly corroborating theories around in science and technology (compare ¶2.2); of course they may be falsified by the next test, but they deserve more philosophical attention (§3). Indeed, it is rational even to discuss reliability science without abandoning any fallibilist principles.

4.4 One of Popper's principal epistemological claims was to have solved 'the problem of induction' by showing that it cannot happen in science. However, others hotly disputed this claim, especially during the time when I was a student in the Department. Reflection on the problem years later led me to realize that both Popper and his critics *had been discussing the wrong problem*; instead of addressing the test of the theory, they had been disputing over the inventory of theories and sub-theories involved in testing. When the correct problem is addressed, then Popper's claim can be vindicated (§4).

The switch from the inventory problem to the testing problem involves a change of logical form of the propositions associated with testing. Popper was always an advocate of logical monism, advocating the classical two-valued true-or-false logic over all other logics because it delivers *maximal* criticism. However, to logical pluralists such as myself such criticism can be brute, and it is quite consistent to support both pluralism and Popperian philosophy by seeking *a* level of criticism that is *appropriate to the problem at hand* (§2.5.2, and especially §5).

5. Popper and competing philosophies
Part 2 of this book contains three chapters (§7-§9) on Popper's relationship to some contemporary philosophers.

5.1 An important case was *Bertrand Russell*. Popper gave me his side of their correspondence, which I joined up to Russell's side and present here as §7. It is a small but quality collection of documents, including a long letter from Popper written in 1946 straight after his confrontation with Ludwig Wittgenstein over philosophical problems in Russell's presence (the famous game of poker), and a forthright review for Austrian Radio in 1947 of Russell's recently published *History of Western philosophy*. The two men admired each other's work greatly, and held various positions in common, including contempt for Wittgenstein; but they advocated very different philosophies, with Popper's fallibilism strongly contrasted with Russell's positivistic search for secure knowledge (§8).

5.2 Around my time as a student Popper confronted *Thomas Kuhn's* philosophy [1962] of 'normal' science and scientific revolutions, especially at a remarkable session of an International Conference in London in July 1965 that I had the pleasure to attend. For me Kuhn is not clear about the occurrence of scientific revolutions; but his treatment of normal science was very valuable, akin to the importance of corroboration, and Popper's attacks on it were excessive (§2.6.2, §9.3).

Popper's most bitter confrontation was with his colleague *Imre Lakatos*, again especially during my time as a student. One of their disagreements concerned scientific induction discussed in ¶4.3; but a broader issue was the merits of Lakatos's concept of scientific research programmes and that progression or degeneration, which was intended to improve upon Popper's own hit-and-miss philosophy of testing theories. While the principle motivating this approach was understandable, the pretentious outcome deserved Popper's dismissal (§9.3-§9.4).

6. Applications to mathematics and its history

In §9 I also handle some history of mathematics, an interest that Lakatos and I had in common. However, the theme of Lakatos's reaction to me was that of plagiarism, which I deny (§9.2). Part 3 of this book takes up more profitable activities in that area in five chapters (§10-§14).

6.1 Except for his interest in the ancient Greeks and despite his considerable knowledge of logic, Popper rarely wrote on mathematics; indeed, I found it difficult to draw him into discussions of mathematics even when I saw cases of fallibilism. When I published as [Grattan-Guinness 1997] a general history of mathematics I dedicated it to him, not only in his memory but especially because the book was guided throughout by applying fallibilist principles to historiography. For me history *of any kind* is a type of (non-formalised) metatheory about the historical events discussed; for example, the historian should distinguish his own rationality from that of his historical figures, he must find important what his figures found to be important at the time, and so on. There seems to be no ontology as a target for historical theories as there is for scientific ones (§2.9); but fallibilism, ignorance as well as knowledge, and criticism certainly obtain (§2.10, §9.9-§9.10).

These sentiments guide a trio of chapters: §10 on the history of science, §13 on the history of mathematics, and §14 on mathematics education; they complement the essays reprinted in (*ROL*, §3, §5, §6 and §9). In particular, in §13 I distinguish between the history and heritage of older mathematics. The historian tries to reconstruct the problem situations of his historical figures and thereby address the question 'what happened in the past?' (and by implication, 'what did not happen?'). By contrast, the inheritor may modernize that old mathematics in any way suitable, in order to address the entirely different question 'how did we get here?'. Both activities are quite legitimate kinds of (non-formalised) metamathematics; neither is reducible to the other; and the distinction between them is quite fundamental epistemologically, and indeed in some form or other is applicable to history of any kind.

6.2 Two other chapters in this Part address, in the context of mathematics and logics, a surprising lacuna in Popper's philosophy: his silence about how scientific theories are devised in the first place. In §11 I build upon §2.11 to advocate the importance of structure-similarity between one mathematical theory and another, and appraise other ways in a theory can develop in the presence of theories already known. In §12 I extend this theme into an examination of theories are formed in applied mathematics; structure-similarity and analogies play an important role, but so also does the distinction made in phenomenology between the parts and the moments of a multiplicity. The philosophy of applied mathematics has been blocked for half a century by the influential but pathetic doctrine of 'the unreasonable effectiveness' of mathematics in science proposed by the physicist Eugene Wigner; let us hope that now we can do a bit better. Indeed, several of my proposals pertain to pure mathematics also, and maybe to other sciences.

In particular, while preparing this book I realized that the generality of mathematical (and also logical) theories comes in two distinct forms: as 'omnipresent' over some domain of reference D when it applies to almost all circumstances evident in D; and 'multipresent' over D (and maybe over some still larger domain) when it finds there a wide range of uses, each of which, however, is quite specific. The distinction could also be applied to the domains of substantial branches of mathematics, such as mathematical analysis or statistics. I have not modified any of the chapters to accommodate this distinction, but it does bear upon features such as the use of analogies in theory-building, and monism versus pluralism. So I give a summary in §21, including a sketch of a new philosophy of arithmetic; a fuller account is brewing in [Grattan-Guinness 2010].

7. Applications to psychology and psychical research

A quite different but very fruitful application of Popper's philosophy opened up in the 1970s when, partly through working on the history of logics, I came across certain aspects of psychology. Part 4 of this book contains the principal outcomes in six chapters (§15-§20).

7.1 The focus fell principally upon *psychical research*. This is the umbrella name given to the range of studies of phenomena that fall outside the concerns of orthodox science, such as telepathy and psychic healing. The opposition to the field has always been pretty robust, especially with the general rise in popularity of positivism among scientists and some philosophers from the 1830s onwards (§15). To many a professional scientist, proudly sporting his hard nose, it is all complete nonsense and its supporters are liars or tricksters or publicists or charlatans, or dupes or victims of others, or sufferers from inadequate left-brain activity, or some combination thereof. But for someone trained to think of science as guesswork, the situation is not so simple; for me the main attraction was the alleged *non-scientific* character of the field.

In the mid and late 19th century some more congenial lines of thought had developed with sufficient strength to lead to the founding of the Society for Psychical Research (SPR) in Britain in 1882. Two chapters in this Part of the book were prepared for the centenary conference of the SPR, which was held at Cambridge in 1982. Firstly, I edited a general introductory book on the subject [Grattan-Guinness 1982], and contributed my own reflections upon the difference between psychical research and the orthodox sciences. This essay forms the bulk of §16, which begins with Popper stating his own rational view of the field and continues with a list of its principal areas of interest. Secondly, to the conference itself I delivered a historical talk about the intellectual context of the founding of the SPR, and also the influence of positivism upon psychical research itself up to the 1930s as it tried to convert itself to the more "scientific" field of parapsychology (§17).

7.2 Further work in this field included to two questions of general philosophical interest irrespective of the possible involvement of psychical effects. Firstly, while of course we all experience *coincidences* in the course of life, can they *always* be attributed *only* to chance (§18)? Secondly, are there kinds of mental activity that take place *independent of language* (§19)? This latter question lay close to one of Popper's philosophical interests, namely, the *formation of language* in children: after several fairly unsubtle hints on his part, in 1993 I wrote a paper elaborating his criticisms of Noam Chomsky's theory of the innate deep linguistic structure and in favour of his own advocacy of the step-by-step build-up of sentences (§8.6). He seemed to be unaware that Chomsky's theory had not enjoyed a great press among linguisticians in recent years, so I embodied his alternative view within rather more general considerations. He read the draft in 1993, and I published it two years later, after his death (§20).

7.3 Apart from this last essay and §10, I do not often cite Popper's philosophy in the chapters contained in this and the previous Parts; but it served as a *constant* guide to both epistemological and historiographical issues. Recently I had the opportunity to acknowledge the importance of Popper's philosophy on all my work when I received a medal for my contributions to the history of mathematics; the texts associated with the occasion form the concluding §22.

8. Housework
The original texts have been modified only in modest ways, if at all. The organisation into sub-sections has been altered somewhat in a few chapters, but only substantial overlaps between chapters have been reduced in one chapter or the other. On a few occasions some lines of footnote have been promoted to the main text. Any significant additions are enclosed in curled brackets. The references for each chapter are given in a bibliography at its end, with items cited in the text; the original versions of some chapters used footnotes for this purpose.

Each chapter has an unnumbered footnote at the bottom of its first page. It begins with the details of the original publication of the article, together with indications of significant changes (if any). Copyright information is also provided there when the holder has requested it; I am most grateful to all the holders for the permissions that they have granted.

For consent to reprint Popper holograph I thank the University of Klagenfurt (Austria), Karl Popper Library. The original documents are kept in the Karl Popper Archive, Hoover Institution Archives, Stanford University, USA, with these file numbers: for the letter to Charles Morris (§1.7), file 329.37; for the letters to Russell and my translation of a book review (§7.4), files 45.32 and 345.14; and for his letter to me of 16 February 1984 that I quote at the head of §16, file 301.2.

Bibliography

Grattan-Guinness, I. 1982. (Ed.), *Psychical research. A guide to its history, principles and practices*, Wellingborough: Aquarian Press.

Grattan-Guinness, I. 1997. *The Fontana history of the mathematical sciences. The rainbow of mathematics*, London: Fontana.

Grattan-Guinness, I. 2009. *Routes of learning. Highways, pathways and byways in the history of mathematics*, Baltimore: Johns Hopkins University Press.

Grattan-Guinness, I. 2010. 'Omnipresence, multipresence and ubiquity: types of generality in and around mathematics and logics', in preparation.

Hacohen, M. H. 2000. *Karl Popper. The formative years 1902-1945*, Cambridge: Cambridge University Press.

Kuhn, T. S. 1962. *The structure of scientific revolutions*, 1st. ed., Chicago: University of Chicago Press.

Lakatos, I. and Musgrave, A. E. 1970. (Eds.), *Criticism and the growth of knowledge*, London: Cambridge University Press.

Popper, K. R. 1935. *Logik der Forschung*, Vienna: J. Springer.

Popper, K. R. 1945. *The open society and its enemies*, 2 vols., London: Routledge and Kegan Paul.

Popper, K. R. 1959. *The logic of scientific discovery*, London: Hutchinson.

Popper, K. R. 1963. *Conjectures and refutations. The growth of scientific knowledge*, London: Routledge and Kegan Paul.

Popper, K. R. 1966. *Logik der Forschung*, 2nd ed., Tübingen: Mohr.

Popper, K. R. 1972. *Objective knowledge. An evolutionary approach*, Oxford: at the Clarendon Press.

Popper, K. R. 1976. *Unended quest*, London: Fontana.

Popper, K. R. 1982a. *Quantum theory and the schism in physics* (ed. W. W. Bartley, III), London: Hutchinson.

Popper, K. R. 1982b. *The open universe. An argument for indeterminism* (ed. W. W. Bartley, III), London: Hutchinson.

Popper, K. R. 1983. *Realism and the aim of science* (ed. W. W. Bartley, III), London: Hutchinson.

Popper, K. R. and Eccles, J. C. 1977. *The self and its brain*, Berlin: Springer.

Schilpp, P. A. 1974. (Ed.), *The philosophy of Karl Popper*, 2 vols., La Salle, Illinois: Open Court.

PART 1

ELABORATIONS

Es irrt der Mensch so lang er strebt.
(Man errs as long as he strives.)

Johann Wolfgang von Goethe, Der Herr in the Prologue,
Faust. Eine Tragodie, Tübingen: Cotta, 1808, 22

1 Truths and Contradictions about Karl Popper

This fine biography describes in detail the remarkable contexts in which Popper formed his philosophy and met his first friends and foes.

MALACHI HAIM HACOHEN, *Karl Popper. The formative years 1902-1945*. Cambridge: Cambridge University Press, 2000. xiii + 610 pp.

1. Introduction

The philosophy of Karl Popper (1902-1994) has gained a range of interest and reaction far wider than that normally received by professional philosophers; in recent times only Bertrand Russell (1872-1970) gained comparable (probably still greater) attention. Convinced that philosophical problems and issues came from outside philosophy itself, especially science, he addressed a broad audience. But he also entered the professionals' field, and indeed attacked some major epistemological tenets held there, such as the assumption that knowledge was accreted by the inductive accumulation and classification of facts.

The response to Popper's work has created interest in his life, which was known a falling in three periods: birth and early career in Vienna, followed by nearly a decade in New Zealand and finally a rather reclusive life-style in Britain after the Second World War. This book, devoted to the first two periods, is the outcome of a long research effort by the author started well before Popper's death and incorporating a doctorate received in 1993. His book shows the incompleteness of Popper's autobiography [1974], which hitherto has been our main source.

The author has used many archives, including the Popper Papers kept at Stanford University and available on microfilm in some other institutions. However, the only photograph is a portrait, on the dust-jacket. Popper's library, which has been purchased by Klagenfurt University in Austria, seems not to have been used; maybe some annotation or marks await the finding.

Footnotes welcomely appear as such, although the font chosen by the publishers may be rather small for some readers. Typographical errors are rare, although the computing cut-and paste technique is evident in a repetition on pp. xi and 5. A separate bibliography is included; however, the items under each author are listed by alphabetical order of titles instead of chronological order of publication or writing, so that the progress of work cannot be scanned. The main casualty here is Popper himself, for whom dozens of items are listed.

This essay review was first published in *Annals of science*, 59 (2002), 89-96; it is reprinted by permission of Taylor & Francis (http://www.informaworld.com).

2. Forming the isolated philosopher

Popper was born in Vienna to a well-placed Jewish family, the last child after two sisters. He set on his lonely track in his late teens, when he moved for few years into a former barracks in the suburb of Grinzing occupied by various groups of organisations. Among his contacts of that time were future major musicians such as Hanns Eisler and Rudolf Serkin (pp. 78, 85). He took a strong interest in music himself, composing on occasion and joining but soon leaving Arnold Schoenberg's society for the promotion of modern music (p. 100). Life was held to limb by various jobs, most notably as a (trainee) carpenter. Taking courses on several subjects in and around Vienna University, his roving intellectual curiosity homed onto education, school reform and cognitive psychology.

Chapters 3 and 4 provide much information on various somewhat neglected figures influential upon Popper, such as Julius Kraft, Viktor Kraft, Karl Polanyi, Edgar Zilsel, Karl Bühler, Heinrich Gomperz and Leonard Nelson.[1] One factor common to some of them and to Popper was socialism, with a desire to create a less class-rigid society and provide a better quality of education for the populus. A thesis on '"habit" and "lawful experience" in education' was written at the University's Pedagogic Institute under Bühler in 1927 (pp. 142-149 but missing from the bibliography), to be followed in the following year by a University dissertation on cognitive psychology (pp. 156-163), and then a further thesis on the foundations of geometry (pp. 172-178) to provide qualification to teach physics and mathematics at school.

Such a posting, duly obtained in 1930 (p. 178), was the limit of Popper's professional ambition at this time. (Very little information on his teaching seems to survive.) But his philosophical horizons were widening, for he saw his various studies as examples of two more general problems: the status of induction in the formation of knowledge, and the demarcation between scientific and non-scientific knowledge. Between 1930 and 1932 he gradually grasped the full generality of these questions and convoluted his position into a non-foundationalist epistemology, in which the assumption that (scientific) knowledge was certain had to be abandoned. It is easy now to write such short sentences; Popper had to struggle hard even to understand his own stance. Chapter 5 describes in great detail the twists and turns that he took in formulating his position, which finally asserted that the key to the demarcation lay in falsifiability, using deductive reasoning, and that it was fundamentally opposed to inductive epistemology. Both here and elsewhere the author correctly emphasises the central place of Kantian philosophy in Popper's development, including the active role of the mind and the need for metaphysics — far removed from characterisation of Popper as a positivist that is still regularly made.

3. Circling the Vienna Circle

Another difficulty for Popper, largely of his own making, lay in his personal situation; even though he was living in one of the intellectual

---

[1] Among other background figures mentioned, Hugo Dingler was a German philosopher of science and of mathematics, not a 'Viennese physicist' (pp. 204, 249).

centres of the world, he worked largely on his own. The main product was a large manuscript on 'the two problems of the theory of knowledge', of which the first volume was completed in 1932 with a second left in fragments; there was little chance for publication (pp. 195-208).

By then Popper was known to the 'Vienna Circle' (hereafter 'VC') of philosophers, which had formed in 1924 around Moritz Schlick (1882-1936) and was given its name late in the decade (by Otto Neurath, to little enthusiasm from other members).[2] Popper's relationship with this circle, a major feature of his early career, receives much attention here in chapters 6-7. However, the author's characterisation of the circle as a united team with Wittgenstein's *Tractatus* (1921, 1922) as its bible (p. 192) is too simple, and indeed is contradicted three pages earlier. This was true for Schlick for some time, and always so for Felix Kaufmann (who however also adopted phenomenology) and Friedrich Waismann; but among others the adhesion was much less marked. In particular for Popper, Rudolf Carnap (1891-1970) rejected some of the seer's chief pronouncements, such as the meaninglessness of tautologies, and *emphasised* the distinction between theory and metatheory (the word 'metalogic' in the modern sense is his) that Wittgenstein rejected.[3]

Popper rewrote his manuscript into a shorter one called 'Logic of investigation. Towards the epistemology of modern science' ('Logik der Forschung. Zur Erkenntnistheorie der modernen Naturwissenschaft'). The developments included a detailed evaluation of the scope and limits of conventions, and examination of the role of probability theory. The place of corroboration of theories was rather marginalised, and the *interaction* of theory with experiment (where, for example, an experiment may suggest the scientific problem in the first place) was reduced to a one-way influence from theory.[4]

Thanks to further editing work by an uncle, the manuscript was reduced to a length commercially suitable for Julius Springer in Vienna,[5] and appeared late in 1934 in a VC book series edited by Philipp Frank and Schlick. Thus the book was in the Circle's own repertoire; and moreover it received much warm praise, especially from Carnap,[6] even though a

---

[2] The chief single source on the VC, including for this book, is [Stadler 1997] {English edition [Stadler 2001]}.

[3] For more on this topic see [Grattan-Guinness 1997]. Pace p. 191 of this book, I see the decisive influence upon Carnap as Russell rather than Frege.

[4] The bearing of Popper's philosophy upon technology still needs further study. Apparently during the Great War young Popper worked in a factory (p. 91); I remember him mentioning that he had worked in one that made lifts.

[5] This is a comment on the financial situation of the time. Carnap's own *Logische Syntax der Sprache* (1934) had had to lose about 50 pages for the same reason; they soon appeared as two separate papers but were restored to their original locations in the English translation *The logical syntax of language* (1937).

[6] As well as a positive review, Carnap praised Popper in various manuscript notes: for example, to Schlick in 1934 (Schlick Papers, State Archives of North Holland, Haarlem, file 95), Morris in 1934 (Morris Papers, Peirce Project, Indiana University at Indianapolis), Kaufmann in 1935 (Carnap Papers, University of Pittsburgh, files 28-20-08 and -21-01), and at a Chicago research seminar 1937-1938 (C. K..

thought experiment proposed for quantum mechanics was soon to be shown by Einstein and others to be faulty (pp. 257-259). It sold 414 copies by 1937,[7] quite reasonable for such a volume.

However, Popper was still an outsider to the VC, because Schlick never invited him to join. Popper explained the difficulty as due to his different philosophy, especially over the clash between falsificationism and verificationism and the status allowed (or denied) to metaphysics. However, the author shows that another factor was Popper's own personality: a clever but arrogant and rather immature man. Popper still felt his exclusion in later life,[8] thus compounding difficulties already in place.

4. Exile in New Zealand, 1937-1945

For all Jews and intellectuals in general, life in Central Europe was becoming steadily more precarious. To members of the VC the situation became especially clear when Schlick was murdered by a former student in September 1936 and the authorities did not work hard at finding the assassin of a professor who was not even Jewish (p. 190). In 1937 Popper secured a lectureship at the Canterbury College in the University of New Zealand, where he remained throughout the War years. The author provides in chapters 8-9 a vivid account of this period, which hitherto has been little known; Colin Simkin, Popper's former colleague there, has provided much valuable information.

This phase of Popper's career is best known for the writing of a book finally entitled *The open society and its enemies* (hereafter '*OS* '), a sustained attack on the role of philosophies underlying social determinism and the totalitarianisms of the time. The approach was based upon a meta-epistemological analogy with the clash between falsificationism and verificationism in science, with falsification now generalised to criticism of theories of all kinds, even metaphysical ones. Plato, Hegel and Marx were the main targets; the author assesses each part of the book very carefully, and in particular sustains the criticism made after its publication that Popper did not include a proper account of Hegel's positions in his attack (pp. 436-438).

Popper saw his book as his 'war work';[9] but also on his mind was the early part of the century (when New Zealand had been noted for its

---

Ogden Papers, McMaster University, Hamilton, box 7). The defence of Popper by Carl Hempel in a review [1939] of a critical paper on probability theory by Hans Reichenbach (p. 280) in the *Jahrbuch über die Fortschritte der Mathematik* is not quite correctly cited: it followed immediately Hempel's warm review of the *Logik*. The set-theorist Erich Kamke [1935] also wrote very positively about that book in the *Jahresbericht der Deutschen Mathematiker-Vereinigung*. Several other reviews are listed on p. 275 of this book.

[7] Carnap Papers, file 27-60-106; for other sales figures, see p. 275 of this book.

[8] Popper pointedly contrasted his own position around 1930 with that of his exact contemporary Herbert Feigl (compare pp. 185-186) at the beginning of an article [1966] for a *Festschrift* for Feigl.

[9] The author describes the writing of *OS* mainly from 1939 onwards. In the preface to the second edition (1959) Popper stated that the decision to write it was

socialist practices, incidentally). A strong memory from his Vienna childhood was of the *Dienstmädchen* system. The woman worked as servant to a family for 13 days per fortnight, from a Sunday to the following Saturday week; then the head of the household decided whether or not she be sacked at the end of the next fortnight. Around 1911 Popper's father accused their servant of stealing some money and dismissed her under this convention; upon asking about the woman's prospects, young Karl 'did not receive a satisfactory reply', as he put it to me. He confirmed that *OS* had been written to oppose that sort of system as well as the Nazis and the Stalinists.

The author also sheds light on one of Popper's most perplexing pieces, allied to *OS:* a 'stodgy' (K. R. Popper) essay on 'the poverty of historicism', which was started in the mid 1930s but was completed only (shortly) after *OS* was finished (pp. 352-382). This spread of time and of influence on either side of the writing of *OS* explains its rather inconsistent character [Popper 1944, 1945].

5. Exile in Britain, 1946-1994

OS had a protracted publication history from its completion in 1943 (pp. 450-459). After some American and British rejections it was taken by Routledge, thanks to reader Herbert Read; parts were rewritten and readied for publication late in 1945. Admirers in Britain, especially F. A. von Hayek and Ernst Gombrich, already wanted Popper employed there (pp. 496-499); and a readership was secured at the London School of Economics, which he took up in 1946 and where he was to remain for the rest of his career. This is the best known sector of his life, 'enjoying bad health to the ripe old age of 92' (p. 460): explicitly not treated in this biography, the author supplies a fine succinct summary in an epilogue with an assessment in ch. 10 of some of Popper's influence on social philosophy and economics.

The very first part of Popper's British phase involved positive involvement with Russell and negative reaction to Ludwig Wittgenstein. It might have borne a little more detail than is given here, as much territory in British philosophy was marked out; three mutually different philosophers, with Russell siding for Popper. For example, *OS and* Russell's *History of Western philosophy* were published within months of each other, and the famous confrontation between Popper and Wittgenstein in 1946 in Russell's presence (p. 523) occurred over an anti-Wittgensteinian topic that Russell had encouraged Popper to present (§7.2).

6. 'A difficult man'

This biography is excellently objective throughout. While the author clearly admires Popper's philosophy (pp. 288, 551), he records the failings of the man in many places.[10] My own contact with Popper

---

taken in March 1938, after he heard that Austria had been invaded (compare pp. 326 and 347 of this book).

[10] The title of this section of the review alludes to p. 179, and especially to an essay [Bartley 1980] by one of the Popperians (footnote 12 below). A favourite quip at the London School of Economics, credited to Ernst Gellner, was 'The open society by one of its enemies'.

suggested to me that one cause, or at least epiphenomenon, was his short height; for he exhibited more than anyone else of my acquaintance the aggressive body and speech language to others taller than himself.

A main cause of Popper's mediocre social skills must be his chosen isolation, from his youth. Although he seems to have recalled his father with affection, apart from anti-Zionism he did not follow him in many ways;[11] on servant girls, quite the opposite, as we saw. His mother and two elder sisters do not appear to have made substantial impact on him (pp. 62-64). At a rather young age he left home for the barracks. His wife, a schoolteacher before their marriage in 1930 and up to their departure from Austria (which left her with permanent homesickness for Vienna), seems not to have influenced the content of his thought, though she advised him on topics to study and spent much time as his typist and contact with publishers and others (pp. 179-180, 460). They had no children, and may not have indulged much in the necessary preliminaries. Life was work and little else, '360 [sic] days a year, day and night' (p. 222); however, periods of intense activity alternated with those of longueur and sleeplessness, a pattern sometimes held to manifest manic depression. Various acolytes emerged during his London phase, but many were struck off the roster at some stage.[12]

The author expressly denies the intention of writing a volume 2; but any such work will benefit massively from this volume 1, which must rank among the best of its genre. An especially nice feature is its publication by the house that rejected *OS* in 1943 for its disrespect for Plato (p. 457).

7. Popper to Charles Morris, 1936

Much material remains in the Popper Archives and elsewhere to fill out still more details of his life and career. It is appropriate to include here a letter written on 9 September 1936 by Popper to Charles Morris (1901-1979), and mentioned on p. 321. A philosopher in the American pragmatist tradition, Morris was also associated with the VC and an important colleague of Carnap at the University of Chicago, especially over the International Encyclopaedia of Unified Science that Neurath was (sort-of) launching from Amsterdam [Rossi-Landi 1953].

This letter must be one of the first that Popper wrote in English (compare p. 312), and shows that his command was already not bad. He wrote this letter right at the end of his Vienna time — consciously so, as he indicates his hopes of emigration and hopes for advice about work

---

[11] For example, Popper did not become a Freemason, a movement within which his father was master of a lodge in Pressburg (p. 41); his [1909] was a tribute on the death of his predecessor, and in [1910] he wrote on the possible influence of the movement upon schools. These items are cited from [Wolfstieg 1911-1926, vol. 1, 950 and vol. 2, 348].

[12] These are 'the 'Popperians', usually professional philosophers, including Feyerabend, B. Magee, J. Agassi, I. Lakatos, J. W. N. Watkins, I. Jarvie, Bartley and D. W. Miller (pp. 537-538). They are to be distinguished from 'the Popperian knights', eminent beknighted (naturalised) British scientists and humanists who publicly supported his philosophy (P. B. Medawar, J. C. Eccles, H. Bondi and E. Gombrich), and of whom he appears to have been much more tolerant.

possibilities in the USA. In his reply of 10 October 1936 Morris recommended that Popper publish in American journals (p. 321).

Popper begins by recalling meeting Morris a year earlier at the first of the VC congresses of philosophy, held in Paris. He describes some features of the succeeding gathering in Copenhagen, including the news of the murder of Schlick. He also recalls some events during his recent visits to England (when, among other contacts, he met Russell), and summarises his indeterminacy interpretation of quantum mechanics following the failure of his thought experiment.

The letter is transcribed from the top copy, held in the Morris Papers, Peirce Project, Indiana University at Indianapolis; a carbon copy is kept in the Popper Papers (file 329.37). It is typewritten; orthography is followed exactly, and a few penned corrections have been silently followed. Editorial footnotes fill out some details.

September 9th, 1936,
Neukräftengasse 8,
Vienna XIII.

Dear Professor Morris,
it is nearly one year now that I have seen you the last, in Paris. I would like to know how you are getting on. Are you now back in Chicago, and do you see often Prof. Carnap?

I have been in England the main part of this year, delivering there some lectures, under them: On the Method of Soziology (London School of Economics), and a course, On Probability (the Mathematical Department of the Imperial College of Science).[13] Then I have been in Copenhagen, visiting there the 'Second Congress'. I had the ocasion to speak to Nils Bohr; I was very much impressed by his extraordinary personality.[14] I discussed with him the Problem of Interpretation of the New Quantum Theory.

Having become clear already some times ago that I have to modify something of my former position I explained to Prof. Bohr a standpoint — something like a compromise — the main features of it beeing the following theses, maintained already in my elder position: (1) Heisenbergs Formula, looked at as to be deduced from the New Quantum Theory (viz. its deductions by Pauli, Dirac, and others) must not be interpreted in an other way than only as an <u>statistical</u> statement, saying itself nothing about 'indeterminacy'. (2) The non-statistical Thesis of Indeterminacy is no statement of the physical theory and has not got the character of a 'law of nature'. (Nevertheless — this I had to admit — it plais a certain role in the interpretation of the theory with help of 'classical' concepts.) (3) It is not possible to solve the problems of Einstein's New Paradox by means of the so called 'Heisenberg'scher Schnitt'.[15]

---

[13] [On this period, see this book, pp. 311-319.]

[14] [For a photograph taken at the Congress showing both Popper and Bohr, see [Stadler 1997, 415].]

[15] [These are allusions to the Einstein-Podolsky-Rosen paradox (a more ssuccessful thought experiment than Popper's own), and to Heisenberg's indeterminacy principle (this book, pp. 253-259).]

I was very happy to get Bohr's consent to this standpoint. He encouraged me in his kind way to a new publication on the matter I am preparing just now.[16]

During the Congress the message of Schlick's death reached Copenhagen. Returning to Vienna I read the newspapers there and at first I had the impression that one must not see as political murdering in this case although the whole athmosphere of shooting and killing which poisons the brains of the younger generation of Central Europe certainly must held as to be responsible of this case.

But now the situation begins to change and I am rather certain you will take an interest in the fact, that there is now something like a press-campaign against the dead Schlick. Especially a periodical with the charming title "Schönere Zukunft" (The Fairer Future) representing the now dominating (so called 'christian') course, published a paper the essenc of which can be formulated as: Not the murder is guilty but the murdered.[17] The whole 'mental' armament of this 'Fair Presence' we are living in, is used to prove this thesis; Antisemitsm, as always, is playing the main role: It was the Jewish-destructive Philosophy of Schlick which poisoned the brains of his pupils, especially the brains of his 'arisch' pupils, who are not enough prepared to bear this poison. So it seems to be something like a reaction of self-defendence that one of the pupils killed the man who poisoned him first. Therefore: Men like Schlick must not be allowed to teach at a Christian and German University, — they only may be allowed to teach Jews. (As a matter of fact Schlick was not of Jewish origin but as I learn of very ancient nobility his ancestors having been Counts, very famous in Austrian and Bohemian History.)

It is a pity, but this happening is characterising the 'Fair Presence' of Austria. Antisemitism and nacism are leading the fashion under Austrian teachers unfortunally also of the school at which I am working, making it impossible for me to practise my post. I don't feel myself any longer to listen day by day to allusions and affronts. I am decided to leave this country as soon as possible. But it is rather difficult. Thirring (who is Professor of Theoretical Physics at the University of Vienna) has tried to get me a Rockefeller-fellowship, but he got the answer, the foundation is not in the situation any longer to give fellowship to physicists;[18] and as a philosopher I cannot get it either.

I am prepared to go to America without a fellowship, but only if I could get some invitations to deliver lectures there (as I did in England) or, if possible, a temporary lectureship, perhaps as a substitute of a lecturer who has gone abroad. Prof. Nils Bohr is willing to help me by writing letters about me to people who could do something in this matter (and in a similar way Einstein) — but I don't know whom I should name him.

---

[16] [No item in the Popper bibliography in [Schilpp 1974] seems to correspond.]

[17] [On this article, see this book, p. 190. A large file on the murder is provided in [Stadler 1997, 920-961].]

[18] [A Rockefeller fellowship had enabled Feigl to move to the USA in 1930 (compare footnote 8).]

Have you ever read in my book? I would be interested to hear your opinion about it. I hope to publish very soon two papers in English.[19]
With my best regards,
Yours very sincerely,
Dr. Karl Popper. [signed]

Bibliography

Bartley, W. W. 1980. 'Ein schwieriger Mensch. Eine Porträtsskizze von Sir Karl Popper', in E. Nordhofen (ed.), *Physiognomien: Philosophie des 20. Jahrhunderts in Portraits,* Königstein: Athenäum, 43-69.
Grattan-Guinness, I. 1997. 'A retreat from holisms: Carnap's logical course, 1921-1943', *Annals of science, 54*, 407-421.
Hempel, C. 1939. Review, *Jahrbuch über die Fortschritte der Mathematik, 61*(1935). 979.
Kamke, E. 1935. Review of *Logik der Forschung, Jahresbericht der Deutschen Mathematiker-Vereinigung, 45*, pt. 2, 91-92.
Popper, K. R. 1938. 'A set of independent axioms for probability', *Mind, new ser., 47*, 275-277 [errata on pp. 415 and 552].
Popper, K. R. 1944; 1945. 'The poverty of historicism', *Economica, 11,* 86-103, 119-137; *12,* 69-89. [Repr. in book form London: Routledge and Kegan Paul, 1957.]
Popper, K. R. 1966. 'A theorem on truth-content', in P. K. Feyerabend and G. Maxwell (eds.), *Mind, matter and method,* Minneapolis: University of Minnesota Press, 343-353.
Popper, K. R. 1974. 'Autobiography' in P.A. Schilpp (ed.), *The philosophy of Karl Popper,* 2 vols., La Salle, Ill.: Open Court, 1-181. [Repr. as *Unended quest,* London: Fontana, 1976.]
Popper, S. C. S. 1909. 'Ansprache in der Warmholtz-Trauerfeier der L.[oge] „Humanitas"', *Der Zirkel, 40,* 35-37.
Popper, S. C. S. 1910. 'Von unser neuen Generation. Deutsche Festrede, gehalten in der Grosslogenversammlung von 24. April 1910', *Orient, 35,* 120-126.
Rossi-Landi, F. 1953. *Charles Morris,* Rome and Milan: Bocca.
Stadler, F. 1997. *Studien zum Wiener Kreis. Ursprung, Entwicklung und Wirkung des logischen Empirismus im Kontext*, Frankfurt am Main: Suhrkamp.
Stadler, F. 2001. *The Vienna Circle. Studies in the origins, development, and influence of logical empiricism.* Vienna and New York: Springer.
Wolfstieg, A. 1911-1926. *Bibliographie der freimaurerischen Literatur,* 3 vols., Leipzig: Verein Deutscher Freimaurer.

---

[19] [Only [Popper 1938] seems to correspond, but maybe it was written in New Zealand.]

2 What do Theories Talk About?
A Critique of Popperian Fallibilism, with Especial Reference to Ontology

To the loving memory of Anita Gregory

Popper's philosophy of fallibilism is considered with especial emphasis on the referents and reference of scientific theories. After some preparatory remarks in Part One, several preliminaries are outlined in Part Two, dealing with the limits of experience, theories as explanations, and the character of facts. The main proposals are presented in Part Three: they concern the importance of reification and the significance of ontological issues. Part Four considers certain prolongations of the proposed version of Popper's philosophy outside science, including history, mathematics, education and Popper's own theory of Worlds 1, 2 and 3.

> I avoid, or tried to avoid, the term 'ontology' in this book, and also in my other books; especially because of the fuss made by some philosophers over 'ontology'.
> Popper [1982b, 7]

PART ONE   PROLOGUE

> Thus practical hydraulics today is still preponderantly a power-domain of coefficients and many times its working methods [are] only an interpretation of empirical data. [...] A theoretically unsatisfactory solution, even if it shows itself useful only on the boundaries within which the technology deploys it, is therefore always better than nothing at all.
> Philipp Forchheimer [1905, 327]

1. The interpretations of corroborations
1.1 *Testing theories.* According to Popper, a scientific theory can never be proved to be true, but its falsity is (or appears to be) established when a prediction is refuted by a test. Should the prediction be upheld by

---

This article was first published in *Fundamenta scientiae*, 7 (1986), 177-221; it is reprinted by kind permission of Springer Science+Business Media. I have modified the division of some sections into subsections, and added all their titles. A quotation from W. B. Yeats at the head of ¶1 has been "promoted" to the head of Part 4 of this book, and replaced by the passage from Forchheimer.

[1] While speaking of corroboration here, I am not concerned with the degree of corroborability of the theory. On verisimilitude, see footnote 21.

the test, then the theory is said to be corroborated, although a later test might lead to a refutation.

Although the theory cannot be proved to be true, the aim of science is to search for the truth. Furthermore, truth is defined as the property of a statement in the theory that it corresponds to the facts. Therefore, the aim of science is toward theories that correspond better and better to the facts, and so exhibit ever greater measures of verisimilitude. The competition between theories is to be appraised in terms not only of verisimilitude but also refutability and corroborability of the problems that they were designed to solve in the first place.

1.2 *A very mundane phenomenon.* Scientific theories are used quite often after the manner of the following simple example. I throw a pebble from my desk to land at a point P on the floor a few yards away. I seek a scientific theory that will predict the landing of the pebble at P given the details of the initial launch and location of the floor. In this case, in the current state of development of dynamics at least four theories will suffice (with the pebble treated as a particle): Newtonian mechanics, variational dynamics, special relativity and general relativity. The first two, being logically equivalent for those parts of mechanics to which they both apply, will give exactly the same prediction, while the differences between their prediction and those of the other two theories are too small to be detected in this particular case. Thus all four theories are corroborated by this test. However, their methods of explaining the motion of the pebble are quite different; even the logically equivalent variational dynamics differs from Newtonian mechanics, for it is based in an assumed teleology of 'least action', and gives epistemic priority to potentials over forces. Thus in terms of explaining how the pebble actually did get to P, at least three of these theories are false, quite possibly all four of them.[1]

What then, are false theories talking about? Not the phenomena that they attempt to explain, not even those that furnish corroborations; but not about anything else either. Moreover, the refutation of a theory only discloses to us this state of affairs; it obtained before, and whether or not a competing theory was available. Hence any refutable theory may not talk about anything at all — a puzzling conclusion to draw concerning the mighty and successful engine of science.

Alternatively, one might say: it does not matter what a theory talks about (apart from avoiding logical inconsistencies, of course); for it is required only to deliver predictions, which can then turn out to be corroborated or refuted. Further, the verisimilitudes of rival theories would be defined only in terms of truth- and falsehood-contents, without reference to the objects and processes of which they speak (forces, potentials, or whatever, in the above case). But this view is of an instrumentalist kind, or else carries an empiricist preponderance, views that Popper has criticised.[2]

---

[2] Popper states his anti-instrumentalist position in many places; for example, [1963, 107-119; 1983a, 103-146]. However, my impression of contemporary scientists is that, if they adopt Popper's position, they construe it in something like the

Of course, the considerations just presented are loosely formulated; 'talk about' will itself need some talking about, and in fact a good deal of preparatory work will be needed on 'theory', 'explanation', 'fact', and 'truth'. This chapter sketches out a treatment of such questions, and considers some of the consequences.

2. Intentions

> In studying a philosopher, the right attitude is never reverence nor contempt, but first a kind of hypothetical sympathy, until it is possible to know what it feels like to believe in his theories, and only then a revival of the critical attitude, which should resemble, as far as possible, the state of mind of a person abandoning opinions which he has hitherto held.
>
> Bertrand Russell [1946, 58]

2.1 *The organisation of this chapter*. In the next section I emphasise the apparent feature of this universe that some of its substances and process lie beyond our means of sensory detection, and examine its consequences for empirical testing. ¶4 deals with two preparatory issues concerning scientific theories in general (their supposed universality and their modes of explanation), while ¶5 contains a similar discussion of facts on which, for Popper, true theories correspond. Then in ¶6 the notion of reification is developed, and is related to a process that I call 'desimplification'. ¶7 contrasts the appraisal of scientific theories in terms of their empirical record with their prowess as accounts of 'what happens'; it is here that ontological issues come to the fore. ¶8 serves as a brief summary.

The rest of the chapter deals with the prolongation of Popper's philosophy outside science. ¶9 considers the possibility that other areas of human thought have ontological potential. As case studies, ignorance is proposed for this role in history and education in ¶10, and the special status of mathematical knowledge is noted in ¶11. Finally, in ¶12 Popper's theory of Worlds 1, 2 and 3 is appraised in the light of earlier discussions.

The chapter is written within the broad framework of Popper's philosophy. Thus I accept without discussion the main features of his views on the testing, refutability, and corroborability of theories; the irrefutability of metaphysical theories; and his indeterminism, and fallibilistic methodology and epistemology. Further, I have little or nothing to say on his attitudes to probability and induction, or on his evolutionary epistemology. I also follow his principal omissions and silences, so that there is little or nothing

---

neo-instrumentalist version outlined in the text here, which is doubtless the 'somewhat rigid and over-simplified form' of it mentioned by Eccles [1974, 349]. Further, even then they may not be able to effect refutations anyway; for, according to the dismal news from Kern, Mirels and Hinshaw [1983], many scientists cannot use the *modus tollens* rule of inference correctly.

It is rather interesting to note the (nervous?) assurance that scientists often give, that they know 'the [sic] scientific method', pace Medawar's caution that 'There is no such thing as the Scientific Method' [1967, 275]; compare footnote 5.

here on the psychology of creating theories (although, as with Popper, much on the appraisal of those theories as solvers of the problems set), on the economics of testing theories, or on the social aspects of science (for example, the pressures of career-making on a scientist's modes of working). Finally, I have not attempted to compare his philosophy in detail with competing philosophies.

I have tried to speak in as general terms as possible, and to be sparing in the use of examples. Those that are chosen usually come from the physical sciences, as these are the areas of science with which I happen to be most familiar. But I have tried to ensure that no special features of these areas are crucial for the examples given, and I am sure that comparable ones could be found in geology, biology, psychology, and medicine. For similar reasons, the examples are historical; but the doctrine is, I trust, applicable to modern science.

2.2 *The references* are confined to specific passages in Popper's (or others') writings, and to items in which themes under discussion are compactly treated in more detail. Many other sources have been of value to me over the years, and I am most grateful for them; some are cited in [Grattan-Guinness 1979 = §9].

Some remarks on terms need to be made. 'Prediction' includes also retrodiction. 'Refutation' is preferred to 'falsification', partly because the latter word has the connotation of fraud (which is not of concern here), and partly for the special character of truth and falsehood adopted in ¶7.1. 'Universe' is used rather than 'world', which is confined here to Popper's theory of the three Worlds. As I shall not be dealing with ontogenetic questions, I take 'empirical' as synonymous with 'experiential', and avoid the latter because of its similarity to 'experimental'. 'Statement', the content of a sentence, is synonymous with 'proposition' (and I pass over the fact that scientists often do not express themselves in statements). 'Theory' refers to a theory as such and not any particular person's view of it, unless this alternative is explicitly stated. A few other terms are discussed *in situ*.

PART TWO   PROPAEDEUTIC

3. On the limits of experience

But several of the phenomena which accompany the motion of fluids escaping observation, others not being able to be calculated due to lack of knowing exactly the laws in virtue of which they take place, one must seek to draw near [to them] as maximally as possible, in the current state of our knowledge, despite the small hope that one has of attaining it completely.

Benoit Fourneyron [Crozet-Fourneyron, 1924, 2]

3.1 *Facing the complications.* To us the universe is a maze of complexity, a miracle of action, and a mystery of purpose. One reason is that many of its substances and processes lie beyond the limits of our sensory means, at both ends: not only the 'micro' cases, where the fundamental constituents seem to be too tiny for our experience, but also in the 'very large', where parts of our universe are beyond our powers of sensory experience. Their range can of course be extended by the use of apparatus such as microscopes, telescopes, remote sensing equipment and the like (at the intellectual price of accommodating the theories of their modes of operation into our work[3]); but even then limits to our experience will still be found.

Further, they are not the only kind of limits that we face, even though once again technology can help. For some phenomena happen too quickly for our direct observation (as in aspects of optics or radio) while others occur too slowly (as in geology or usually in evolution, for example); certain phenomena reach us only as effects of some apparent cause (such as the result of some supposed force), or hide their mode of action in transparency (as Fourneyron noted above with water-flow); for some, it is not even clear what *type(s)* of phenomena are involved (as in aspects of mind and brain research; Eccles [1974, 350] gives a nice example).

In addition, some phenomena seem to change character with scale. Breaking a stick of chalk into pieces needs only strong fingers; breaking the atoms of which it is held to be composed is another matter (as it were), with serious consequences; and breaking up a galaxy is doubtless different again. Or, take as an example from life sciences, people eat food and (may) love the music of Mozart, unlike the chromosomes that are said to contain their genetic makeup.

3.2 Modes of experience. Yes, the brown desk is as brown as it was yesterday, and the visual *Times* newspaper accompanies the tactile and auditory *Times* (to use A. N. Whitehead's example); but it seems clear that human sensory and mental processes are enormously complicated. (How many senses are we held to have, by the way? Still the classical five?) Philosophers tend to take the reliability of these processes far too much for granted (perhaps because of the emphasis on sight). As far as fallibilism is concerned, it is no tiresome quibble to insist that if we follow Popper in adopting metaphysical realism (that the universe exists independently of our experience of it) then we must regard it as referring to the noumenal stratum. Its components gain phenomenal attributes when we experience them, and enter our phenomenal stratum: this is composed of the subliminal, empirical, and supraliminal realms (where 'supraliminal' is taken from [Myers 1903, xxi]) as an antonym to 'subliminal', on the other side of the empirical). As was noted in ¶3.1, the boundaries between these realms can be altered by the use of apparatus, without the empirical realm knowingly absorbing the other two.

---

[3] Note that instruments are, of course, developed in the first place in response to theoretical and observational needs. Hacking [1983, ch. 11] gives nice examples from microscopes, but refutes them *en courant* by stating that 'One needs theory to make a microscope. You do not need theory to use one' [p. 191]. I do not pursue further the role of instruments, although my general points apply to them; see, in particular, ¶6.2 on more accurate measurements.

The phenomenal processes may be no more problematic than (say) reading some dials (correctly) and checking the experimental set-up; but they can come to the fore — as with the colour-blind John Dalton, for example, when he came to study the composition of colours [Shearman 1981, 121-125]. More generally, our theorising is saturated by the character of our sensory means. Colours are a good example: the tree may be in the quad when no one looks at it, but are its leaves green when no one looks at them? Again, is not human understanding of the universe much dependent on the forward mounting of our eyes, as opposed to the sideways position evident in many other types of animal? Can we take phenomenal aspects of experience so much for granted when, for example, the variation in auditory sensitivity is so wide — every level from people who can barely distinguish two notes apart, up to those who can hear every player in a symphony orchestra at once (and also the hearing aid in row K that is turned up too high?). Indeed, *all these phenomenal processes themselves exist in the noumenal stratum*, and are subject to phenomenal attributes of their own; when studied in the appropriate branches of science, for example.

3.3 *The variety of human experience.* I regard all the above features of our universe, and of us humans living within it, as *contingent*. In other universes, the changes of property with scale may not apply, say, or the sensory means of inhabiting sentient beings could be sufficiently refined to experience directly the fundamental constituents and/or processes operative there, and perhaps also the activities involved in their version of thinking. (Of course, they may not know that they are so equipped.) However, we are not in such a situation, and seem unlikely to evolve into it, not for a while anyway.[4] Thus the bearing of these features is likely to be long lasting.

Science, including those parts of it that study the human being himself, is concerned with all these levels of experience, and its theories form a spectrum across them. For example, an astrophysicist may ponder on the motion of distant stars without much concern for the details of their shape; questions of that kind concern the geodesist, who in turn is not likely to share the civil engineer's worry over the stability of embankments; he in turn is unlikely to have to deal with quantum mechanics (compare [Popper 1959, 207-209]). Again, a physiologist's interest in histology will not necessarily penetrate to the level of the constituents of tissue that interest the molecular biologist or the geneticist. Of course, scientists may be mistaken in ignoring factors in these ways; on occasion they attempt unifications across different parts of the spectrum (¶7.4): my point here is to emphasise the spread of scientific concerns in toto.

---

[4] Some of the so-called 'paranormal' phenomena suggest that our sensory means are greater, and the modes of interaction between physical and mental phenomena more complicated, than orthodox science allows (see [Grattan-Guinness 1982, part 2] for a survey of these phenomena; also §16-§17). However, psychical research does not suggest that we have the sensory means suggested in the text here, so I shall not consider these phenomena further in this chapter (apart from a trace in footnote 38).

3.4 *The need to theorise.* In view of the limited sensory means of man, it is not surprising that he has (evolved the ability to) transcend the empirical realm in the formation of theories. The 'mind's eye' is capable of thinking up objects, processes and relationships that the head's eyes have never seen, and perhaps never could see: fundamental particles and 'invisible' processes, and also theories themselves as objects for metatheoretical considerations. What, then, do scientific theories talk about?

Several highly influential traditions in philosophy, to which names like 'empiricism' and 'positivism' are often attached, assign legitimacy only to theorising about the empirical realm; the rest is at best *façon de parler*, and preferably should be avoided altogether. Now one can certainly operate philosophically to some extent in this direction: for example, one may put forward (refutable) laws that appear to be rooted only in the empirical realm, and thus offer (say) Huygens's construction of double refraction of light in certain crystals, the sine law of radiation of heat, or Gestalt perception, without making any assertions as to the constitution of light or heat or the mode of working of perception. But, as with Popper, I grant a central place to mankind's trans-empirical capacities just mentioned, and so regard as inadequate or incomplete philosophies that are so constructed.

Thus I also seek an alternative to the neo-instrumentalist interpretation of his philosophy suggested in ¶1.2. Conventionalism is stale fare, although Popper's attitude to conventions in general seems rather glib [1974b, 1115-1117] (compare ¶9.2). Although discontented with conventionalism ([1959, 77-81], and the longer discussion in the 1930s manuscript published as [1979b, 177-219]), Popper himself nudges towards it in [1959, 136-145] in that he gives a short account of the relationship between the simplicity and the refutability of theories. However, it is not very efficacious, and is not much developed in his later writings, though [1959, 378-386] contains a new appendix on 'content simplicity and dimension'. Indeed, it would be a formidably complicated task to produce a general metatheory of simplicity — a good reason not to be a conventionalist.[5]

Thus a central place needs to be found in our thought-up objects. As a first approximation I say that in forming and developing a theory we reify a (sub-)universe in which the states of affairs are held to occur,[6] and then try to appraise that universe against the appropriate parts of this one. Part III of this chapter is devoted to a second approximation; the rest of this

---

[5] Bachelard [1927] anticipated Popper's philosophy in stressing the approximative character of theorising [chs. 1-3], and the complications involved in articulating a philosophy of simplicity [ch. 7]; however, his fundamental principle of 'the rectification of concepts' [ch. 2] was directed largely to the 'approximate verification' of theories [ch. 16]. I have never seen Popper cite this book; perhaps he does not know it.

[6] The phrase 'reification of a (sub-)universe' passes over, with rather pat glibness, a large collection of complicated questions concerning the parsing of statements, distinctions between sense and naming, and so on (see [Quine 1961] for an important study of such questions). It also covers what scientists often call 'making a model' (§21.3.1), although they often muddle up the model with theories based upon it.

one is concerned with important preliminaries. I begin with some cautionary tales on 'real' and 'ideal'.

3.5 *The 'reality' of theories.* In view of our capacity to think up theories beyond the empirical realm, and also to think about those theories, Popper has broadened his use of the term 'real' beyond the empirical realm [1963, 118]:

> [...] there is no point in denying reality to dispositions, not even if we deny reality to all universals and to all states of affairs, including incidents, and confine ourselves to using that sense of the word 'real' which, from the point of view of ordinary usage, is the narrowest and safest: to call only physical bodies 'real', and only those which are neither too small nor too big nor too distant to be easily seen and handled.

He has since cautioned that 'theories in themselves [...] are not quite as "real" as tables and chairs' [1974a, 184], and even that 'a spurious fact is simply not real' [1972, 46]; but the scale of 'reality' is still somewhat overwhelming. Further, it makes his reluctance to talk ontology all the more surprising.

I do not regard the question of 'real' as a mere quibble over a word; for it has been used in numerous ways by philosophers over the centuries. As John Blackmore shows well in his paper [1979], and even better in unpublished work that he kindly sent to me, the variations in usage even incorporate inversions; for example, a positivist's 'realism' would be phenomenalism to a Kantian, whose 'realism' is metaphysics to a positivist. (A source of this confusion seems to be that realism is a Good Thing; so all philosophers are realists.) If the word is to be used, it should be explained in context and/or qualified with adjectives, such as Popper's 'metaphysical realism' (¶3.3) on the existence of the universe independently of ourselves.

As for 'reality' of theories and their component notions, I take Popper's position to be that of a conceptual realist — we think up these notions and then use them — as opposed, for example, to the Platonic realist who claims that they exist independently of the mind and that we discover them. (My criticism of the Platonic realist is: how does he know that they exist independently of the mind until it 'discovers' them?) Conceptual realism is intimately linked with reification; but if we regard an aim of science as concerned with comparing reified universes with this one, the sense of realism needs to be carefully indicated. To treat, for example, muons or collective unconsciousness as 'real' just because they occur in influential theories strikes me as too glib. To take the pebble-throwing example of ¶1.2, if we can reify four different theories about its flight and thereby create four universes, all 'real' but each differing from the others in some basic respects, then we are in danger of populating our minds with Quine's 'slum of possibles' [1961, 4] in jumbo proportions. In the rest of this essay I shall try to be sparing in the use of 'real', and provide it with qualifiers when used.

Companion remarks are called for concerning 'idealism', a word that, as Blackmore [1979] again shows, has suffered from over-use as much as has 'realism'. In a sense its fate has been worse; for as 'idea-lism' it denotes solipsistic and related philosophies that usually admit the existence only of ideas in the mind and of appearances, while as 'ideal-ism' it refers to ideal-ised (that is, certain types of reified) states of affairs, whether the attendant underlying philosophy is idea-list or not. For example, Popper's metaphysical realism opposes idea-lism, but his philosophy of science admits the ideal-istic character of theories. Again, when using 'idealism' I shall try to establish the sense intended. Let us turn, now, to the theories themselves.

4. Some introductory remarks on theories
CERTAIN SCIENTISTS

Man's law is precision, God's is chaotic. Man's wisdom is offensive to God, therefore He shows His displeasure in complications. To man the complications are chaos, thereby is man deceived. To God, man's precision is the fretfulness of a babe, aye, and man at his wilful deceiving is undone. Then to God, man is precisively chaotic; to man, God is the disruption of precision.
Patience Worth [Prince 1927, 185]

Popper has made clear that theories are scientific if they are refutable by empirical test, but the same can be said of theories outside science, too; impractical legislation, for example. The question of demarcating science from other areas of human thought is taken up in ¶9; here I point out some "obvious" but important features of scientific theories.

4.1 *Corroborability and refutability* exhibit a symmetry in that a refutation of theory T is a corroboration of not-T and vice versa. Therefore we need a criterion for preferring to test T rather than not-T (or, again, vice versa); and it is here that the difference between empirical contents and the scales of potential refuters comes into play. We happen normally to assume that T has 'positively' defined predicates, with their negative counterparts in not-T; but as [Frege 1918b] pointed out, there are many cases where the positive/negative distinction is not obvious, or is a matter of convention.

4.2 *The universality of scientific theories.* 'I consider the distinction between universal and individual concepts or names to be of fundamental importance', writes Popper [1959, 64]. The distinction is explained 'with the help of examples of the following kind: 'dictator', 'planet', 'H_2O' are universal concepts or universal names; 'Napoleon', 'the Earth', 'the Atlantic' are singular or individual concepts or names' [p. 64]. Then 'Scientific theories are universal statements' [p. 59], containing only universal names, which 'can be defined without the use of proper names' [p. 64]. 'It is the *strictly universal statements* which I have had in mind so far when speaking of universal statements - and theories or natural laws', as opposed to 'numerically universal statements' that are at most 'conjunctions of singular statements' [p. 62].

Popper's restriction of scientific theories to (refutable) universal statements has surprising consequences. For example,

> All heavenly bodies contain atmospheres suffused with oxygen (A)

is scientific, but not

> All heavenly bodies of this universe have atmospheres suffused with oxygen (B)

since it contains the proper name 'this universe',[7] nor

> The atmosphere of this planet is suffused with oxygen (C)

for similarly containing 'this planet', nor

> There exists a planet with an atmosphere suffused with oxygen (D)

for being existential, and thus irrefutable [p. 70].

The scientist interested in (C) could claim that it was universal enough for his purpose: it is 'numerically universal' for Popper, in the form

> All regions, however tiny, of the atmosphere of this planet contain some oxygen. (E)

It is also refutable: some very tiny regions, perhaps at the periphery of the atmosphere, may not contain oxygen, or they may be created artificially by enclosure and burning off, and so on. Yet in Popper's view the scientist should take theory (C) as an instantiation of (B) (or other forms like it), and point to the growth of knowledge more in terms of refutation by the case of the Moon rather than corroboration by the Earth.

This is a most unusual view of scientific theories, and I do not think it deserves, or needs, to become popular: the ineffable power granted by Popper to proper naming should be ignored. On the contrary, the scope of empirical reference of a scientific theory can be as small as circumstances impose; for example, this sheet of paper in this room in ... in this universe is sufficient for theorising about the constitution of paper, for we need only to adjoin two meta-epistemological principles in order to avoid inductive epistemology.

The first principle is *trans-empiricism*: that the theory transcends the empirical base, however great (and therefore, *a fortiori*, however small) it is. The second principle is modality: that the theory is drawn from, and tested within, a certain range of space-time but it might possibly apply in other cases — indeed, we would want to find that out. (In a few cases, such as in cosmology, the very nature of the theory may demand the whole of the universe as the range from the start.) Of course, the content of the theory, and the severity of the testing, is markedly limited if a small range is imposed; but this is a methodological and scientific question, not an epistemological one. Further, after all, while the paper theorist above is currently deprived of other paper, he still has access to other theories, from which to make analogies and contrasts. There is nothing un-Popperian in adopting the view just described — and worries about proper names play no role in it. I shall adopt this position in the rest of this chapter.

4.3 *Scientific theories as explicanda*. Philosophers and scientists are (largely) agreed that scientific theories are explanatory (or try to be: their

---

[7] If 'this planet' is a proper name, why not also 'this planetary system', 'this galaxy', ... even 'this universe'? There is some point in distinguishing (B) from (A) anyway, since (A) is testable (in principle!) in other universes.

success is not at issue here). Popper is one of the few to recognise that, epistemologically speaking, a theory explains the known in terms of the unknown [1963, 63, 89] — or, in the simple case of the known pebble in ¶1.2, several different and competing classes of unknowns. It is all too common for explanations to be misunderstood as passing from the unknown to the known. This reading can only apply in a methodological context, when a theory known to be well corroborated by some groups of phenomena is then tried out on some other phenomena, where it is as yet untested and so unknown. This is a good example of a situation where epistemological and methodological issues should not be confused.

While Popper is very clear on this point, a related one needs discussion: namely that theories often appear to be *descriptions*, not explanations of any kind. Take one of his examples:

All orbits of planets are ellipses   (F)

[1959, 122]: it *describes* the motions of planets, without offering any explanation ('because ...'). Yet it seems to be perfectly satisfactory as a scientific theory. Can this be so?

(F) explains in various ways. Firstly, it explains by exclusion; why, for example,

No orbits of planets are squares (or helices or ...).   (G)

It also explains by *instantiation*, such as

The orbit of Venus is an ellipse.   (H)

With the help of initial data it explains specific cases, such as

At time t_0 Venus is to be seen in direction D_0 from place P_0 on the Earth.   (I)

Conversely, (I) may be regarded as an empirical prediction concerning the calibre of (F) as an explanation, as is noted in ¶5.1.

(F) can be subject to explanations of its own, in terms of other theories. Again, different kinds need to be distinguished. The explanation of (F) from

All heavenly bodies move in ellipses,   (J)

which Popper terms as explanation in terms of levels of universality [1959, 75-77, 121-123], depends on the meanings assigned to the words 'planet' and 'heavenly bodies'. (This was not a trivial point between Ptolemy and Copernicus concerning the status of the Sun.) Newton's explanation of (F) from

All planets obey the inverse-square law   (K)

is achieved as a deduction (using mathematics, though mathematics itself is not at issue here). Then there may be causal explanations such as

Plena of force-fields sweep the planets round in ellipses,   (L)

theological explanations, entelechic explanations (as are used in life sciences sometimes), and so on. Note that these explanations compete as explanations of (F), and possibly also of each other.

All contentual statements explain by exclusion: for example, even (I) explains, in this mode, why Venus is not seen in direction D_1 ($\neq D_0$) from P_0 at time t_0. I shall assume that a *statement is theoretical only if it explains in at least one of the other modes described above*. Further, when referring to theories as explanations, the modes of explanation need to be borne in mind.

4.4 *On induction.* While Popper is not concerned with the psychology of creating theories (and, I recall from ¶2.1, neither am I in this chapter), he stresses that theories should be appraised, in an intellectual sense, as solutions of the problems against which they are set. However, as emerges well from [Hacking 1983, part B] (among other writers), it is not easy to distinguish a problem from a theory in the first place. In Popperian terms, a theory qua fallible conjecture, is problematic. For example, is determining the boiling point of olive oil a problem, or a theory under test? It is too easy to concoct theories of which the experimental work is a test: at random,

The boiling point of olive oil is 88.1° Centigrade. (M)
Further, the most "obvious" theory,
There exists a boiling point for olive oil, (N)
will *not* do, as it is existential, and thus irrefutable. Determining the boiling point is best regarded as a *problem*, to which a (M) is a proposed solution.

Suppose for sake of argument, that (M) is the theory that we propose. How did we come to it, rather than, say, the value 90.8° C? Usually we draw on experimental work done over a few samples, and perhaps also cross-checks against several thermometers; and this is a type of induction. The endless polemics over the status of induction seem to me somewhat pointless, since different questions are involved. As method we may come to theory (M) by some sort of inductive strategy (usually far more complicated than this particular case, of course: compare [Rescher 1980]): we may also use such strategies to help isolate one problem from another in the first place. Such questions, belong to inductive *methodology*. By contrast, theory (M) as such is trans-empirical, in the sense of ¶4.2: inductivist epistemology does not work, for the usual reasons stressed by Popper and others;[8] further instances of 88.1° C do not make (N) any more probable, and so on. Popper's claim to have solved '*the* problem of induction' [1974, 1, italics inserted] seems excessive; indeed, he has himself pointed out that there is more than one problem of induction [Popper 1959, 281-282 (1972 addition)], where inductive epistemology is called 'logical and methodological'.

In what sense is (M) a theory? Firstly, the value 88.1° C will have been decided as a (possibly weighted) mean from data, and so is open to revision such as an increase in precision (¶6.2) and distinctions between hitherto unnoticed different kinds of olive oil. Again, it asserts the constancy of the temperature in question, which can be tested against further samples. Indeed, the 'constant' may turn out to vary with other factors (such as environmental pressure in this case), which therefore have to be

---

[8] Bhaskar rejects Popper's refutation of inductivist epistemology on the grounds of conflating the question of the conformity of nature with the truth of a theory [1978, 217-218)]; I have not found this conflation in Popper's writings. Rescher rejects Popper's position for inefficacy, in that refutation does 'no more than to eliminate one possibility' [1980, 217]; he does not discuss either corroborability or verisimilitude.

specified in an expanded version of (M). It also draws on the calibration of thermometers, a procedure that lends special status to statements such as

    The boiling point of pure water at atmospheric
    pressure is 100°C.   (O)

I stress theoryhood here, for it is common to regard statements such as (M) as facts. To that complicated category I now turn.

5. On the character of facts

[...] I do not believe in the existence of a line which separates two kinds of statements: simple statements of fact, and sophisticated explanatory theories. I believe that apparent statements of fact are theory-impregnated, and that whether or not a theory is explanatory depends entirely upon the question whether it is being used to explain, that is, to solve an intellectual problem, a problem of understanding.
    Popper [1974b, 1094]

[...] the question whether our man-made theories are true or not depends upon the real facts; real facts, which are, with very few exceptions, not man-made.
    Popper [1972, 328]

5.1 *Facts as theories.* For Popper 'Our main concern in science, and in philosophy is, or ought to be, the search for truth' [1972, 319], and he draws on Tarski's theory of truth to define the truth of a statement by its correspondence to the facts. Therefore facts are a main target for scientific theories. However, Popper does not give even a working definition of the word 'fact', so that it is not easy to determine its scope and limits. I make a sequence of comments, and adopt a strategy to deal with them in ¶5.3.

In the second quotation Popper declares facts to be, with very few exceptions, not man-made. What are these exceptions? If they are the facts involving man-made materials and artefacts, as studied in engineering, pharmacy, chemical technology, and the like, then 'very few' is a massive under-estimate. I take as irrelevant to the specification of facts the question of whether or not the states of affairs are man-made.

More problematic however, is the place of man-made theories in facts, to which Popper alludes in the first quotation. Since factual statements incorporate (implicitly, perhaps) theoretical ones, the test of a theory T is dependent upon their corroborability also. They belong to what Popper calls the 'background knowledge' B of T and its test [1983a, 236-244]. Now B itself can be tested; but one must watch out for vicious circles, in that T might be part of the background knowledge of some part of B. To take the oxygen example (C) from ¶4.2, we may test for the presence of oxygen by using fractional distillation, but also test out distillation using oxygen. Procedures involving measurement can give such examples, too; one could be constructed from (M) of ¶4.3. In any given case the fact involved in the test should be clearly distinguishable from the background knowledge; but in a large ensemble of cases, circularity would arise, and the 'empirical basis' of theories [Popper 1959, 44-48, 93-111] may be more complicated than it appears.

There is also the question of the range of statements regarded as 'factual': not only those drawn from the empirical basis but also ones taken from logic and language, or mathematics or even philosophy itself. These have to be considered, since a theory can be tested from such points of view as well as empirically: for logical consistency, for example, or a candidate for a reduction programme. How is the range of facts to be controlled here?

For Popper, facts are 'reality pinned down by *descriptive* statements' [1963, 214, italics inserted], whereas theories appear to be explanations. However, as was pointed out in ¶4.3, even the most straightforward statement can explain by exclusion, so that theories and facts cannot be distinguished as explanations and descriptions respectively.

5.2 *Truth as correspondence to the facts.* As Susan Haack [1976] points out, Tarski advocates his correspondence theory of truth only for formalised languages and regards it as epistemologically neutral; and Popper's claim that the theory is applicable 'even to a "natural" language' [1963, 223] finds little evidence in Tarski. In his reply to Haack [Popper 1979a] quotes some supporting passages from Tarski, but they lack conviction when set in context (as is done in [Grattan-Guinness 1984 = §6], from which this section is drawn). A favourite example with Popper, 'snow is white', is a particularly clear source of one difficulty: the bivalency of the underlying logic (true or false, and no other possibilities). To draw on ¶5.1, this statement could be part of a theory under test; indeed it was refuted by the rather grubby stuff in my garden the other day.[9] Popper's insistence on the monism of bivalent logic [1972, 304-307] strikes me as a contentious part of his philosophy, both here and in other contexts (§6):[10] I regard it as a refutable theory of applied logic (§5.4).[11]

Another aspect concerns the correspondence specified between a fact and a statement (its truth-value is not at issue here), as established via the notion of satisfaction. The specification could involve beliefs and other

---

[9] These remarks may constitute parts of an answer to J. Bronowski's complaint to Popper that 'Snow is white' is not the kind of statement used in science [Bronowski 1974, 618]. Popper's answer, that the statement is theory-laden [1974b, 1094], does not meet the point raised.

[10] Henry Margenau expresses well the artificiality of saying that a theory is, or is not, corroborated when experimental tolerances are borne in mind; but he speaks of accuracy and tolerance as a matter of 'convention' [1974, 755]. I prefer to regard it as an aspect of the sub-theories involved in the perceptions, experiments and instruments involved in the test of the theory. Popper demurs over Margenau's use of conventions, but does not really tackle this point in his reply [1974b, 1121-1125].

[11] Popper expresses his monistic philosophy for bivalent logic in [1972, 139-140, 304-308]: it constitutes an unusual point of agreement with Quine's philosophy [1970, ch. 6]. Popper's argument is that science demands the severest testing and thus the strongest logic, which bivalency furnishes. I find the strength of bivalency to be rather brute in many cases, 'Snow is white' being one (§6), and prefer logical pluralism (on which see [Haack 1978] and §5.4).

subjective elements; and also perhaps some parsing of the statement (which Popper disparages as language analysis, while also taking certain of its features for granted[12]).

5.3 *Handling facts.* Somewhat tentatively, I adopt the following strategy. I use the correspondence theory of truth, while accepting the possibility that in some walks of life and perhaps science itself, other theories of truth would be more relevant; and for now I accept also bivalency, although with much reluctance. In addition, I emphasise Popper's tri-distinction between truth, possession of truth, and proof of its possession — that 'the task of science is the search for truth, that is, for true theories (even though[,] as Xenophanes pointed out[,] we may never get them, or know them as true if we get them)' — as it is essential for Popper's view that theories cannot be proved to be true.[13]

To help distinguish facts from theories, I make a contrast with ¶4.3 and say that factual statements can explain only by exclusion. Further, it seems necessary to qualify the word 'fact' with adjectives like 'empirical', 'mathematical', 'linguistic', and so on, as circumstances require, in order to avoid misunderstandings (and also essentialism). In particular, empirical facts include the singular existential statements used in the empirical testing of a theory [Popper 1959, 102]. They refer to states of affairs in any part of the (expandable) empirical realm. Since sensory mediation plays some role in forming this type of fact (¶3.2) and also in theories involved in the use of instruments such as microscopes that enter the empirical realm (¶3.1), the corresponding statements relate to the phenomenal stratum.

I exclude from empirical facthood statements involving reifications outside the empirical realm. Thus, for example, 'sub-atomic particle A releases such-and-such energy on impact with sub-atomic particle B' or 'the collective unconscious caused Mr. X's hatred of Mr. Y' are not empirical facts, although they may lead to them (such as an accurately predicted trace in a cloud chamber, or the circumstances of reconciliation between Mr. X and Mr. Y). I shall also avoid the common use of 'fact' to refer to familiar, highly corroborated, theories such as the oxygen theory (C) of ¶4.2, and phenomenological theories such as Huygens's law and examples given in ¶3.4; for they can explain by means other than exclusion.

---

[12] For example, in connection with truth and in various other contexts, Popper asserts the primacy of statements over concepts (see especially the table in [1963, 19] or [1974a, art. 7)}. Thus he has a philosophy of verbs (or, at least, of recognitions of statements). What is it?

For an interesting advocacy of moments as truth-bearers, see [Mulligan, Simons and Smith 1984].

[13] [Popper 1963], 229; see also p. 225. However, for some reason that I cannot determine, he wrote later that '*We are seekers after truth but not its possessors*' [1972, 47, italics in original]. His tri-distinction, which includes theories which happen to be true (and which by luck have been thought up), refutes the view sometimes held that all theories are born refuted (and note that metaphysical theories are unfalsifiable anyway).

PART THREE   PROPOSALS

6. Reification and desimplification

> In this section a mathematical model of the growing embryo will be described. This model will be a simplification and an idealization, and consequently a falsification. It is to be hoped that the features retained for discussion are those of the greatest importance in the present state of knowledge.
>
> Alan Turing [1954, 37]

6.1 *Reification.* Scientific research often proceeds as follows. A scientist looks at some examples of the phenomena that interest him and finds that they are much too complicated to be handled theoretically. So he simplifies them by omitting what he regards as small or peripheral effects; for example, he takes the temperature of the component bodies to be constant, or some particular property of the materials to be isotropic, or ignores air resistance and the rotation of the Earth. (In the case of the pebble in ¶1.2 he would be wise to ignore its own rotation while in flight.) These decisions contribute to the reification of a sub-universe mentioned in ¶3.2; a universe is set up in which such states of affairs hold, whereas in this universe they rarely, if ever, do.

The theory T_1 that he constructs refers to this reified universe, U_1; but the experiments and observations are carried out in this universe, U_0. Thus, if he experiences a refutation of his theory, he may well conclude that his simplification went too far, and that the discarded and/or overlooked effects were not so small or peripheral after all. He may then follow a line of thought that I call 'desimplification', in which he tries to enrich T_1 by allowing for some of the effects discarded from U_1. From a sequence of such moves he reifies one or many new universes U_2, U_3, ... to which his new theories T_2, T_3... respectively relate, and tests again. Corroboration of the predictions means that the theory T_n in question reifies a universe U_n in which the pertinent types of phenomena are similar enough to those experienced in U_n to escape discrimination from them, given the current state of experimental design and observational technique. Two possibilities arise.

6.2 *Desimplification.* One possibility was exemplified in ¶1.2, where four theories predict the (apparently) same point of landing of the pebble on the floor. When theories give predictions that are indistinguishable relative to a given empirical arrangement, or when one theory contains a prediction that is too small for empirical detection, we try to desimplify that arrangement in order to effect a more severe test. This is what underlies the notion of a more accurate measurement.

Another possibility is that one desimplifies the theory to check on the size and/or scope of a knowingly omitted factor and see whether or not it is worth allowing for in the empirical test: if so, the theory will need desimplifying.[14] In a variant form, several such factors may turn out to make more or less equal but opposite modifications to the arrangement (at least, within the measures currently set on its parameters).

The notion of 'desimplification' is intended to serve more or less as a synonym for Popper's phrase 'ad hoc hypotheses' [1959, 42, 80-82], although there are two related differences. Popper admits the use of such hypotheses only with reluctance, and when the main theory has been refuted. By contrast, desimplification need not be executed with reluctance when a factor has knowingly been omitted; and it can be carried out whether or not the main theory has been refuted. Further, desimplified theories can be compared for testability, in contrast to Popper's claim that 'Ad hoc hypotheses cannot be compared in this way' [1959, 145 (1972 addition)]. However, I am not proposing any calculus, or measure, of desimplification.

Desimplification is also a way of describing aspects of Kuhn's theory (1962) of normal science; namely, the basic lines of the theory T have been laid down but all kinds of desimplification are carried out to cope with small (or not so small) effects, extend the detailing, apply to special cases, and so on. While much normal science is routine to a fault, and creates much of the conservative, trade-unionist atmosphere of science, it can involve the creation of difficult new (sub-)theories and experimental techniques; so I find Popper's strictures against normal science ([1974b, 1144-1148], for example) rather over-stated (§9.3.2).

Under the view proposed here the occurrence of so-called 'revolutionary' change[15] is now seen as arising when the moves to desimplify are patently unsuccessful, or grossly contrived (or even impossible), and a more radical change of theory is needed; or when a desimplification turns out to have major repercussions on a theory to which it was originally offered as a special adornment.[16] It is worth noting that some scientific 'revolutions' have started out from specific details of the theory that later was to be overthrown, rather than from an attack on its basic principles. Popper [1975] rightly defends the role of rational criticism in so-called revolutionary situations in science, distinguishing them also from ideological revolutions. However, elsewhere he seems to grant 'revolution' a surprisingly wide brief

---

[14] For example, during the 19th century the theory of the so-called 'simple' pendulum was raised to a very high level of desimplification, largely because of the use of instruments in geodesy. Studies included the stretching and torsion of the suspending wire, the effect of pivoting on the chape, and the angular momentum caused at the change of upswing into downswing [Wolf 1889].

[15] My jib at the word 'revolutionary' arises partly from its excessive use by historians and philosophers of science, but also because its meaning as a change in state of affairs clashes with its origins as a return to a given state (as in the 'revolution' of the heavenly bodies). In Elizabethan times the word was understood in this latter sense even for philosophical contexts, but it changed its meaning later [Cohen (B.) 1976]. {See now my advocacy of the notion of convolution in ROL, §8.}

[16] It is a pity that Kuhn entitled his fine book [1962] *The structure of scientific revolutions*, for its explanation of their occurrence is rather weak. If thought of as a study of 'The structure of normal science', then it makes much better sense (§9.3.2).

in the remark 'There is much less accumulation of knowledge in science than there is revolutionary changing of scientific theories' [1963, 129].

6.3 *Decision or discovery with theories?* In ¶5.3 I objected to calling 'facts' statements drawn (for example) from scientific theories about the interaction of sub-atomic particles or the action of the collective unconscious. Here I offer a companion objection to calling such statements 'discoveries'. I find the use of this word excessive[17] (akin to the over-use of 'fact' noted in ¶5.1), for it takes reification to excess to regard as discoveries the objects and states of affairs so 'named'. These supposed discoveries usually constitute reifications, or other assumptions, made for the benefit of the theory concerned.

The distinction between decision and discovery operates also within the empirical realm; for example, the 'discovery' of a new planet hinges not only on its observation but also on the decision that it is a new planet. One of Popper's favourite examples of the profitable employment of ad hocness in science (and for me of desimplification) is the discovery of Neptune in 1846 [1974a, art. 9; 1974b, 946]. Now it is instructive to note that the planet was observed several times before 1846 [Hoyt 1980, 32-35]. In other words, in 1846 a certain physical object was seen again (that is, (re)discovered); but a decision was taken that it was a new planet of our system. 'Thus it is decisions which settle the fate of theories' [Popper 1959, 108]. Our decision is backed up by a well corroborated, but refutable theory: and if the object actually is such a planet, then the decision turns out to be a discovery. This brings us up against ontology.

7. On ontology and facthood

I believe that ideas such as absolute certitude, absolute exactness, final truth, etc., are figments of the imagination which should not be admissible in any field of science. On the other hand, any assertion of probability is either right or wrong from the standpoint of the theory on which it is based. This looseness of thinking seems to me to be the greatest blessing which modern science has given to us. For the belief in a single truth and in being the possessor thereof is the root cause of all evil in the world.

Max Born [1978, 298-299]

7.1 *Ontology as a target.* Popper's aversion from discussing ontology, exhibited in the quotation at the head of this chapter, seems to arise from his mis-identification of ontology with essentialist 'what is?' questions instead of 'What is there?' questions [1977, 4]; interestingly, he correctly appraises teleological questions as of the type 'What is it for?' [p. 105]. I

---

[17] Note that Popper [1959] is called *The logic of scientific discovery*, whereas 'The logic of scientific investigation' would have been a preferable translation of *Logik der Forschung* (§9.7.4). In places in his writings he uses the word 'discovery' to refer to reifications; a particularly interesting example occurs in [1982a, xviii], where a 'proposal' by Einstein becomes a 'discovery' eight lines later.

find this aversion rather a pity, since clearly ontological questions are prominent in his philosophy. In any case, why should a 'tottering old metaphysician' [Popper 1974b, 993] not discuss ontology?

I take Popper's view of the aim of science as a search for truth (¶5.1) as an attempt to answer the question 'What happens?', or, more precisely, 'What is there, what are they doing to each other, and how are they doing it?'.[18] This mixture of ontology and teleology is hard to disentangle (compare [Bhaskar 1978]); indeed, the distinction between substance and process is not clear-cut and theories about a given class of noumena may well differ in their views about the places and roles of each.[19]

So I shall take ontology to embrace both substance and process in the noumenal stratum. Further, in the context of a particular theory 'ontology' refers to *whatever that theory asserts or assumes that there is*.

On this view the aim of science is to search for theories that are ontologically correct; that is, theories whose own ontologies are identical with the objects and processes that they talk about, and so correctly explain what happens within the class of noumena to which they relate. (I assume that we can develop languages sufficiently sophisticated to express such theories in the first place; after all, what better motivation could there be?) Theories are said to be (empirically) factually true when they correspond to the empirical facts expressed by the singular statements involved in the corroborating tests, and factually false otherwise. Truth and falsehood are now defined rather strictly relative to the empirical realm; and since theories can be tested in ways additional to empirical test, I draw on the decision reported in ¶2.2 to speak of the 'refutation' of theories in this chapter, rather than their 'falsification'.[20] Here are some features of the distinction between ontological (in)correctness and factual truth and falsehood.

7.2 *Ontology versus facts*. My reluctance to adopt a bivalent logic in connection with factual truth and falsehood and corroborations was recorded in ¶5.2; see also §5.4. By contrast, an ontologically correct theory can display such bivalency: for it is either correct or incorrect. More pre-

---

[18] There is also the 'why?' question of Ultimate Purpose; I can offer here only the weak response, 'Why not?'. Explanatory theories 'X because Y' offer answers to "local" questions concerning levels of universality noted in ¶4.2.

[19] For example, the teleological character of variational dynamics in such principles as that of least action was contrasted with Newtonian mechanics in ¶2.2. Again, some theories of heat spoke of a heat-bearing substance such as caloric entering or leaving bodies (basically a substance-oriented theory with some attention to processes) while others talked only about excitation of 'molecules' of these bodies (basically process theories with few or no commitments on substance beyond those concerning the molecules themselves).

[20] Usually Popper takes 'refutable' and 'falsifiable' as synonymous, although in [1959, 100] he says 'falsifiable, i.e. empirical'. One could propose the following distinction: between 'intrinsically irrefutable' theories such as metaphysical positions, which are irrefutable by their very nature; and 'empirically irrefutable' theories, which arise in the context of ¶6.2, for example, when more accurate measurements cannot be made by means of the state of the technological art then available.

cisely, it can be correct in some components and not in others; for example, it may propose the correct kinds of substance but the wrong types of process connecting them.

A theory that is ontologically correct can also be factually true; but the pebble example of ¶1.2 shows that the converse relationship need not hold, for a theory can both be ontologically incorrect and factually true. In other words, it does not describe or explain what actually happens in the noumenal stratum but nevertheless can be corroborated in empirical tests.

Further, since experimental work is often subject to empirical approximation, as was stressed in ¶3.1, the corroboration or refutation of a theory may be appraised in some statistical way, within a range of variations in the experimental arrangement. By contrast, ontological (in)correctness is not susceptible to such handling, not even in the case where the ontologically correct theory is statistical; the statistical part of a competing theory would be ontologically correct statistics or not (compare ¶11.2).

7.3 *Ontology and corroboration.* Assume that theory T_1 was more highly corroborated than T_2 by a given class of phenomena (their performance over other phenomena is irrelevant). Now, it does not follow that T_1 is more ontologically correct than T_2: it could even be less correct in that it may not even talk about the actual substances and processes involved. For example, if bodies actually do more under the action of forces on particles, then Newtonian mechanics is more ontologically correct than special relativity, although it embraces fewer factual truths (and, the refutations tell us, it is incorrect in the type of force and/or particle involved and/or the states they may be in). The same applies to desimplifications (¶6.1); while they may help in proving the severity of the empirical tests, the reified universes U_r involved in the desimplified theories T_r are not necessarily closer to the (relevant parts of) the actual universe U_0.

When a theory is ontologically correct, it literally refers to the noumena that it treats: there is an identity relation between its reified universe U_n and the parts of the actual universe U_0 to which it relates. Further, if it is correct in all its components, it is the ultimate explanation of those noumena (modulo forms of linguistic expression and translation, of course, which may pose some non-trivial problems). Popper claims that ultimate explanations cannot be found in science [1945, vol. 2, 9-11; 1972, 194-196; 1983a, 134-136]; but his argument is based on the mis-identification of ultimate explanations with essentialist definitions, in tune with his view of ontology noted in ¶7.1. I suspect that a confusion of necessary and sufficient conditions is involved here: an (alleged) essentialist explanation may (claim to be) the ultimate explanation, but the converse relation does not have to hold.

Two further related points need to be stressed. Firstly, the existence of such a theory is another contingency to add to the one noted in ¶3.3; while it is ontologically correct in this universe U_0 it may well not be in another one. Secondly, due to the possibility of later refutation in U_0 the theory cannot be proved to be the ultimate explanation, even if it is. In other words, we imitate the tri-distinction of ¶5.6 between truth, its possession, and proof of its possession and note the difference between ontological correctness (that is presumed to exist), its possession (that might be gained by luckily happening to think up such a theory) and proof of its possession (that cannot be established). In these terms noumena cease to be unknowable, as it is customary to say, but instead are unprovably knowable.

7.4 *Corroboration and verisimilitude.* This tri-distinction relates to Popper's remark that corroboration is not (necessarily) a measure of verisimilitude (or truth-likeness): in the example of ¶7.3, T_1 has passed more tests than has T_2, but nevertheless it may contain fewer true consequences and/or more false ones than T_2, these still awaiting to be exhibited.[21] I extend this result as follows: corroborability is not necessarily a measure of truth-likeness, which in turn is not necessarily a measure of ontology-likeness; thus corroborability is also not necessarily a measure of ontology-likeness. For example, the comparison between T_1 and T_2 just made is valid even if T_1 is ontologically correct, because, for example, for some reason the tests that it calls for cannot be executed. Thus corroborability, verisimilitude and ontology can lie in three different directions: in a given range of phenomena, one theory is the best tested (to date); another contains more factual truths; and yet another explains what actually happens.

Figure 1 presents a schema of the principal relationships involved. It is based on an analogy with a theory containing basic principles (or 'hard core') and auxiliary factors (or, for me at ¶6.1, desimplifications). The question marks are a shorthand way of saying that the relationship in question may not apply.

Figure 1

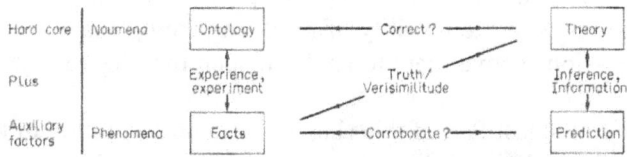

7.5 *The remit of theories.* Even an ontologically correct theory cannot be expected to refer to more than a proper subset of the total noumenal stratum; for example, the correct theory of the path of the pebble (¶1.2) is likely to be entirely uninformative about the digestion of fat by locusts or even the flow of water in a river. Thus there is no reason to expect that the universe acts under one uniting principle (apart from invoking the logical ruse of forming this principle as the conjunction of all the ontologically correct theories). However, in no way does this situation deny the benefit to

---

[21] See [Popper 1963, 235; or 1974b, 1010]. He defines verisimilitude relative to all kinds of true statements, whereas I am concerned with the scope and limits of truthhood. I shall not pursue the question here, however, since I mistrust verisimilitude on other grounds; namely its assumption of a bivalent logic (compare footnote 11). I take this view independently of the difficulties that the theory faces anyway (on which see Andersson [1978], for example, and §9.7.5).

theorising of unifying (some group of) scientific theories; for, among other reasons, the unifying theory may be factually true for many of the phenomena against which it is tested, without gaining the additional accolade of ontological correctness.

For example, during the early years of the 19th century P. S. Laplace reified a universe of which "all" its actions were to be expressed in terms of central forces acting between the fundamental 'molecules' of substance [Fox 1974]. But his programme was not an ontological claim; on the contrary, he stated quite explicitly that we do not know the shape of the 'molecules' and thus not the law of attraction between them, and as a general principle he stated that the final causes and sources were unknowable to us, and so was agnostic on the question of unity at the noumenal stratum. Reductions are to be distinguished from unifications, although they can work in harness.[22]

Laplace's programme was a genuine attempt at unity within physics. I suspect that many claims for 'the unity of science' (or at least some of its branches) are more rhetoric than philosophy; for they are based upon confusing unity with interconnectedness, which is a necessary condition for unity but not a sufficient one. Reductionists sometimes express the hope that the problems that interest their anti-reductionist opponents will "go away". The anti-reductionist can reply, in a sort of intellectual special relativity: which does the going away; the problem, or the reductionist?

The points made here relate to what are sometimes called 'natural kinds', a special type of theorising that tries to divide up the universe into sorts. My examples above were chosen deliberately; there are not only (proposed) separations such as mechanics from biology but also within mechanics itself (such as solid things like pebbles from fluid things like rivers, on the grounds of the inability of fluids to take up shape on their own). All the discussions above seem to apply to such theories refutability, desimplification, ontological correctness, and so on.[23]

In the last Part of this chapter I extend the notion of kinds to what I call 'intellectual kinds'. First, however, a little summing up is in order.

8. The philosophy of natural philosophy

> One can work with something like a world "behind" the world of appearance without committing oneself to essentialism (especially if one assumes that we can never know whether there may not be a further world behind that world).
>
> Popper [1963, 173]

---

[22] On the possible benefits of specific reductions, together with criticisms of the Ockham's Razor Chapel of Philosophies, see [Popper 1972, 289-297]. On the difficulties in effecting reductions, see [Popper 1982b, 163-174], and [Yoshida 1982].

[23] Questions of natural kinds should not be confused with the alleged uniformity of nature (on which see [Popper 1959, 252-254]). The former topic is concerned with the stuffs of the universe, as it were; the latter claims a certain constancy in their inter-relations.

According to the view adopted in this chapter, a theory is put forward as an explanation (of some type) of certain classes of noumena; it may also (partly) embrace other theories, and so serve as explanation of them also. It is expressed in terms of a reified (sub-)universe, relating somewhere within the subliminal and/or empirical and/or supraliminal realms. Empirical testing has to fall within the empirical realm (even though perhaps extended by the use of apparatus); it is also subject to possibly non-trivial phenomenal adumbrations upon the noumena in question. Corroboration or refutation of the theory is a decision taken upon the measure of empirical proximity of theoretical prediction and experimental or observational result. Whichever decision is taken, the theory can be revised by desimplification, or even by 'revolutionary' change of line. Theories can be compared for measures of corroboration (and corroborability) and of verisimilitude, which are to be distinguished from each other and also from ontological correctness. Truth correlates with the correspondence (or proximity) to the facts, while ontology is concerned with 'what happens (etc.)' in those realm(s) of the noumenal stratum to which the theory relates.

In several respects Popper's philosophy is Kantian, as he himself has acknowledged in places (the preface of [Popper 1966], for example); the thrust of the considerations outlined in previous sections is to make it perhaps more Kantian. Indeed, I have characterised it in the title of this section as 'the philosophy of natural philosophy', since it belongs more closely to the days when science was that richer study, the contemplation of nature in all its forms. Now this neo-Kantian philosophy is not necessarily tuned into the *Naturphilosophie* of those days [Gower 1973]; for example, several commentators have pointed to similarities with certain philosophies of the 18th and 19th centuries in English-speaking countries.[24] However, this does not consign Popper's philosophy to the historical archives; for at all times, including ours, the philosophically refined practitioners of science have aspired to the condition of studying nature, and some of them have even concerned themselves with philosophical questions pertaining to their interests, with implications for philosophy in more general terms.

At the time when natural philosophy was at its prime, philosophy itself maintained close connections with it, and also with related fields such as mathematics and psychology. However, over the decades these links have largely been severed, and now philosophy tends to be dominated either by reductionisms and language analysis (although, regrettably, relating but little to linguistics) or by efforts to make something out of nothingness. Further, in some way that is connected with these changes it has become 'very much a self-contained professional activity, self-reproducing by training up the next generations via 'the *prima facie* method of teaching philosophy' [Popper 1963, 72]. Thus the practice of philosophy today splits into two largely disjoint groups; these professionals, and the amateurs

---

[24] Some articles in [Schilpp 1974] point to this feature, most valuably in the reprint there of [Medawar 1967]; Popper confesses ignorance of the cited sources in his reply [1974b, 1036]. Among recent sources see [Niiniluoto 1978] on Whewell and Peirce, and [Elkana 1984] on Whewell.

above who profess something else as their chief activity. Popper is one of the few philosophers whose work attracts the interest of both groups.[25] Let us see how his philosophy, in the form outlined in this chapter, may be applied to areas of thought outside science.

PART FOUR   PROLONGATIONS

9. Intellectual kinds: the scope of refutability

> What is the world that science reveals to us as the reality of the world we see? A world dark as the grave, silent as a stone, and shaking like a jelly. [Is that] the ultimate fact of this glorious world? Why you might as well say that the ultimate fact of one of Beethoven's violin quartettes [sic] is the scraping of the tails of horses on the intestines of cats.
>
> James Hinton [Hopkins 1878, 13]

This final Part generalises upon the remarks on natural kinds made in ¶7.5 into a foray into what might be called 'intellectual kinds', where science stands as one kind among others. How, and to what extent, does Popper's philosophy of science apply to these other kinds? In this section the applicability of the theory of refutation outside science is considered in general terms, and the question of the role of ontology is posed. The other three sections try out these ideas in specific non-scientific areas.

9.1 *Inside and outside science.* Popper points out that his philosophy of refutation in science is itself not refutable since it 'is not empirical, but methodological or philosophical' [1974b, 1010]. Recently he has stated, somewhat nervously, of methodology in general, that 'It is rather, a philosophical — a metaphysical — discipline, perhaps partly even a normative proposal' ([1983a, xxv, italics inserted]; compare [1975, 99] or [1974b, 1036]). I assume that the question of the normativeness of his philosophy belongs to metametaphysics — that is, a metaphysical problem concerning a metaphysical theory.

Questions of this kind take us beyond science; but do they take us beyond refutability? Put another way, is there an effective demarcation between science and non-science? Laudan [1983] has argued well that the demarcation is not important, whatever view is taken of science. Certainly refutation does not provide a criterion, for it occurs often enough in other areas of human activity (as was noted at the beginning of ¶4). However, there may be a difference to be noted at the ontological level. Scientific theories reify, and even refer to this universe when ontologically correct. Can non-scientific theories do the same?

---

[25] Popper notes the place of 'us last laggards of the Enlightenment' with 'an irrational and anti-rational philosophical messianism *à la* Heidegger from the one side, and for a "mathematically exact" philosophical method from the other' [1983a, 177].

9.2 *Rationality outside science.* The question is more interesting than it may seem when the closeness of contact between scientific and non-scientific areas is noted. For example, we can have scientific (specifically, acoustic) theories of the scraping of the tails of horses on the intestines of cats, but can our theories of violin playing (and quartet composition also) also be 'scientific'? Hinton's view was resoundingly negative, but he will have had in mind the reductionist and positivistic philosophies of science alluded to in ¶8.2, that began to gain sway in his day (the late Victorian period). Since then the word 'science' has been stuck on to the names of many activities to make them look good [Truesdell 1968, 75]. But this is often just rhetoric; is there philosophy there, too? Are other intellectual kinds (ethics, law, aesthetics, and so on) autonomous of each other? Popper's advocacy of norms, and their non-reducibility to (what he calls) facts, suggests such pluralism.[26] But how do we demarcate between these kinds, especially between science and the other kinds? Can there be a logic of sociological discovery, say? Does axiology serve in ethics as the correlate to ontology in science?

One negative answer would argue that scientific theories are concerned with things that are not man-made whereas theories in other areas depend entirely on man-made decisions and conventions; thus no ontology awaits the non-scientist. There is much force in this point; but we recall from ¶5.1 that several parts of science deal with man-made artefacts and materials. Continuing the theme of links between science and non-science, take the case of the sociology of science; under the view presented here, qua sociology it is not scientific, even though its subject matter is science (or more precisely, scientists and perhaps their science too).

Another negative answer would point to the difference between the universality of scientific theories and the particularistic character of theories outside science. However, a riposte was given in ¶4.1: scientific theories can be put forward in terms of trans-empiricism and modality without any claims of universality; and so can theories outside science.[27]

A third negative answer would stress that rationality is a common desideratum both in science and in other fields. Popper stresses the bearing

---

[26] See especially [Popper 1945], vol. 2, 234-237, 383-388 (this latter an addendum of 1961) and vol. 1, 292 on differences between public ethics and private aesthetics; and [McIntyre and Popper 1983] on medical ethics, which is more of a preliminary, metatheoretic outline. Among examples of scientific and non-scientific theories mingling together, [Nervi 1966] is a beautiful essay on mechanics, engineering and aesthetics as found in architecture.

[27] Thus I am dubious of the efficacy of Popper's distinction between the tasks of the 'theoretician' and the historian [1972, 354-357], in that the former tests universal laws against specific cases (as furnished by different sets of initial conditions, for example) while the latter investigates particular situations (such as the place(s) and period(s) of interest) drawing on universal laws, 'usually rather obvious ones' (such as scientific theories themselves). I do not see why historiography must be bereft of 'universal' laws: is not Popper's anti-historicism in *The poverty of historicism* [1957] an example?

of criticism upon rationality: indeed, his advocacy of it as a norm depends much upon this feature [1974b, 1036]. Further, another similarity with science arises in this context: while the rationality principle is false, it still serves as an efficacious guide to persons' rationality [Popper 1967], just as a knowingly refuted scientific theory can still be used for convenience if it is sufficiently accurate (or corroborable) for the purpose at hand.

I am somewhat reserved of the emphasis laid on rationality, since so many conventions operate in life that, while not arbitrarily chosen, are somewhat arational in character. In fact, I suspect that rationality itself is an intellectual kind — Bartley calls it 'metacontext' [Popper 1982, 122] — that needs its own criterion of demarcation, together with desimplification by study of its arational aspects. For example, the demarcation should relate to human capacities; after all, the behaviour of extra-terrestrials may be 'rational' to them but not to us (compare ¶12.4). Again, in sociology — but, interestingly, not in science — it has to cope with antinomies such as that of the one *vis-à-vis* the many (for example, it is rational for a society to conserve fuel resources, but also rational for any member not to bother, as he is only one among so many).

9.3 *Searching for criteria.* Now the negative answer may indeed be what is required: science is demarcated from non-science by the meta-property that ontological correctness is predicable only of scientific theories. However, the similarities and connections between scientific and non-scientific areas are marked enough that a positive answer feels worth seeking, especially as the negative one does not demarcate one non-scientific area from another.

Whatever form a positive answer might take, it would have to be general, and meta-epistemological in character. One possibility is the following. Popper has generalised the notion of refutation to a self-referring methodology of criticism [1945, vol. 2, 386-393 (1961 addition)], which has become known as 'pancritical' (or 'comprehensive critical') rationalism [Bartley 1983b]. This self-referring criticism applies both inside and outside science, and contrasts with the non-self-referring character of refutation in science; so maybe non-scientific areas need non-self-referring criteria of demarcation of their own. As I regard the search for such criteria as a particularly important prolongation of Popper's philosophy of science, and one in which ontological questions seem to be prominent, I try out a trio of cases in the next three sections.

10. On ignorance in historiography

> Only now could a real history of thermodynamics be written, since only in the last twenty years have the expressed aims of the creators of thermodynamics been achieved. In blunt terms, only now do we know a decent theory of the scope the creators sought, so only now can we see just where the old authors stopped short or even went wrong.
>
> Clifford Truesdell [1980, 3-4]

10.1 *Fallibilist historiography.* In many ways parallels between scientific and historical activity are clear, and need only summary notice here. The historian too has his empirical realm, with lost and destroyed materials lying beyond them; he also has to sort out his facts from his theories, and he may offer explanations of various kinds. Some of those may go beyond the empirical; for example, into the (controversial) area of psychohistory. He can indulge in desimplification when bringing in, say, family ties in his currently unsatisfactory study of political figures. The analogies extend even as far as the use of statistics, where the claims of cliometrics as "scientific" history have been debated in recent years [Fogel 1982].

Agassi [1963] has applied several of Popper's ideas to historiography, in criticism of inductivist, conventionalist and continuist types of history{; see also [Schupp 1975]}. My concern here is to focus on the status, if any, of ontology. Can there be some category to serve for history the role suggested in ¶7 for ontology in science? In this section I outline the candidacy of ignorance for the historiography of ideas, using the history of science as a test case; in a sequence of allied remarks constituting ¶10.5, I apply it to education also.

In philosophical questions in general, ignorance serves as the companion to knowledge, with both growing together (for example, a corroborated prediction of a fresh kind brings us not only its new knowledge but also awareness of our ignorance about other possibilities). Popper brings out well the intimate and complicated links between knowledge and ignorance; for example, in the introductory chapter 'On the sources of knowledge and ignorance' of [Popper 1963], where he criticises as simple-minded the empiricist view that the growth of the former necessarily causes reduction in the latter. What more can ignorance offer to historiography of ideas?

10.2 *History and heritage.* (§13) One of the most potent traps on the historiography of science is the view that the main historical task is to pose the question 'How did the present come to be?' and then answer it by tracing the history as a Royal Road to Here. Only the "relevant" parts of the past — problems, solution, methods, the lot — are described; and they are cast in terms of what the historical figures should have done had they followed our approaches. The quotation at the head of this section is an excellent example of this philosophy.

As a means of posing scientific problems, this type of modernising treatment can be excellent; but as an account of what happened, it is highly suspect. It often embodies a version of historical determinism (our arrival at the modern intellectual state of affairs was inevitable) and verificationism (our views must be right, because our predecessors tried to obtain them). In conversation H. J. M. Bos coined the excellent term 'unveilism' to capture these properties; it is a version of historicism, criticised by Popper in [1957].

Unveilism is also a form of relativism; for if our history reveals the Royal Road to Here, then history for our descendants is the Royal Road to There, and the two cannot be compared, even when they track parts of our common past. A similar remark applies to the history-is-bunk historiography, widely supported by scientists; for it follows that their own work will rapidly become bunk too. (But who are mere historians to comment?)

Unveilism also leads to a form of attack that I have sometimes called, rather disparagingly, 'the methodology of feedback criticism'. To speak in the most general terms, an historical figure initiates some new kind of intellectual theorising, but he develops it only to a very limited extent. It is enriched and extended over generations, and may even become a reasonably well-known doctrine. Then the unveilist uses that doctrine to attack the achievements of the historical figure, and thus to bite back at the hand that fed him.

10.3 *The status of ignorance.* In opposition to such views I stress the need to reconstruct the 'ignorance situation' of our historical figures, in which we emphasise not only what they achieved relative to their own prehistory but also what they did not achieve relative to what happened after their period but before ours, and of which they were wholly or partially ignorant. In particular, we need to reconstruct their unclarities, and to do so clearly.[28] When an ignorance situation is exposed, then new historical questions can be posed, of the type: how, when, and why was a particular part of that situation filled in?

In this approach the emphasis is laid on the historical figures' problems, not ours, and the character of the solutions that they found (compare [Popper 1972, 170-180]). We are freed from having to like, or justify, the past; on the contrary, one frequently dislikes what is found. We can see what options were available at a given time other than those that were actually taken up; we may even understand why, and not just how, certain opportunities were missed. We also have no need to accept that the present must judge the past, or that we have the correct yardstick by which the regrettable blunders of the past must be judged. In line with our aim in science to answer 'what happens?' (¶7.1), in history we give priority to the question 'what happened?' over the often preferred alternative 'how did the present come to be?'.

10.4 *History as metatheories.* The reconstruction of what happened (with, of course, its corollary of what did not happen) can be represented, as a task for the historian, as follows. Popper gives this schema for sequences of problem-solving in science:[29]

Problem (P) $\Rightarrow$ Theory (T) $\Rightarrow$ Error Elimination (EE) $\Rightarrow$ New Problem (P*) $\Rightarrow$ ....

---

[28] In this and many other contexts it is important to distinguish vagueness from ambiguity. Vagueness is often unavoidable: it attends the imprecision of experience discussed in ¶3, for example. Ambiguity concerns the muddling of distinctions, and its avoidance can be a valuable form of advance in theorising. The distinction between vagueness and ambiguity is itself such an example; people are often vague or ambiguous about it, or both.

[29] See, for example, [Popper 1972, 164, 176-177]; or [1974a, art. 29]. On the allied theme of his evolutionary epistemology, see ¶12.3.

It generalises for history to

Historian(s) H: $\quad\quad\quad P_H \Rightarrow \quad T_H \Rightarrow \quad EE_H \Rightarrow \quad P^*_H \Rightarrow \quad T^*_H \Rightarrow \ldots$
$\quad\quad\quad\quad\quad\quad\quad\quad\quad\quad\quad\quad\quad\quad\Downarrow$
Historical figure(s) F: $\quad\quad\quad P_F \Rightarrow \quad T_F \Rightarrow \quad EE_F \Rightarrow \quad P^*_F \Rightarrow \ldots$

The construction of the ignorance situation at stage P_F involves, among other things, appraising the difference between T_F and T^*_F and P_F and P^*_F. There may also be philosophical differences between H and F themselves: while H may work with fallibilism, he must allow for the possibility that F did not do so, but instead was (say) an inductivist. Indeed, when handling a group of Fs, H must note the variety of philosophies that they collectively exhibit. Indeed, he can turn it to good advantage by using a technique that I call 'the method of mutual illumination', where he reads some F_1 with the eyes of F_2 (and vice versa): clashes of view thereby arise, and can be a useful source of historical questions, especially concerning the reconstruction of each F's position.

The hierarchy illustrated above, and the attendant consequences, can be extended to the history of the history of science; to education, where the teacher formulates theories about the student's understanding of the subject matter (§13-§14); to the history of education; and so on. As mentioned above, different levels may employ different philosophies and regulatory principles, especially when the mixture of ¶9.2 of scientific and non-scientific areas is involved.

10.4 *'History-satire' in education.* The example of education, lightly raised above, needs further account, since ignorance has a status there of importance comparable to that in history. For a student starts out with an impressive measure of ignorance of the material that, by definition as a student, he is supposed to be learning. Indeed, one can point easily to cases where education imitates history, in that the historical record can be imitated in the classroom. For example, a curriculum design over several years of mechanics teaching could contemplate, say, this analogy:
Greeks $\Rightarrow\quad\quad$ Merton school $\Rightarrow\quad$ Galileo $\Rightarrow\quad$ Newton $\Rightarrow \ldots$
upper primary $\Rightarrow\quad$ lower secondary $\Rightarrow\quad$ sixth form $\Rightarrow\quad$ university $\Rightarrow \ldots$
Often such analogies are very profitable, although appeal has to be made to 'history-satire' [Grattan-Guinness 1973 = §14.4] in order to avoid excessive time being spent on local details of the past. For educational purposes the past is a bank to raid, not a church to worship in; but it is also not a useless museum. It can help transform the usual form of science education, with its premature and excessive imitation of formalised research-level theories, full of answers but devoid of questions.

The direction of these considerations is that one includes ignorance in forming an analogue to ontology for history and education. In this view histories are appraised, among other ways, in terms of their accounts of what did not happen, and educational strategies are judged by what they keep back from the students until certain stages. One would not treat scientific theories this way, so that a demarcation based on the place of ignorance may be put forward: historical and educational theories contain ignorance integrally in their content, whereas (for example) scientific theories do not.[30] This criterion is not self-referring, since ignorance is not an essential to it; thus the requirement of ¶9.4 is satisfied.

A criticism of this proposal is that an ontologically correct theory in the history of ideas, or in education, would have to allow for an indefinitely extended range of knowledge of which the historical figure (or the student) was (is) ignorant. But the same is true of ontologically correct scientific theories also, as follows from ¶7.5 on natural kinds: such a theory of the path of the pebble in ¶1.2 is silent about fat digestion in locusts (and so on). So indefinite ignorance is no handicap: indeed, it furnishes a way of understanding the old adage that 'historians are the only people who can change what has happened'.

11. On mathematical theories

On the other hand, pure mathematics and logic, which permit of proofs, give us no information about the world, but only develop the means of describing it.
Popper [1945, vol. 2, 13]

Several of the features of theories discussed earlier seem to apply to mathematics without too much difficulty: mathematical theories can be refuted by counter-examples (among other means); desimplification can be carried out, normal theories in Kuhn's sense (¶6.2) are developed; and in addition, the historical matters discussed in ¶10 are well illustrated by the history of mathematics. On the other hand, some unease arises when appending the word 'mathematical' to words such as 'truth', 'reality', 'fact' and 'discovery', for the considerations of ¶4 and 5 do not seem to carry over easily. My approach will be somewhat Hegelian: in ¶11.1 and ¶11.2 I shall exploit the similarities but face the differences in ¶11.3, and see if something "in the middle" can be worked out, in the cause of demarcating mathematical from non-mathematical theories.

11.1 *Structure-similarity.* Let us start the mathematicians' way and think of mathematics as covering the full range of theories with which they are collectively concerned: foundational and abstract stuff, general theories such as mathematical analysis and specific numerical methods, mathematics divorced from empirical interpretations and mathematics involved with them,[31] 'the lot'. I take up the 'neutral Platonist' position often favoured by mathematicians, and assume that the object and processes under study exist in some sense. However, mathematicians usually regard the matter as dealt with: that is, if these things do not exist, then too bad. I use ¶6.1 to interpret

---

[30] The metatheoretic character of this statement is very important; it has nothing to do with (say) scientific (specifically, psychological) theories of ignorance.

[31] I avoid the standard adjectives 'pure' and 'applied' mathematics, since the term 'applied' is too broad: one can apply arithmetic to hydrodynamics, say, but also to abstract algebra. The preferable phrase 'mixed mathematics', referring to mathematical theories mixed with empirical ones, was current for part of the 18th and 19th centuries.

the position as one concerning reification in a mathematical theory and ask: in which ways do the considerations of ¶6 and 7 apply, or not apply?

Consider the use of a mathematical theory M in a non-mathematical theory P that reifies a certain sub-universe in the way(s) outlined in ¶6.1. I call 'structure-(dis)similarity' the meta-properties that M and P do (do not) share the same ontology and structure (§11). For example, there have been many cases of dissimilarity in mathematical physics, in which P drew on some atomistic philosophy of matter but where the calculus was used in its Eulerian form in which that matter was reified as a collection of slices of infinitesimal thickness. Again, a solution to a differential equation may take the form of a sum (say); but whereas the sum is interpretable in P, its individual members may not be.[32] To take another form, one may accept with ¶7.8 that liquids and solids are different natural kinds and still use the same differential equations or expressions in mathematised studies of each. The use of approximative and numerical methods [Bachelard 1927, ch. 12] raises similar issues, in that the approximating theories may or may not display structure-similarity with the corresponding physical interpretations.

I stress cases of dissimilarity here, because they show that *double reification* can be involved in mathematicised theories — one for P, a different one for M — in a way from which non-mathematical theories are free. This situation is often overlooked; for example, it seems to be the cause of the physicists' habit of referring to a mathematised theory in physics as a (mere!) mathematical abstraction as if a non-mathematical theory in physics was not an abstraction also.

11.2 *Testing mathematical theories.* Whether or not the worlds reified by P and its associated M are the same, there is an important difference between them; that if P is refutable by testing, but only the use of the world for M in P is refutable also. As Popper says above, mathematics gives us no information about the world; for example, if two glasses of water are poured together into a third, it does not follow that $1 + 1 = 1$ and therefore that arithmetic is refuted, but only that arithmetic was not the mathematics appropriate to study that phenomenon.[33]

Similar points attend what might be called 'statisticised' theories. In many cases one uses statistics because of lack of details of some state of affairs, or at least the impracticability of obtaining them: thus one reifies a simplified universe for the (scientific or non-scientific) theory, and uses

---

[32] Fourier series provide particularly interesting examples (§11.6). Joseph Fourier himself came across them in connection with heat diffusion, but he did not regard their trigonometric form as an obligation to adopt a wave theory of heat. By contrast, the series can be well interpreted in acoustics theory, as super-harmonics.

[33] Popper gives a similar example in *Conjectures and refutations* [1963, 211]; but in the previous paragraph he 'applies' $2 + 2 = 4$ to apples in a series in which he holds that '"2 apples + 2 apples = 4 apples" is taken to be irrefutable and logically true'. I do not understand this sense; perhaps it hinges on a particular method of defining 'apple'. Grattan-Guinness [1987] {and now [1990 *passim*} contains an historical survey of using mathematics in physics, using a scenario in which structure-similarity (§11), and also desimplification (¶6.1), play important roles.

statistics, among other reasons, to appraise its degree of approximation to those actual states. But there may be situations in the world: according to [Popper 1982a] the behaviour of fundamental particles of matter is actually statistical. In these cases structure-similarity between the statistics and the theory can obtain; indeed the statistical theory may even be ontologically correct, and so refer (compare ¶7.3). But the statistics, as such, still remain empirically irrefutable.

11.3 *The place of nominalism.* This irrefutability of mathematics grounds the claim that mathematics is a non-scientific area (so that the possibility of demarcating it from science is a genuine question to ask). It is also the reason why mathematics needs metamathematics, whether formulated in a strict Hilbertian style or in looser terms. Now the questions raised in this section are often regarded as belonging only to proof theory, mathematical logic and set theory. But this activity, usually called 'philosophy of mathematics' since its eruption in the early decades of this century, is rather narrow in its range (while, of course, fascinating in its details[34]); in particular, it is largely silent about almost all the mathematics envisioned in ¶11.1.

However, one approach within this tradition has broken through the usual confines to offer a most unusual and stimulating attempt to demarcate mathematics. Field [1980] tries to articulate a nominalistic philosophy of mathematics (potentially "all" mathematics, indeed) in which, in a spirit compatible with Popper's quotation at the head of this section, questions of truth in mathematics are rejected, the empirical interpretation of mathematical theories is taken as a prime issue, and the view is advocated that no specifically mathematical objects exist at all. Thus questions of 'mathematical truth', facts, and discovery do not have to carry the weight that they bear in science, so that the words are used purely conventionally. Further, a demarcation can be suggested: if a mathematical theory M is added to some body of nominalistic knowledge B, no nominalistically expressible consequences can be drawn that could not be obtained from B itself alone, whereas when a scientific theory is thrown into an intellectual discourse of which it is not already part, it will be essential to the production of some new consequences [Field 1980, 9-12]. As required by ¶9.3, this criterion of demarcation is not self-referring.

Field sketches out a treatment of mathematics from set theory through arithmetic, geometry, and space to various aspects of Newtonian dynamics and related topics. The details are very clever, too intricate to be summarised here. My reservations lie in the reliance that seems to be made on axiomatisation, of both mathematics and physical theories: for most mathematics is not like that, and the cautions of Kurt Gödel's incomplet-

---

[34] For example, [Quine 1969] provides a splendid survey of the rings changed on various axiomatic set theories and their logics. Mathematicians have some awareness of these studies, but often only enough to muddle the distinction between unity and inter-connectedness stressed in ¶7.5 and assert mathematics is 'unified' (somewhere in this foundational stuff). Even the most 'rigorous' modern mathematics falls far short of the requirements of modern logical theory [Corcoran 1973].

ability theorem are not the only reason to wonder if it could be captured in nominalistic terms. Further, the demarcation of mathematics from logic (usually first-order predicate logic with identity) is not clearly made (and the concerns of ¶5.2 with logics are not raised).[35] Nevertheless, his efforts are greatly to be applauded, and maybe a union of some form of nominalism and an enriched and general theory of structure-(dis)similarity will indicate how it is that mathematics 'only develop the means of describing' the universe (§11-§12, §21).

Finally, the assumption that arithmetic is a basic component of mathematics can be questioned. Presumably it comes from the generally stable character of the Earth (as was noted in ¶3.2); by contrast, a form of sentient being who "lived" in the photosphere of the Sun, say, would take something like topology as basic and use arithmetic as a rather erudite branch of mathematics interpretable only in special circumstances such as counting the heavenly bodies. But are they no less 'rational' (¶9.3) about their mathematics than us?

12. Popper's three real worlds

> I am not in the fortunate situation of a mineralogist, who shows his listeners a mountain crystal. I cannot give a thought into the hands of my readers, with the request that they consider it quite precisely from all sides.
> Gottlob Frege [1918a, 66]

Popper's Kantian emphasis on man's capacity to think up theories has led him to concentrate in recent years on the mind and its activities, and propose his theory of Worlds 1, 2 and 3. Interesting criticisms of this theory have been made, from various points of view, in [Cohen (L. J.) 1980] and [Place 1979]. After some discussion of the various forms that Popper has granted it, I relate it to certain of the considerations presented earlier in this chapter, and its status as a prolongation of his philosophy of science.

12.1 *Popper's definitions of his Worlds* vary somewhat. The most frequent form states that World 1 is that of physical objects, World 2 of subjective experiences, and World 3 of 'the products of the human mind' [1972, 106; 1974a, 181; 1977, 16, 38]. But these clauses need some elaboration; and when it arrives, unclarity or even contradiction arises. For example, World 1 is 'the world of physics [...] chemistry and biology' rather than the referents of theories within these disciplines [1982b, 114]; World 2 contains 'all kinds of subjective experiences, including sub-conscious and unconscious experiences' [*ibidem*] but elsewhere is 'the world of our con-

---

[35] A propos of the demarcation of logic from non-logic, see Schroeder-Heister's reconstrual [1984] of Popper's studies of the foundations of deduction in the 1940s, which were unsuccessful for the purposes posed [Popper 1974, art. 27]. Apart from the question of logics, the demarcation itself involves positions over whether logic deals with truth and/or information and/or deduction and/or logical consequence ([Corcoran 1969]: compare [Bartley 1982, 174-183] on 'checking' logic).

scious experiences' [1972, 74]; and World 3 is both 'the world of objective contents of thought, especially of scientific and poetic thoughts and works of art' and 'the world of the logical [sic] content of books' and of similar artefacts [1972, 106, 74]. Further, Worlds 1 and 3 overlap, since 'Many World 3 objects exist in the form of material bodies, and belong in a sense to World 1 and World 3' [1977, 38].

Popper's rather liberal use of the word 'real', noted in ¶3.4, extends still further in connection with his theory of Worlds 1, 2 and 3; for 'not only are the physical World 1 and the psychological World 2 real but so also is the abstract World 3; real in exactly that sense in which the physical World 1 of rocks and trees is real' ([1982a, 116]: note that World 3 is now abstract). He describes this statement as a 'truth'; thus, by virtue of his use of Tarski's theory of truth described in ¶5.2, it corresponds to a fact. In tune with the cautions over the use of 'fact' uttered in ¶5.3, I would prefer to regard it as a theory awaiting test.

Popper emphasises the ubiquity of reality by defining it in the passage just cited: 'I propose to say that something exists, or is real, if and only if it can interact with members of World 1, with hard [sic], physical bodies'. However, under this definition, all objects of any kind are real; for any member of any of his three Worlds can — that is, has the potential to — interact with physical bodies (I ignore his qualifier 'hard'), whether for the refutation or the corroboration of a theory.[36] But by his own philosophy, if a theory says everything then it says nothing: so here, if every thing is real, then reality is not worth having.

12.2 *World 2* poses the special question: whose world is it? I have my own such experiences, and you have yours; but yours could be my World 1 objects of study (as your psychologist, for example). Hence World(s) 2 (mine, yours, everybody's) is a subset of World 1, characterised from its complement by its special role as the sole means of communication between members of that complement and members of World 3.

This view of World(s) 2 surely squares with the character of subjective experiences, even those such as hallucinations of a purely subjective type: that one can try to objectify them, even though, of course, special difficulties are involved (as with Dalton's colour-blindness, noted in ¶3.2). For example, there are various cases in medicine of scientists testing out drugs on themselves. Popper's interpretation of 'the testing of experiments [...] as a fundamentally objective procedure in which subjective experiences play a

---

[36] In response to this remark J. Shearmur made the interesting suggestion that reality be denied to the ultimate reified substances of a theory, even though all the subsequent components and the objects (of all kinds) attending its testing would be real under Popper's definition; he gave the example of phlogiston as a non-real type of object. However, let X be the substance and/or processes in the ontologically correct theory T of combustion to which phlogiston theory was a refuted guess. Then it follows from this suggestion that X is not real either; for the corroborations involved with T, like the corroborations and refutations of phlogiston theory, involve 'interaction with members of World 1'. But one can hardly deny reality to X, when 'X' refers.

totally inferior role' [1974b, 969] can admit of some exceptions. Indeed, they can be of at least two kinds, those linked with what Eccles calls the 'inner sense' and purely mental activity such as thoughts, feelings and imaginings and those involved in the 'outer sense' of perception, which draw on the normal sensory means and also secondary qualities such as colour and sound [Eccles 1977, 359-360].

12.3 *The Worlds in Popper's epistemology.* In addition, Popper's theory of three Worlds relates closely to his evolutionary epistemology, which has its genesis (as it were) in his book manuscript of the early 1930s [1979, esp. ch. 4] but received fuller exposition only much later (especially [1972, ch. 7; 1983b; and *passim* elsewhere]. Campbell [1974] provides a splendid critique of this theory and its history before Popper; Bartley [1976, 480-490] suggests that Popper was ignorant of much of this background, but elsewhere advocates its general utility for philosophy [Bartley 1983a, 846-853]; by contrast, Ruse [1977] criticises Popper's understanding of evolutionary theory itself. Possessing such ignorance myself in impressive proportions, I confine my reservations to these two remarks that search for the limits to the analogy between evolution and epistemology. Firstly, are there not features of knowledge, such as radical changes of theory, which have no image in evolutionary processes? Secondly, what in evolution corresponds to the search — and even finding — of truth in Popper's version of fallibilism (or to ontological correctness in mine)?

12.4 *Sites for the Worlds.* For me World 1 consists of the universe of physical objects and states, at all levels of phenomenal accessibility, inanimate and animate. The latter includes the biologically and biopsychologically animate,[37] and also the physical components involved in what is called 'consciousness'.[38] The former include (for example) books as physical objects; as carriers of intellectual content, they have only potential World 3 status, requiring actuation by means of minds, through linguistic appreciation and (in case of paintings, sculpture, and music) semiotic or phenomenal apprehension also.

World 3 still comprises the products of human minds;[39] the World of problems, theories as such, languages (including mathematics, and neo-languages such as music), meditation, philosophy, metaphysics, Popper's theory itself, and so on. However, there seems to be nothing to gain from

---

[37] I assume here, fallibly but inessentially, that there are objects that are biologically animate but psychologically inanimate. This anti-panpsychistic position receives support from Popper [1977, 67-71].
[38] I support the claims for the existence of consciousness — indeed, of consciousness in various states — partly for the reasons advocated by Popper (see, for example, [1974a, art. 39] or [1977 *passim*]), for certain features of psychical phenomena [Grattan-Guinness 1982, especially chs. 3, 6, 20 and 23-25; and §16-§19].
[39] For Popper the capacity to develop and retain language is a necessary condition for entry into World 3, so that animals cannot access it [1977, 438-442]. On the amazing variety of forms of sensory perception and awareness that are being found in animals see [Griffin 1981], a book notable also for its attacks on behaviourism.

following Popper and admitting into this World objects that have not yet been thought up; indeed, this clashes with his conceptual realism noted in ¶3.5, sets up further difficulties for the analogies with evolutionary theory noted in ¶12.3, and trivialises (or avoids) questions concerning ideation.

Much more attention needs to be given to the structure of the World. For example, the extent to which it has to copy World 1 cognitively needs study. Again, on its linguistic side, theories of denoting are required, in order to distinguish reference (or reification) from naming. In addition, the distinctions of theory from metatheory, epistemology from metatheory, and so on, will supply much wanted structure. And the world 'real' needs careful taming, broadly along the lines of ¶3.5.

These two Worlds are disjoint, but they are brought together by the activity of minds. Popper [1982c] claims as the principal novelty of his theory the role of minds to act as traffic gates between Worlds 1 and 3, and there seems to me to be no need to claim more and raise these minds into a collective World 2.

The modes of interaction between Worlds 1 and 3 seem to be extremely varied. For example, to take Eccles's distinction of ¶12.2, processes involving the outer sense pertain to World 1 while possibly also provoking World 3 ideas, while those from the inner sense relate largely to World 3 (although the mental activity itself belongs to World 2, as usual).

Theories of the mind form a particularly important group of examples of interaction between World 1 and World 3 objects. These theories can be factually true or false, ontologically correct or incorrect, and so on. (In particular, if we think up the ontologically correct theory T of thinking by means M, then $T = M$; this is how self-reference works here.) Theories of other minds have the same character, although the interaction is much more complicated than the usual hierarchy of theory, metatheory, and so on; for when several thinkers think about each other, they might also think of each others' thoughts of each other, so that their individual hierarchies embrace. Testing of such theories is a rather complicated process, especially when it involves the impossibility of self-prediction [Popper 1982b, 62-77]. Theories of interaction between human and extra-terrestrial intelligences are still more tricky, as they raise questions of detecting (the equivalents of) languages, comprehending others' 'rationality' (¶9.2), and so on.[40]

Finally, where is *memory*? 'There can be very little doubt that memory is essentially physiological and brain-based', wrote Popper on this last point [1977, 485], presumably on the usual argument that removing parts of the brain causes loss of memory. But Rupert Sheldrake points out

---

[40] L. J. Cohen raises the question of the inter-accessibility of World(s) 3 between humans and humanoid intelligences [1980, 171-177]; Brams [1983] sketches some fascinating scenarios for our interaction with them using game theory. Lem [1983] weaves a delightful tale around the semioticist's problems: if a signal arrives here from outer space, how can we tell if it is a phenomenal process and/or a 'message', and how do we distinguish the content of the 'message' from the 'linguistic' apparatus needed to encode it? As Lem says, the problem resembles the tenor breaking the wine-glass whatever he happens to be singing about ([1983, 95]: compare ¶4.2).

the weakness of this type of argument: according to it, the objects shown on a television screen must be inside the set, since they can be eliminated by tampering with the workings [Sheldrake 1981, 122].

In addition, the boundaries between memory and other mental processes are not sharp. For example, does one remember, or calculate, that 2 x 2 = 4? Or 3 x 7 = 21? Or 7 x 13 = 91? Or ...? Of course, in some cases one might do both without knowing which 'got there first', as it were.

12.5 *Prospects.* Popper's theory of Worlds 1, 2 and 3 is perhaps the most speculative part of his philosophy, elevating a commonplace of life into a rather grandiose metaphysics. In this section I have largely confined myself to a modified presentation of it. While the general lines of intent of the theory are clear, the details are rather difficult to articulate. In particular, it is surprisingly difficult to apply to it other aspects of Popper's philosophy (whether in his form or the variant outlined in earlier sections here): the apparatus of metaphysical research programmes, refutability, corroborability, truth, desimplification and the like stands back a little nonplussed. While irrefutable, the theory may be true; it could even be ontologically correct; but it seems somewhat difficult to talk about.

Acknowledgements

For opportunities to air some of these ideas in lectures, I am indebted to the Northern Association of Philosophy, the Society for Psychical Research and the British Society for the Philosophy of Science. I have also enjoyed profitable correspondence and/or discussion at various times with J. T. Blackmore, the late Mrs. Anita Gregory, and Sir Karl Popper.

Bibliography

Agassi, J. 1963. *Towards an historiography of science*, The Hague: Mouton.

Andersson, G. 1978. 'The problem of verisimilitude', in G. H. Radnitzky and Andersson (eds.), *Progress and rationality in science*, Dordrecht: Reidel, 291-310.

Bachelard, G. 1927. *Essai sur la connaissance approchée*, Paris: Vrin. [Repr. 1968, 1969.]

Bartley, W. W. 1976. 'The philosophy of Karl Popper. Part I. Biology and evolutionary theory', *Philosophia, 6*, 463-495.

Bartley, W. W. 1982. 'The philosophy of Karl Popper. Part III. Rationality, criticism and logic', *Philosophia, 11*, 121-221.

Bartley, W. W. 1983a. 'The challenge of evolutionary epistemology', in *Absolute values and the creation of the new world*, New York: International Cultural Foundation Press, 835-880.

Bartley, W. W. 1983b. 'The alleged refutation of pancritical rationalism', in *Ibidem*, 1139-1179.

Bhaskar, R. 1978. *A realist theory of science*, 2nd ed., Hassocks: Harvester.

Blackmore, J. T. 1979. 'On the inverted use of the terms "realism" and "idealism" among scientists and historians of science', *British journal for the philosophy of science, 30*, 125-134.

Born, M. 1978. *My life. Recollections of a Nobel Laureate*, London: Taylor and Francis; New York: Scribners.

Brams, S. J. 1983. *Superior beings. If they exist, how would we know?*, New York: Springer.

Bronowski, J. 1974. 'Humanism and the growth of knowledge', in [Schilpp 1974], 606-631.

Cohen, I. B. 1976. 'The eighteenth-century origins of the concept of scientific revolution'. *J. hist. ideas, 37*, 257-288.

Cohen, L. J. 1980. 'Some comments on third world epistemology', *British journal for the philosophy of science, 31*, 175-180.

Corcoran, J. 1969. 'Three logical theories', *Philosophy of science, 36*, 153-177.

Corcoran, J. 1973. 'Gaps between logical theory and mathematical practice', in M. Bunge (ed.) *The methodological unity of science*, Dordrecht: Reidel, 23-50.

Crozet-Fourneyron, M. 1924. *Invention de la turbine ...* , Paris: Beranger.

Eccles, J. C. 1974. 'The world of objective knowledge', in [Schilpp 1974], 349-370.

Eccles, J. C. 1977. Contributions to [Popper and Eccles 1977], 225-421, 425-568.

Elkana, Y., 1984. 'William Whewell, historian', *Revista di storia della scienza, 1*, 149-197.

Field, H. H. 1980. *Science without numbers*, Oxford: Blackwell.

Fogel, R. W. 1982. '"Scientific" history and traditional history', in *Logic, methodology and philosophy of science VI (Hannover, 1979)*, Amsterdam: North Holland, 15-61.

Forchheimer, P. 1905. 'Hydraulik', in *Encyklopädie der mathematischen Wissenschaften mit Einschluss Ihrer Anwendungen*, vol. 4, pt. C, Leipzig: Teubner, 324-372.

Fox, R. 1974. 'The rise and fall of Laplacian physics', *Historical studies in the physical sciences, 4*, 81-136.

Frege, F. L. G. 1918a. 'Der Gedanke. Eine logische Untersuchung', *Beiträge zur Deutschen Idealismus, 1*, 58-77. [Repr. in [Frege 1977], 342-362. English trans. *Mind, 65*(1956), 289-311; repr. in P. F. Strawson (ed.), *Philosophical logic*, London: Oxford University Press, 1967, 17-38.]

Frege, F. L. G. 1918b. 'Die Verneinung. Eine logische Untersuchung', *Beiträge zur Deutschen Idealismus, 1*, 143-157. [Repr. in [Frege 1977], 362-378. English trans. *Translations from the philosophical writings of Gottlob Frege*, 2nd ed., Oxford: Blackwell, 1960, 117-135.]

Frege, F. L. G. 1977. *Kleine Schriften* (ed. I. Angelelli), Hildesheim: Olms.

Gower, B. 1973. 'Speculation in physics: the history and practice of *Naturphilosophie*', *Studies in the history and philosophy science, 3*, 301-356.

Grattan-Guinness, I. 1973. 'Not from nowhere. History and philosophy behind mathematical education', *International journal of mathematical education in science and technology, 4*, 321-353. [Parts in §14.]

Grattan-Guinness, I. 1979. 'On Popper's philosophy and its prospects', *British journal for the history of science, 12*, 317-337. [§9]

Grattan-Guinness, I. 1982. (Ed.), *Psychical research. A guide to its history, principles and practices*, Wellingborough: Aquarian.

Grattan-Guinness, I. 1984. 'On Popper's use of Tarski's theory of truth', *Philosophia, 14*, 129-135. [§6.]

Grattan-Guinness, I. 1987. How it means: mathematical theories in physical theories. With examples from French mathematical physics of the early 19th century', *Rendiconti dell'Accademia del XL*, (5)9, pt. 2 (1985), 89-119.

Grattan-Guinness, I. 1990. *Convolutions in French mathematics, 1800-1840. From the calculus and mechanics to mathematical analysis and mathematical physics*, 3 vols., Basel: Birkhäuser; Berlin: Deutscher Verlag der Wissenschaften.

Griffin, R. 1981. *The question of animal awareness. Evolutionary continuity of mental experience*, rev. ed., New York: Rockefeller University Press.

Haack, S. 1976. 'Is it true what they say about Tarski?', *Philosophy, 51*, 323-336.

Haack, S. 1978. *Philosophy of logics*, Cambridge: Cambridge University Press.

Hacking, I. 1983. *Representing and intervening. Introductory topics in the philosophy of natural science*, Cambridge: Cambridge University Press.

Hopkins, E. 1878. (Ed.) *Life and letters of James Hinton*, London: Kegan Paul.

Hoyt, W. G. 1980. *Planets X and Pluto*, Tucson, Arizona: University of Arizona Press.

Kern, L. H., Mirels, H. L. and Hinshaw, V. G. 1983. 'Scientists' understanding of propositional logic: an experimental investigation', *Social studies of science, 13*, 131-146.

Kuhn, T. S. 1962. *The structure of scientific revolutions*, 1st ed., Chicago: University of Chicago Press.

Laudan, L. 1983. 'The demise of the demarcation problem', in *The demarcation between science and pseudo-science* (ed. R. Laudan), Blacksburg, Virginia: Virginia Tech., 7-35.

Lem, S., 1983. *His Master's voice*, London: Secker and Warburg. [Polish original 1968.]

Margenau, H. 1974. 'On Popper's philosophy of science', in [Schilpp 1974], 750-759.

Medawar, P. 1967. 'Hypothesis and imagination', in *The art of the soluble ...*, London: Methuen, [ch. 9]. [Repr. in [Schilpp 1974], 274-291.]

Mulligan, K., Simons, P. and Smith, B. 1984, 'Truth-makers', *Philosophy and phenomenological research, 33*, 287-321.

Myers, F. W. H. 1903. *Human personality and its survival of bodily death*, vol. 1, London: Longmans, Green. [Cited from the 1954 printing.]

McIntyre, N. and Popper, K. R. 1983. 'The critical attitude in medicine: the need for a new ethics', *British medical journal, 287*, 1919-1923.

Nervi, P. L. 1968. *Aesthetics and technology in building*, Cambridge, Mass.: Harvard University Press.
Niiniluoto, I. 1978. 'Notes on Popper as a follower of Whewell and Peirce', *Ajatus, 37*, 272-327. [Repr. in Niiniluoto, *Is science progressive?*, Dordrecht: Reidel, 1984, 18-60.]
Place, U. 1979. Review of [Popper and Eccles 1977], *Annals of science, 36*, 403-408.
Popper, K. R. 1945. *The open society and its enemies*, 2 vols., London: Routledge and Kegan Paul.
Popper, K. R. 1957. *The poverty of historicism*, London: Routledge and Kegan Paul.
Popper, K. R. 1959. *The logic of scientific discovery*, London: Hutchinson.
Popper, K. R. 1963. *Conjectures and refutations*, London: Routledge and Kegan Paul.
Popper, K. R. 1966. *Logik der Forschung*, 2nd ed., Tubingen: Mohr.
Popper, K. R. 1967. 'La rationnalité et le statut du principe de rationnalité', in E. M. Chassen (ed.), *Les fondements philosophiques des systèmes classiques*, Paris: Payot, 142-150.
Popper, K. R. 1972. *Objective knowledge. An evolutionary apporach*, Oxford: at the Clarendon Press.
Popper, K. R. 1974a. 'Autobiography of Karl Popper', in [Schilpp 1974], 1-181. [Repr. as *Unended quest*, London: Fontana, 1976.]
Popper, K. R. 1974b. 'Replies to my critics', in [Schilpp 1974], 961-1197.
Popper, K. R. 1975. 'The rationality of scientific revolutions', in R. Harré (ed.), *Problems of scientific revolution*, Oxford: Oxford University Press, 72-101.
Popper, K. R. 1977. Contributions to [Popper and Eccles 1977], 1-223, 425-568.
Popper, K. R. 1979a. 'Is it true what she says about Tarski?', *Philosophy, 54*, 98.
Popper, K. R. 1979b. *Die beiden Grundprobleme der Erkenntnistheorie*, Tübingen: Mohr. [Manuscript of 1930-1933, edited for publication by T. E. Hansen.]
Popper, K. R. 1982a. *Quantum theory and the schism in physics* (ed. W. W. Bartley, III), London: Hutchinson.
Popper, K. R. 1982b. *The open universe. An argument for indeterminism* (ed. W. W. Bartley, III), London: Hutchinson.
Popper, K. R. 1982c. *Bücher und Gedanken*, Vienna. [Lecture delivered to the Austrian Book-trade Organisation. {English trans. in *In search of a better world*, London and New York: Routledge, 1992, 99-106.}]
Popper, K. R. 1983a. *Realism and the aim of science* (ed. W. W. Bartley, III), London: Hutchinson.
Popper, K. R. 1983b. 'Evolutionary epistemology', manuscript [kindly made available to me by Popper].
Popper, K. R. and Eccles, J. C. 1977. *The self and its brain*, Berlin: Springer.

Prince, M. 1927. *The case of Patience Worth*, Boston: Society for Psychic Research, 1927. [Repr. New York: University Books, 1964.]

Quine, W. V. O. 1961. *From a logical point of view*, 2nd ed., Cambridge, Mass.: Harvard University Press.

Quine, W. V. O. 1969. *Set theory and its logic*, 2nd ed., Cambridge, Mass.: Harvard University Press.

Quine, W. V. O. 1970. *Philosophy of logic*, Englewood Cliffs: Prentice Hall.

Rescher, N. E. 1980. *Induction. An essay on the justification of inductive reasoning*, Oxford: Blackwell.

Ruse, M. 1977. 'Karl Popper's philosophy of biology', *Philosophy of science, 44*, 638-661.

Russell, B. A. W. 1946. *History of Western philosophy*, London: Allen and Unwin.

Schilpp, P. A. 1974. (Ed.), *The philosophy of Karl Popper*, 2 vols., La Salle, Illinois: Open Court.

Schroeder-Heister, P. 1984. 'Popper's theory of deductive inference and the concept of a logical constant', *History and philosophy of logic, 5*, 79-110.

Schupp, F. 1975. *Poppers Methodologie der Geschichtswissenschaft. Historische Erklärung und Interpretation*, Bonn: Bouvier.

Shearman, P. D. 1981. *Colour vision in the nineteenth century* ..., Bristol: Hilger.

Sheldrake, R. 1981. *A new science of life* ... , London: Blond and Briggs.

Truesdell, C. A., III. 1968. *Essays in the history of mechanics*, Berlin: Springer.

Truesdell, C. A., III. 1980. *The tragicomical history of thermodynamics, 1822-1854*, New York: Springer.

Turing, A. M. 1954. 'The chemical basis of morphogenesis', *Philosophical transactions of the Royal Society of London, B237*, 37-72.

Wolf, C. J. E. 1889. 'Introduction historique', in (ed.), *Mémoires sur le pendule*, pt. 1, Paris: Gauthier-Villars, i-xliii.

Yoshida, R. 1982. 'Reduction as replacement', *British journal for the philosophy of science, 32*, 400-410.

3 The Place of the Notion of Corroboration in Karl Popper's Philosophy of Science

While Popper's philosophy of science is dominated by falsification and criticism of theories, he does allow that a theory may pass a test and thereby be 'corroborated', although not confirmed, by the evidence associated with the test. However, his treatment of corroboration is rather cursory; an elaboration is attempted here, to cover situations in which, for example, small effects are crucial and so the boundary between corroboration and falsification is not sharp.

1. Corroboration in Popper's philosophy

The main text of Popper's *Logik der Forschung* (1935) consists of seven chapters outlining the main features of his falsificationist philosophy of science, followed by two involving probability theory and quantum mechanics, and finally one on corroboration, which is the main concern here. I shall cite his English translation of the book [Popper 1959], noting its newer material when appropriate.

The chapter begins with a section on non-verifiability of theories that world have been better placed in chapter 6 on testability, and then two sections on the probability of events and of hypotheses and on probability logic, which belong rather in chapter 8. Only then come four short sections on corroboration, the property of a theory that it passes its tests. The chapter is placed last because of a link between corroboration and probability; Popper even claimed that there was no difference between the two in that both are appraisals [1959, 265]. In a later appendix *ix, and in more detail in other work [1982a, 240-255] he offered quantitative measures of corroboration involving probabilities, but at the end of the latter text he almost dismissed the enterprise; and in similar vein I shall not consider it further, since the measures depend upon the tests that happen to have been carried out and so are not very significant. Otherwise he did not discuss corroboration very much in his writings, and it is not well handled by commentators;[1] but this rather low status contrasts markedly with the actual practice of science, where scientists and engineers regularly work with (seemingly) highly corroborated theories, even to the extent that 'reliability science' is a subject in its own right.

This article was first published in F. Stadler (ed.), *Induction and deduction in the sciences,* Dordrecht: Kluwer, 2004, 251-262; it is reprinted by kind permission of Springer Science+Business Media.

[1] For example, the four articles in [Schilpp 1974] that treat corroboration seem to be among the weakest in the book; Popper's replies are often unavoidably dismissive [1974, 993-999, 1037-1048].

I shall argue that corroboration deserves more attention in falsificationist discussions than it often receives, and that interesting philosophical and educational questions arise beyond its supposed relationship to verification, or contrast with it, in scientific induction. I shall use Newtonian mechanics as my main example, as it is an important theory whose main features and applications should be reasonably familiar to the reader. Thus there is a skew towards the physical sciences and technology; but I am sure that the points made apply also to the natural and human sciences.

2. The content and status of corroboration

The basic tenets of corroboration held by Popper are these. It is a qualitative measure of the success that a theory has achieved in passing tests; that is, that the predictions made from it are borne out by experience or experiment, at least as well can be determined under the conditions and precision of the tests and experimenters and observers. It has no cumulative or inductive character at all; repeated success has no bearing whatsoever on the outcome of the next test, nor is falsifiability compromised. Thus it differs fundamentally from verification as handled in inductivist philosophies of science;[2] indeed, when considerations of the probability calculus are brought to bear upon corroborability, *im*probability increases. It is a methodological notion, to be distinguished from the verisimilitude of a theory, which is a semantic notion that records the extent to which its predictions do (not) correspond to the pertinent facts.[3] As with every other procedure envisioned in this philosophy, including falsification, corroboration can be fallible.

In a later footnote to his book Popper suggested that for a theory T that enjoys a range of domains of applications, corroboration be construed not as a property of T but as a relation between T and a domain within this range [1959, 412]. This was a good modification: to take an example to which I shall soon return, Newtonian mechanics is far more highly corroborated in celestial mechanics than in engineering mechanics. Sadly he rejected his modification at once, but I shall adopt it when it is useful, not only for corroboration but also for testing in general: a theory is (or is not) corroborated in some domain of application. I also adopt the distinction stressed in [Bogen and Woodward 1988] between data needed to provide inferences about the phenomena at hand, and the phenomena themselves

---

[2] Most inductive epistemologies have been cumulative; but some are eliminative, in claiming that when all other theories have failed, the one left must be true. An interesting foray of this kind is [Nicod 1924], who in its course grasped some features of the logic of falsificationism [see esp. ch. 2]; however, his intention was quite different from that of Popper, for whom, of course, all theories could be false.

[3] Popper once wondered if greater corroboration of a theory may go along with its increased verisimilitude [1963, 234]; but he never intended the former to *measure* the latter [1974, 1011].

I note without discussion the well-known difficulties about the definition of verisimilitude, as my main concern lies in its basic difference from corroboration For discussion, see [Fetzer 1981, esp. pt. 3], [Dilworth 1981, esp. ch. 3] and [Oddie 1986]; and for a review of the more recent theories, see [Kuipers 1987]. The distinction is often missed in other philosophies of science.

from which falsifying or corroborating evidence for the theory is drawn. I take the discussion below to apply to theories descriptive of the phenomena under study and those that attempt to offers explanations of them.

The low status of corroborations in falsificationism is well exemplified by a philosophical problem that was very influential on young Popper in the formation of his philosophy. In 1919 Newtonian mechanics was found to provide incorrect calculations of the advance of the perihelion of the planet Mercury and of the refraction of light passing the edge of the Sun, and much of the discrepancies between prediction and calculation from observations were found by using general relativity theory, which thereby was given a great boost [Jeffreys 1919]. But the following feature of the reaction is rarely noted. There was much concern with this discrepancy *only because Newtonian mechanics had already achieved a massive catalogue of corroborations in celestial mechanics; moreover, the two falsifications were irrelevant to almost all of them.* Had this situation not obtained, the discrepancies would have aroused little interest. This point applies even if relativity theory had not had been corroborated by the new data, though the corroboration added greatly to its prestige.

Despite this setback, Newtonian mechanics was and is still very widely used. This shows that *we can still use falsified theories in domains of application away from those where falsification has happened* [Popper 1959, 53-54], in this case in terrestrial and planetary mechanics and even in various other parts of celestial mechanics.

Corroboration also plays an important role in Popper's claimed solution to 'the problem of induction' [1972, ch. 1]. In fact he claims to solve two problems: the epistemological one, where he asserted that no guarantee exists of invariance or regularities from the past to the future; and a methodological problem, where he advised that if competing theories are available for a choice of a course of action we should choose the 'best-tested theory'. But this is a serious slip of the pen; the context makes clear that he meant the best *corroborated* theory. However, even that version of his advice meets difficulties, for critics have pointed out that theories that propose an ad hoc modification of the original theory T at some give future moment of time N are just as well corroborated (and tested) up to N as is T itself, and Popper had to resort to special metaphysical pleading for theories to be invariant over space and time. However, the logical analysis of the propositions involved in testing shows that an alternative logical layout of them can be proffered in which such special pleading is avoided (§4).

3. The status of normal science

Falsification corresponds to the testing *to destruction* of a theory. Now this is very important, just as in the way that, say, a car manufacturer should test his models to destruction in order to assess their robustness and behaviour in accidents. But I buy a car to fulfil much less extreme tasks, and hopefully always safely. In the same way, successful scientific theories gain most of their importance from the corroborations that they (seem to have) achieved, not from their ultimate destruction. Like many other philosophers, Popper tended to focus upon the 'great moments' of science such as that of 1919, when falsification stared a major theory in the eye and a competitor stood waiting in the wings; but this focus is rather naïve, resembling the approach of historians who emphasize only great Events and the Great Men involved in them. Most of the time in science, philosophy and history, far less crucial processes are taking place; but they may well involve major theoretical and/or experimental difficulties.

These remarks are congenial with Thomas Kuhn's characterization of normal science [Kuhn 1962]: paradigms, or several of the many types of them that Kuhn proposed, set the style of theorizing for many ranges of application of the science concerned, and corroboration is normally expected. When some major 'revolution' (or maybe an innovation) takes place, then patterns of basic thinking in this context has to shift. While finding Kuhn's account *least* useful in explaining the structure of revolutionary changes, I accept his advocacy of the *Gestalt* nature of the change: to continue with the above example, after relativity theory the relationship between space-time and velocity and acceleration had to change, and one works either in the old Newtonian mode or in the new one but not with the two together.

The warnings given in [Popper 1970] of the 'dangers' of normal science are well taken, in particular the degeneration of science into mere puzzle-solving, to use Kuhn's particularly unfortunate word in [1963, ch. 2].[4] Other dangers include scientists' elevation of a favourite theory into a quasi-religion in which belief gains an excessive status, and the complacent dismissal of data as 'anomalous' when they are naughty enough to falsify it. Popper's position was reinforced by his view of science as the study of nature, *not* of consensus among the experts about its character, a view that Kuhn also affirmed.

On the other hand, Popper's attack on normal science once again demoted corroborations, and the positive sides of Kuhn's conception are greatly under-estimated. Those who dismiss normal science as robotic exegeses of given principles are profoundly mistaken (they might try to learn some). In cases familiar to me it is hard work in all respects, creative and experimental. Notions and theories subordinate to the main theory can still require deep new notions, leading maybe to new sub-theories normal within their domains, and so on further (for example, from Newtonian mechanics to hydrodynamics to the same for viscous fluids to the same at high temperatures).

4. Science, technology and reliability

The special bearing of corroboration upon technology (including engineering) needs to be emphasized. I make no distinction between theories and tests being applied to natural phenomena or to man-made materials and artefacts. A theory in chemistry is under test whether it involves, say, pure common salt or some artificial compound; likewise, mechanics can be applied equally to the motions of Mercury or to the motions of a pendulum clock.

The main bearing upon corroborations is as follows. In Popper's view science is a risk-taking enterprise, where theories are formed and then

---

[4] The misfortune arises from the fact that puzzles, as normally understood, are exercises within a *finite,* though maybe large, space of possibilities, and solutions may be found by exhaustive inspection. But rarely are even the routine uses of science finite in this sense.

tested as severely as possible. Now science and technology have a very close relationship, and yet technology requires reliability in the performance of its product. *Thus science is risk but technology is safety* — a "paradox" of which the resolution requires careful attention to be paid to corroborations.

An important concern in technology is determining the values of parameters involved in the specification of materials and the design of artefacts; typically, the breaking strength of the types of steel used in a building, and the consequent efforts to ensure that it is not exceeded. Decisions taken here may well be affected by further factors, such as cost [Rescher 1978, esp. pt. D].

Reliability and safety in technology are sufficiently important in technology to have given rise to 'reliability theory'. This subject treats the properties and (in)frequencies of failure of materials, equipment and artefacts, and the strategies for their maintenance and replacement. Largely an interaction between probability with statistics and technology, some of its features were heralded by the introduction in the 1930s of quality production control (see especially [Shewhart 1931]). It was launched in a large scale after the Second World War, being inspired by many technological novelties and production problems that had been faced then [Barlow 1984]. It makes much use of statistical distributions, including some little-known elsewhere; in particular the Weibull, which was originally formed to express the breaking strength of materials but has since been used in many other reliability contexts.

Reliability theory has become a quite wide-ranging subject, taught in many technology courses.[5] However, it seems to be presented as a collection of techniques, without reference to philosophical consequences; and conversely, it seems to be little known to philosophers of science. Yet the distinction between the two concerns is only one of practise, not of knowledge as such: the failure of a light bulb at the 737th switch-on is just as much a falsification of its theory of action as is the motion of Mercury a falsification of Newtonian mechanics. Further, the successes of the previous 736 switch-ons are corroborations of its theory of action just as much as those enjoyed by Newtonian principles elsewhere in mechanics. Other points of contact with the philosophy of science include the use in both subjects (though not usually by falsificationists) of Bayesian techniques; and the reliability expected, but maybe not delivered, of instruments and machines used in testing scientific theories.

The role of probability in reliability theory seems to be philosophically neutral — hence the lack of philosophical discussion in its literature. The distributions are usually applied to assessing repetitions of action or behaviour, sometimes using large runs of data; thus the most natural philosophical interpretation of its theories is frequentist. But its theorems are equally intelligible to, say, a Popperian advocate of probabilities as propensities. Reliability, strong corroborations (up to now, anyway); none of these

---

[5] For a good survey of the main statistical techniques deployed, see [Ebeling 1997]. The influence of [Shewhart 1931] has been substantial in its own right, especially for its role in the ideas on production reliability and management promoted after the War by W. Edwards Deming.

features of theories in science or technology lies outside the philosophical remit [Ziman 1978].

Reliability theory also takes due note of unreliability: failures in technology, which involve falsifications of theories. These failures are of different kinds: among others, deliberate failure, as in built-in obsolescence of equipment; vulnerability in design, as in the quick collapse of the towers of the World Trade Centre in New York after an attack that imposed conditions on the buildings far beyond their designed limits; and faults in preparation of materials, as that of the rivets in the sinking of the *Titanic*. The task of determining the cause(s?) of failure can be complicated and even controversial, as in science.

5. The philosophy of small effects

The case of the perihelion of Mercury involves another feature of testing theories that can occur in quantified scientific theories of all kinds and that makes the contrast between corroboration and falsification less straightforward than it appears: the discrepancy was an *extremely small quantity*. Two features are worth noting.

Firstly, as mentioned earlier Newtonian mechanics was already highly corroborated, so that the discrepancies did not *obviously* raise grave issues; maybe some modified details (of perturbation theory for Mercury, say) could resolve the issues. In that test, examining the 'next decimal place' [Richtmyer 1932] led to major consequences; but usually this is not the case, and it hard to see how a scientist could tell in advance whether or not a major difficulty is at hand.

Secondly, while in this case the quantity involved did corroborate one theory and falsify another one,[6] in many cases the magnitude of the discrepancy might fall within the range of experimental, observational and human imprecision pertaining to the test, thus preventing any clear conclusion to be drawn. Imagine a situation in which the precision of observation and experiment is only around 10^{-3} in some unit of relevant measurement, and the discrepancy between a prediction by T and the experimental data is a little larger than 10^{-3}. Has T been refuted or not? It is hard to say, and the usual strategy will be to collect more data — while recognising that *exact* repetition cannot achieved [Popper 1950] — and maybe even effect some statistical assessment of range of data obtained. In situations where the theory under test is itself statistical, Popper advocated assessment involving likelihood [1959, 410-419].

6. The place of truth in Popper's philosophy

An important event for Popper was his meeting with Alfred Tarski in 1935 when he learnt of Tarski's semantic correspondence theory of truth, which took the truthhood of T away from the perplexing epistemologies of belief and assertion and placed it instead in the safe haven of the metatheory of T. The distinction between object theory and metatheory became and remained central to Popper's work; indeed, some of his best contribu-

---

[6] I do not discuss here the non-trivial 'Duhem-Quine' thesis about the underdetermination of theories by observations.

tions are metaphilosophical, such as arguing for indeterminism and against determinism [Popper 1982b].

Popper adopted this theory of truth in harness with his emphasis upon the truth-*content* of theories. However, in thrusting it to the centre of his concerns he gave it an epistemological weight that Tarski always eschewed ([Grattan-Guinness 1984 = §6], with special reference to [Popper 1972, ch. 9]). Popper seems to have thought it to be necessary and sufficient for his purposes; but in fact further assumptions are required for sufficiency to be achieved. The assignment of correspondence between fact and proposition is itself a convention. Again, the restriction of truth to bivalency somewhat restricts the applicability; pace Tarski's own example 'snow is white', perception of colours is hardly a yes-or-no matter, and bivalency imposes a discontinuity upon a continuous spectrum.

More seriously, Popper's emphasis on the facts to which true propositions correspond sounds like empiricism on the rise: theories are appraised *merely* on the resemblance of their predictions to the pertinent facts, without concern for their contents, which may differ radically (such as in our example of Newtonian mechanics versus relativity theory). This restriction does not square with Popper's important point that scientists *think up* theories, which may invoke notions that transcend experience. (Ever felt a Newtonian central force?) It also underrates the unavoidable role of theories both in the expression of facts themselves and in the decision to regard them as relevant to the theory in the first place. Again, it mitigates against his advocacy of common-sense realism but rejection of common-sense epistemology [Popper 1972, ch. 2]. It is consistent with philosophies of science to which Popper does not subscribe, such as proposing that science in general is progressive, and perhaps inevitably so.[7]

In my view, there is *no* contradiction in the situation where a theory T_1 has all the corroborations of T_2 and more of them and yet is also further away from the actual state of affairs (whatever they are). Suppose, for example, that light actually is a waval phenomenon: then particulate theories are more distant from actuality than waval ones. Yet before the arrival of A. J. Fresnel in the mid 1810s, particulate theories had provided the more elaborate and testable theories in optics for a century.

Popper's scheme

Problem — tentative theory — error elimination — new problem [1972, esp. pp. 242-244] seems vulnerable to the same doubts. Error is to be eliminated at the level of correction of falsifications by new theories that (hopefully) gain fresh corroborations; but the theoretical structure and assumptions of the old and new theories are not considered. Part of the difficulty here arises from his regrettable silence, in all of his philosophical writings on science, over the *formation of scientific problems in the first place* — in particular, that it is subject to meta-problematic criticism. In particular, it led to him to an overly narrow conception of the role of experiments and observations in science as being subservient to theory in merely providing the means of determining corroboration or falsification

---

[7] For discussions of cumulation see [Bachelard 1928, esp. bk. 3], [Niiniluoto 1984, esp. chs. 4-6] and [Dilworth 1981, esp. ch. 5].

[1959, 106-111]. In real scientific practise, however, experimental work may set up the problem in the first place, and the relationship with theory becomes much more dynamic [Franklin 1986]. For example, why did a certain experiment on an acid not work when the temperature of the acid was very low? Does this failure also obtain with other acids? In such a way a problem is created.

A preferable approach distances Popper's falsificationism from the empiricist elements just noted. It preserves Tarski's correspondence theory of truth, but starts from Tarski's own appraisal of it rather than Popper's: that it requires a realist ontology but otherwise is *neutral* over epistemological assumptions.[8] Truth and falsity are confined to the factual record, of passing or failing tests. Theories as such are held to be 'ontologically (in)correct' when they do (not) correctly describe the way (whatever it is) in which the phenomena actually happen [Grattan-Guinness 1986 = §2.7]. Of course we shall never know if a theory is correct, because the next test could lead to a falsification; this is the mate of Popper's distinction between truth, possession of truth, and proof of possession of truth.

Concerning the formation of scientific theories I lay emphasis upon *reification*. The scientist is often faced with a vastly complicated situation S in which the phenomena are taking place; so in order to set up a soluble problem [Medawar 1967] he simplifies S to s and treats it *as if*, for example, the environment is at constant temperature, the Earth is not rotating, the phenomena are taking place in a vacuum, and so on — factors that hopefully have only small effects upon S in the sense discussed above.[9] Then (let us hope) he formulates a theory T about testable s, which is falsifiable at least within the current apparent limits of experimental accuracy. If success follows, in the form of corroborations and/or resolvable falsifications of T, then he puts back some of these factors in a sequence of *desimplifications*: allow first for, say, air resistance somehow, then the rotation of the Earth, then maybe its precession and nutation, then.... Thereby s gradually returns towards S via successively more developed versions of T, with the simplifications reduced though probably not eliminated.[10] This sequence of theories could be *locally* progressive in the sense of achieving comparably higher measures or precisions of corroboration, at least in some domains(s) of application; but a *general* progressionist philosophy of science is not

---

[8] Thus the confirmationist philosopher Rudolf Carnap was just as impressed with Tarski's theory as was Popper; on the historical context see [Wolenski and Köhler 1999, esp. pts. 1-3].

[9] There is no necessary link between this 'as if' philosophy and that advocated by Hans Vaihinger from the 1910s, especially not his relativism [Vaihinger 1911]. However, exploration of similarities may be worth undertaking.

[10] An outstanding example of successive desimplifications is the history of the so-called 'simple pendulum', which is anything but simple when many decimal places are required from the observations. Numerous small(ish) factors were considered: not only air resistance and rotation of the Earth but also flexure and torsion of the suspending wire, the non-planarity of the swing, possible differences in time of the up- and down-swings, possible angular momentum at the end of each swing, flexure of the supports of a large pendulum, ... [Wolf 1889-1891].

entailed. Indeed, different desimplifications may work better in different domains of application of T.

Another benefit of working with reification is the role needed for Popper's 'World 3' of abstract objects such as problems, theories as such, mistakes, and so on [1972, esp. ch. 4; 1994, chs. 2-3]. The notion itself is quite legitimate — Popper was not the first to advocate it — but his version has been justly criticised (for example, in [Cohen 1980]). The doors of World 3 seem to be too wide open, for anything can enter it. One reason for this liberality may be the construal of truth (and also verisimilitude) to the predictions of theories without regard for their respective contents. But when reification is admitted, then one grants membership to notions (of all kinds) only when they are required by a problem, theory or test; then its population can be reined in to some extent.

7. Some remarks on science education

In order to be a scientist, almost certainly education will be required. As is well known, science education can be a very off-putting experience to the student, not only or necessarily because of the inherent theoretical and/or experimental difficulties but also due to the theological way in which scientific theories are presented: corroboration raised to the level of unquestioned truth, normal science becoming factory-like science, science portrayed as a non-human activity. This manner of presentation seems particularly regrettable but partly unavoidable at school level, when the science conveyed is inevitably rather elementary, and therefore less likely to be falsifiable. There is a great problem of balancing the need for skills with the excitement of original creation, even at a modest level [Ravetz 1972]. Popper's philosophy may sometimes be mentioned, but the normal reaction to the failure of a student to corroborate a theoretical prediction is to claim that he made a mistake somewhere — and indeed the claim is almost certainly correct!

The philosophy of science education is a vast topic, and I make only two points here. Firstly, the notion of corroboration should be emphasised, and distinguished from verification: at least some basic points are conveyable, especially that theories can be well corroborated in at least some domains of application, and so are well worth learning! Secondly, there have been quite strong initiatives in many countries to import the history of science into teaching, where the difficulties and even blunders of the past can be conveyed sympathetically along with the corroborations, and the human element introduced [Matthews 1994].

Now falsificationism (with Popper's extension, critical rationalism) possesses a nice point of contact here, denied to many other philosophies. For in that philosophy a theory exists at all because it is the outcome of treatment, especially falsifications, of previous theories. (Even in a completely new field of study at least analogies from established theories were probably used, fallibly of course, as sources of inspiration for its theories.) Therefore falsificationism and history fit together well, although the student does not need to learn all the historical details and complications.[11] Popper himself was interested in the history of science to a degree most unusual in a philosopher of his (or our) time; and while this was in part a personal inclination, it was surely also inspired by its natural place in the kind of philosophy that he advocated.

8. Concluding remark

Some major philosophies rest upon platitudes, which the inaugurating philosophers realised contained profound and general insights. Tarski's semantic theory of truth, for example, based upon the distinction between a language and its metalanguage, is grounded upon the platitude that we have to stop our work sometimes sand reflect upon its purpose and content. (The logician Kurt Gödel is also to be credited with this distinction, which was also partly grasped by some predecessors.) Popper's philosophy is based upon the platitude that we learn from our mistakes, an assertion of naïve fallibilism that he built into falsificationism and then on to (comprehensively) critical rationalism. The theme of this paper hangs upon another platitude; that we learn from our successes.

Bibliography

The works marked '*' have not been cited above, but they valuably guided the considerations presented.

Bachelard, G. 1928. *Essai sur la connaissance approchée,* Paris: Vrin. [Several reprints.]
Barlow, R. E. 1984. 'Mathematical theory of reliability: a historical perspective', *IEEE Transactions on reliability, R-33,* 16-20. [Followed on pp. 21-133 by articles on the teaching of reliability science, and on its development in various areas of technology.]
Bogen, J. and Woodward, W. 1988. 'Saving the phenomena', *Philosophical review, 97,* 303-352.
Cohen, L. J. 1980. 'Some comments on third-world epistemology', *British journal in the philosophy of science, 31,* 175-180.
Dilworth, C. 1981. *Scientific progress,* Dordrecht: Reidel.
Ebeling, C. E. 1997. *An introduction to reliability and maintainability engineering,* New York: McGraw Hill.
Fetzer, J. 1981. *Scientific knowledge. Causation, explanation, and corroboration,* Dordrecht: Reidel.
Fleck, L. 1935. *Entstehung und Entwicklung einer wissenschaftliche Tatsache: Einführung in die Lehre vom Denkstil und Denkkollectiv,* Basel: Schwabe. [English trans.: *Genesis and development of a scientific fact,* Chicago: University of Chicago Press, 1979.]

---

[11] In advocating roles for history in mathematics education I advocate the method of 'history-satire', where the broad historical record is respected but most of the barnacles are avoided or only treated lightly [Grattan-Guinness 1973 = §14]. The same kind of procedure could work for the various sciences, though maybe in less detail than is possible for mathematics.

Franklin, A. 1986. *The neglect of experiment*, Cambridge: Cambridge University Press.
Gjertsen, D. 1989. *Science and philosophy. Past and present*, London: Penguin.*
Grattan-Guinness, I. 1973. 'Not from nowhere. History and philosophy behind mathematical education', *International journal of mathematical education in science and technology*, 4, 421-453. [Parts in §14.]
Grattan-Guinness, I. 1984. 'On Popper's use of Tarski's theory of truth', *Philosophia*, 14, 129-135. [§6]
Grattan-Guinness, I. 1986. 'What do theories talk about? A critique of Popperian fallibilism, with especial reference to ontology', *Fundamenta scientiae*, 7, 177-221. [§2]
Hacking, I. 1983. *Representing and intervening. Introductory topics in the philosophy of science*, Cambridge: Cambridge University Press.*
Jeffreys, H. 1919. 'On the crucial tests of Einstein's theory of gravitation', *Monthly notices of the Royal Astronomical Society*, 30, 138-154. [Repr. in *Collected papers*, vol. 6, London: Gordon and Breach, 1977, 176-192.]
Kitcher, P. 1993. *The advancement of science. Science without legend, objectivity without illusions*, New York and Oxford: Oxford University Press.*
Kuhn, T. S. 1962. *The structure of scientific revolutions*, Chicago: University of Chicago Press]. [2nd ed. 1972.]
Kuipers, T. van 1987. (Ed.) *What is closer-to-the-truth?*, Amsterdam: Rodopi.
Matthews, M. R. 1994. *Science teaching. The role of history and philosophy of science*, New York and London: Routledge.
Maxwell, N. 1998. *The comprehensibility of the universe*, Oxford: Oxford University Press.
Medawar, P. B. 1967. *The art of the soluble*, London: Methuen. [Paperback reprints.]
Medawar, P. B. 1969. *Induction and intuition in scientific thought*, London: Methuen.*
Nicod, J. 1924. *Le problème logique de l'induction*, Paris: Alcan. [English transs.: *Foundations of geometry and induction* (1930) and *Geometry and induction* (1969), London: Allen and Unwin.]
Niiniluoto, I. 1984. *Is science progressive?*, Dordrecht: Reidel.
Oddie, G. 1986. *Likeness to truth*, Dordrecht: Reidel.
Perutz, M. 1991. *Is science necessary?*, New York and Oxford: Oxford University Press.*
Popper, K. R. 1950. 'Indeterminism in quantum physics and in classical physics', *British journal in the philosophy of science*, 1, 117-133, 173-195.
Popper, K. R. 1959. *The logic of scientific discovery*, London: Hutchinson.
Popper, K. R. 1963. *Conjectures and refutations*, London: Routledge and Kegan Paul.
Popper, K. R. 1970. 'Normal science and its dangers', in *Criticism and the growth of knowledge* (ed. I. Lakatos and A. Musgrave), Cambridge: Cambridge University Press, 51-58.

Popper, K. R. 1972. *Objective knowledge. An evolutionary approach.* Oxford: at the Clarendon Press.
Popper, K. R. 1974. 'Replies to my critics', in [Schilpp 1974], 961-1197.
Popper, K. R. 1982a. *Realism and the aim of science,* London: Hutchinson.
Popper, K. R. 1982b. *The open universe. An argument for indeterminism,* London: Hutchinson.
Popper, K. R. 1994. *Knowledge and the body-mind problem* (ed. M.A. Notturno), London: Routledge.
Ravetz, J. R. 1972. *Scientific knowledge and its social problems,* Oxford: Clarendon Press.
Rescher, N. 1978. *Scientific progress. A philosophical essay on the economics of research in natural science,* Oxford: Blackwell.
Richards, S. 1987. *Philosophy and sociology of science. An introduction,* 2nd ed., Oxford: Blackwell.*
Richtmyer, F. K. 1932. 'The romance of the next decimal place', *Science, 75,* 1-5.
Schilpp, P. A. 1974. (Ed.), *The philosophy of Karl Popper,* 2 vols., La Salle, Illinois: Open Court.
Shewhart, W. A. 1931. *Economic quality control of manufactured product,* London: MacMillan.
Vaihinger, H. 1911. *Die Philosophie des Als Ob. System der theoretischen, praktischen und religiösen Fiktionen der Menschheit auf Grund eines idealistischen Positivismus. Mit einem Anhang über Kant und Nietzsche,* 1st ed., Berlin: Reuther & Reichard. [English trans.: *The philosophy of 'as if': a system of the theoretical, practical and religious fictions of mankind* (trans. C. K. Ogden), London: Routledge and Kegan Paul, 1924.]
Waddington, C. H. 1977. *Tools for thought,* London: Cape. [Paperback reprints.]*
Wolenski, J. and Köhler, E. 1999. (Eds.) *Alfred Tarski and the Vienna Circle,* Dordrecht: Kluwer.
Wolf, C. 1889-1891. (Ed.) *Mémoires sur la pendule, précédés d'une bibliographie,* 2 vols., Paris: Gauthiers-Villars.
Worrall, J. 1989. 'Why both Popper and Watkins fail to solve the problem of induction', in F. d'Agostino and I. C. Jarvie (eds.), *Freedom and rationality,* Dordrecht: Kluwer, 257-296.
Ziman, J. 1978. *Reliable knowledge. An exploration of ground for belief in science,* Cambridge: Cambridge University Press.

4 Karl Popper and the 'Problem of Induction': a Fresh Look at the Logic of Testing Scientific Theories

For the centenary of Sir Karl Popper (1902-1994)

A major source of controversy surrounding Popper's philosophy of science is his claim to have solved the problem of induction by showing negatively that inductive inference is impossible; for critics felt that his assumptions about the circumstances surrounding the testing of theories admitted induction by the back door. I argue that both Popper and his critics were addressing the wrong problem, and that when the correct problem is handled, Popper's position is vindicated.

1. Problems of induction, and corroboration

Popper formulated his philosophy of science in the late 1920s largely concerned with two problems [Popper 1930]. One was to demarcate science from non-science, and there he proposed falsifiability (synonymously, refutability) as the necessary and sufficient condition for a theory to be scientific. The other problem was that of induction, where the traditional view held that theories are proven to be true by the accumulation of 'confirming' instances, or at least that the achievement of this happy state becomes steadily more probable. Popper rejected this view entirely, claiming that no such certitude could be guaranteed: instead, falsification was the key, and theories remained conjectural, however well confirmed. (Thus science is the activity of writing wrongs.) Popper's claim has met much opposition; I support it here, although in a modified form. To do so I focus upon the logical form of statements involved in the testing of theories, and also on the notion of corroboration (Popper's replacement for confirmation), two rather neglected aspects of his philosophy.

There is no single problem of induction; as we shall soon see, Popper himself claimed to have solved two. I note quickly a third one, of great importance to scientists, concerning strategies for framing problems and theories in the first place. These strategies may resemble induction in, for example, wondering if similarities between, and/or statistical inference across, collections of data or observations suggest common causes or properties among their kind; but of course no inductive *import* attaches to these strategies or proposals. Indeed, scientists aware of Popper's philosophy

---

This article was first published in *Erkenntnis, 60* (2004), 109-122; it is reprinted by kind permission of Springer Science+Business Media. In view of §3, the account of corroboration has been reduced.

sometimes invoke it in such contexts, where it fits nicely. Ironically, Popper himself did not stress them; in his scheme

Problem — tentative theory — error elimination — new problem (1)

he took the theory to be *already formed,* and concentrated upon the interpretation of tests [Popper 1972, 119, 242-244]. It is a pity that he did not consider these creative sides of science; as one consequence he took observations and experiment to be subservient to theory, whereas the interaction between these three categories is much more dynamic (see, for example, [Franklin 1990] in the case of modern physics; and §12).

Popper's problems of induction lay elsewhere; and his solutions of them drew upon corroboration, discussed in detail in §3. If the prediction of a theory T is found actually to obtain, then T has 'passed its test' and been 'corroborated' [Popper 1959, 33, 262-275]. But his criticism of induction remains in force; to it we now turn.

2. Popper's solution to 'the problem of induction'

Classical inductive theory reveals no clear grasp of the *critical* function of experimentation.
     Peter Medawar [1969, 34]

Predecessors on this problem, such as David Hume, usually cast it in subjective terms, concerning past experience perpetuating (or not) into the witnessed future; but an objective interpretation was surely intended, and Popper emphasised this aspect by framing the problem thus: 'Can the claim that an explanatory universal theory is true be justified by "empirical reasons"; that is, by assuming the truth of certain test statements or observation statements (which, it may be said, are "based upon experience")?' [Popper 1972, 7; see also 1974, 1013-1020]. Thus the question was clearly epistemological; and he answered it in the negative, for reasons analogous to Hume's: no finitude of confirming instances, however great, can jump across the divide to achieve universality [1959, ch. 1].[1] Further, inductive epistemology is not itself founded on induction (as some enthusiasts may have thought) but is metaphysical [Popper 1959, 253-253; 1974, 1020-1027]. As for probability theory, increasing the content of a theory *decreases* its probability in the sense of a calculus [1963, 218-221]. By contrast to all this, a false test statement reveals the falsity of the attendant universal theory, thanks to the *modus tollens* rule of inference [1959, 75-77].

Popper also formulated a second problem of induction, which is methodological in character: 'Can a *preference*, with respect to truth or falsity, for some competing universal theories [sic] over others ever be

---

[1] Popper was attacking the *enumerative* version of inductive epistemology; there have been a few advocates of *eliminative* induction, who assert in Sherlock Holmes style that, when all competitors have been eliminated, the lone survivor theory must (ought to?) be true. The processes involved in (allegedly) finding it resemble falsification in striking ways (see, for example, [Nicod 1924, esp. ch. 2]). The Hume-Popper answer can be modified to criticise this version also.

justified by such "empirical reasons"?' [1972, 8; see also p. 21]. He stressed that *'there can be no good reasons'* in the sense of a guarantee underlying any choice; success could just be lucky. With this caveat, his answer was to choose 'the best-tested theory' [p. 22, repeated in 1974, 1025]; but this must be a slip, and the surrounding discussion shows that he intended to choose the best *corroborated* theory. The notion of 'best-tested theory' is methodological, dependent upon the time at which the various competing theories are compared, and at a given time the best-tested theory is not necessarily also the best corroborated one; an old but much falsified theory could have a much larger test record than a very promising but new theory. Popper also assumed that one theory would be better tested than all the others; as we shall soon see, this assumption was challenged.

Popper has found few supporters for his claims; one is [Miller 1982, 2002], who has offered rebuttals to several criticisms. One group of them states that Popper admits induction by the back door in various ways; we shall note an important possible case below. Another criticism is based upon the remark that theories created by modifying a general theory in certain ad hoc ways are just as corroborated as the original, so that his solution to the methodological problem fails. I shall take the example cast in [Worrall 1989], for it captures the issues involved.

Consider two persons on the top of the Eiffel Tower who plan to descend to the ground. Galileo's daughter Virginia (my character) believes theory G, which asserts that gravitation is working as usual according to father's law of fall; so she plans to use the lift, although she is aware of possible failures of the technology. But Professor Floater believes in a modified theory M(N) of G, according to which from some instant N of time soon to come gravitation will weaken considerably; so he plans to wait until then and then jump off the Tower and float down safely to ground. G and M(N) are equally corroborated so far; so which theory should we prefer over the other?[2]

Popper dismissed such modifications as not being falsifiable independently of themselves [1972, 15-16]; but that does not prevent M(N) from being true and G false. To adapt Charles Babbage's argument against Hume's rejection of miracles, ad hocness with respect to one state of theorizing and background knowledge may be a deductive consequence relative to another state.[3] A similar point can be made regarding the principle of

---

[2] The situation is different regarding a spatially modified theory, where G is asserted except that gravitation is always weaker in some (maybe tiny) region around the Earth. Although falsifiable, it very likely hard to falsify.

[3] For Babbage's argument see [1837, esp. chs. 10-12]. In line with his belief in the continuity of nature, Hume claimed that it was more probable that witnesses to a miracle lied about it, or were mistaken in some way, than that it could have happened; but Babbage retorted with examples using his calculating engine. For example (my own), set to generate integers by adding 3 on each turn of the handle, his engine produced sequences such as 2, 5, 8, 11, 45, 48, 51,.... This would be a miracle to the witness — but not to the fiendish Babbage, who had arranged the wheels of his engine so that it jumped to 45 at the fourth turn.

causality, which asserts that events are always effects from causes: a phenomenon that is sudden relative to one theoretical observer or context can be an anticipable effect to another one (for example, a heart attack sudden to the patient but no surprise to his doctor; or, on a larger scale, 11 September 2001 for New Yorkers and for certain other citizens of the world).

Presumably with ad hocness in mind, Popper required of a theory to be scientific not only that it be falsifiable but also that it contains '*a simple, new and powerful idea*' [1963, 241];[4] and any theories proposed after instant N should try to explain the change as well as the preceding regularity [1959, 253]. Some comments are required. Firstly, the demand for newness is irrelevant; it is the novelty of the use that matters. Secondly, as conventionalist philosophers found out long ago, articulating the notion of simplicity in general is a very complicated exercise; in particular, Popper's own identification of simplicity with degree of falsifiability [1959, 140-142] is neither convincing heuristically nor necessary to his philosophy, which is not conventionalism or pragmatism [1959, 79-81].

Thirdly, is the powerful idea still weak enough to escape the clutches of induction? Apparently chosen as an example of a general idea, Popper invoked the 'principle of the continuity of nature', which he construed to be a 'metaphysical interpretation of a methodological rule' that theories 'are to be invariant with respect to space and time; and also if we postulate that they have no exceptions' [1959, 253]. While a methodological rule is not to be confused with a causal law or any kind of inductive rule of inference, this pair of assumptions seems to involve rather reckless metaphysics: the first one, claiming invariance of theories, surely carries a whiff of induction, while the second, on lack of exceptions, may even be falsifiable. Some alternative move is worth seeking; and a closer look at the logic of testing theories may help.

3. The logic and context of testing theories

As stated at the start, part of Popper's attack on inductive epistemology was to insist that any scientific theory be universal in cast: that is, a theory must *start off* universal. If universally quantified, it ought to be laid out as an implication, that is, be conditional in the logical form 'if A, then B', where A and B are statements (or propositions); and his examples made clear that this form drew not only on the statement calculus but especially on the predicate calculus with quantification [1959, 62].[5] Thus, for example, G would start

for all x, t,... if x is a heavy body, then at any instant of time t... (2)

with G stated as a relationship between x, t, and maybe other variables. M(N) commences with the existential quantifier:

---

[4] This requirement distinguishes Popper's 'dressed' from 'bare' falsificationism in [Maxwell 1998], where metaphysical issues in science are well highlighted.

[5] This is one of the few places in *The logic of scientific discovery* where logical form is made manifest. Most of Popper's talk of 'logic' concerns the logic of the situation: while a well-established use of 'logic', it is quite distinct from formal logic where details of the pertaining situations are irrelevant, or almost so.

There exists a specific instant N of time t such that if $t < N$, then G; but if

$t \geq N$, then the same law applies but with weaker gravitation.[6] (3) A quantitative measure or estimate of the weakness might be adjoined to make M(N) more testable.

To analyse the problem of induction, the *full context* of testing a scientific theory needs examination. Further theories of various kinds are thereby involved; some of them are so "obvious" that normally they are taken for granted or assumed implicitly. They comprise the following categories:

Metaphysical assumptions about the universe. These are intrinsically unfalsifiable theories. For example, in an extension of falsification into rational criticism of such theories [Popper 1982a, 1982b] argued well for realism against idealism and for indeterminism against determinism; but if you are a solipsistic determinist, then so be it.

General assumptions about the universe, such as the principle of the continuity of nature and the demand for no exceptions mentioned above; *but in this analysis they are not taken to be metaphysical*. Unless the theory under test is intrinsically statistical in character, another assumption of this kind is the principle of causality, which Popper also took to be metaphysical [1959, 61-62, 246-248].

Any additional desimplifying scientific theories. In this case, as the Eiffel Tower is very high, we might consider the effect of air resistance on G and on M(N).

Simplifying assumptions knowingly made in forming problem and theory. In this case the rotation of the Earth can be ignored, along with all its minor perturbations.

Any *mathematical and/or statistical theories as such* that may have been used: their own reliability regarding rigour of proofs, reasonableness of their empirical interpretations, and so on.

Local assumptions about the circumstances of the test and the attendant predictions. Among many examples, we assume that no meteorite is about to hit the Eiffel Tower, that Paris is not due for an earthquake, that the poles of the Earth are not about to reverse, that the camcorder recording the jump is in working order, that Professor Floater is not endowed with the remarkable but controversial gift of levitation, and that the experimenters shall live long enough to complete the test (the subsequent fate of Professor Floater is another matter).

All these various theories are to be conjoined logically to G; similarly, all of them are conjoined to M(N) except for the principle of the continuity of nature, since we always want the full set of theories to be consistent. (Should any inconsistencies emerge among the various premises come to light, then some modifications will have to be made among the corresponding theories.) In each case the result is a jumbo statement, which I shall call 'Conj', made up of a conjunction of many sub-statements, each

---

[6] Since the value N of t is specified, (2) is falsifiable; were no such specification given, then (2) would be unfalsifiable [Popper 1959, 68-71].

one either in categorical or in conditional form. Then the details of the test are fixed and predictions are made. These latter are singular statements that are obtained by specifying (ranges of) values of variables in the appropriate sub-statements; for example, for G that 'at N + 10 seconds Professor Floater will have descended 50 yards' while for M(N) 'at N + 10 seconds Professor Floater will have descended much less'.[7] At instant N the test will start.

This schema is well enough known, although normally less fuss is made about all the assumptions involved. Now for the alternative formulation: some logical moves are available that better highlight the hypothetical character of testing and also allow change in the status of assumptions such as the principle of the continuity of nature. The details of the moves, and the attendant elaborations, are sketched in the final section of this paper.

Firstly, *the jumbo conjunction of conditional statements can be converted into an implication of jumbo conjunctions*, which takes the conditional form 'if A, then B', where A and B are each a long string of conjunctions. A contains all the antecedents of the conditional statements specifying ranges of the variables such as 'if x is a heavy body' from (2), and maybe also some other preliminary information: B has all the consequents, especially the articulation of a theory such as G in (2). Secondly, any categorical statements are conjoined together to form one categorical statement C that is adjoined both and to A and to B, giving the alternative jumbo statement

$$\text{if (C \& A), then (C \& B)} \qquad (4)$$

(where '&' denotes 'and'), which I shall call 'Imp'. From it (C & A) is asserted, and (C & B) deduced via the *modus ponens* rule of inference.[8] Then the singular test statements are formed and conditions for the test itself specified as indicated above, largely by making specifications on (C & B).

My proposal is that *testing is much better understood in the form Imp of an implication (4) of assertions and conjunctions rather than in the traditional form Conj of a conjunction of assertions and implications*. Two groups of reasons may be put forward.

Firstly, Conj is to be understood as laying out all *the circumstances* under which the test will be carried out: the testing situation itself is grasped by Imp, which captures much more strongly its hypothetical character; that is, *if* (theory) *then* (testable consequence). In Imp all the

---

[7] This example is oriented to quantitative tests, when fixing values to variables is apposite. In many cases, especially but not only in the life sciences and medicine, the test may involve important qualitative and/or indeterminate components, and thereby be less precise (such as, say, when will the monkfish move, and will it approach the bait?). However, the logical moves proposed here still apply.

[8] Also available, as a logical equivalent to (4), are

{if C, then [if A, then (C and B)]} and {if A, then [if C, then (C and B)]}

(*3·3, *3·31). But neither form has the same quality of interpretation for the situation of testing.

premises of the conditional sub-statements are bunched together as part of the antecedent of the conditional, and all the consequences are conjoined in the consequent. Similarly, if falsification should obtain after test, then it would rebound through the negated (C & B) to the negated (C & A); and since each of those statements is a conjunction, then *every* sub-statement in C, A and B is to be questioned, as should be the case.

Secondly, on a point of logic, Imp is a logical consequence of Conj, and so is the logically weaker of the two statements. But Imp is *still strong enough to capture the needs of testing scientific theories*; indeed, the domination of implications shows it to be *appropriate* in logical strength for that purpose. By contrast, the dominance of conjunctions in Conj shows that it is *unnecessarily strong for testing,* and so requires the special pleadings and dubious assumptions discussed earlier. The switch to Imp removes them.

As far as I know, the important distinction between Conj and Imp has not been made before. Of course, the choice between the two is a decision or convention, and not itself true or false; but the arguments just rehearsed give cogent reasons for preferring Imp.

The change from Conj to Imp has basic consequences not only for understanding testing but also for the controversial metaphysical assumptions discussed earlier.[9] Consider in particular the principle of the continuity of nature. It may be summarised as

for all t and T, if t and T are time, and $t \neq T$,

then $S(t)$ if and only if $S(T)$,   (5)

where S is a compound predicate covering all features of the universe that are held to be invariant over (space and) time. (The precise specification of the components of $S(t)$ is a daunting and maybe controversial task; mercifully I need treat here only its logical form.) In the traditional scheme Conj it is one of the conditional sub-statements in the jumbo conjunction, and requires special pleading because it stands there on its own; but in Imp its antecedent forms part of A and its consequent part of B, just like any other conditional sub-statement. Thus it is treated no differently from any other theory, and there is *no* need for it to be given metaphysical status. The same point can also be made for other such principles, such as those of no exceptions, causality, and also over the need for 'simple and powerful ideas'; if you need them, then propose them, but do not demand special metaphysical status for them.

4. Avoiding induction

Let us compare G versus M(N) under the test. Assuming that no unexpected mistake occurred, then there are three possible outcomes. Firstly, the findings will be indecisive; perhaps a change in gravitation will be found, but so slight in measure as to lie within experimental imprecision.

---

[9] This point offers a *specific* avoidance of metaphysical assumptions; it is *not* made in the context of a general hostility towards metaphysics. On the contrary, one of Popper's most impressive feats was to make clear contexts in which metaphysical theories must be entertained, and also their relationship to unfalsifiability.

Secondly, one of the theories will be corroborated. Finally, at least one of the two theories will be falsified, perhaps both (the change in gravitation takes place at N+80 seconds, say); and then much detective work is required, since the negation of a conjunction is a disjunction of negations

not (this premise) or not (that premise) or not ... ,   (6)

and when the disjunction is jumbo as with C, A or B, the culprit theory or theories could take some finding.[10]

If M(N) is proven false again after so many other falsifications for values of t<N, then the modification itself will seem to be at fault; the principle of the continuity of nature is upheld, at least for now. But if G is falsified, then that principle, shorn of metaphysical protection, will be a prime target among the premises for rejection; and, as Popper mentioned above, theorists should then try to seek reasons for its failure. If gravitation is found suddenly to change at N, then the subjective shock that witnesses (literally) will feel is fascinating from many points of view, scientific, psychological and cultural; but the *epistemological and methodological issues* surrounding the principle of the continuity of nature are not affected: they remain in the objective realm.

More generally, it seems that we live our lives with a deeply ingrained habit of taking well-worn hypotheses to be established facts (hence enticing induction), and working always among a *plenum of continuous corroborations*, such as the principle of the continuity of nature. But the unexpected sometimes happen, even if rarely; indeed, this platitude is part of Popper's indeterminism.

Popper's solution to both problems of induction can be vindicated, but shorn of his dubious metaphysical additions. On the epistemological problem, there is no guarantee of continuity from past to future — and now also no need to recourse to special metaphysical pleading to say so. On the methodological problem, G is (fallibly) to be preferred over M(N), since the latter is one of a class of theories of the kind M(t) that has been falsified for all values of the parameter t less than N when G has been corroborated. The only retreat left for the inductivist is to deny deductive arguments in principle;[11] but then he is shorn of deductive tools in *all* contexts, which is his predicament.

5. The logical details

The logical deduction of Imp from Conj is a little cumbersome though simple to effect. In classical logic the existential and universal quantifiers are inter-related, and so are the logical connectives; thus the logical forms chosen below are not unique for their respective parent statements. The leading quantifier(s) determine the forms: categorical, as

---

[10] In scientific contacts where causal relationships are not clear and/or data not too precise, probability theory may be used to compare the candidacies of possible culprits. This is a serious matter, but is independent both of status of the principle of the continuity of nature, and of the role of the probability calculus discussed above.

[11] I think that this point is the burden of [Swann 1988].

when starting with (an) existential quantifier as in (3), or with universal quantifier(s) followed by conjunctions and/or disjunctions; or conditional, starting as in (2) with universal quantifier(s) followed by (bi)implications. Each form may be converted to the other one with suitable use of negations; as we shall see, the matter is not critical. Thus I take no stance over the interesting discussion within logic as to whether or not the material conditional 'if ... then' is regarded as logically equivalent to implication.

The logical moves will be presented first for schematic forms of theories; then the required elaborations will be considered. The logical theorems required are rehearsed in many logic books; I choose to cite them, by starred number, from the Whitehead/Russell *Principia mathematica* [1910].

The jumbo statement Conj laying out the circumstances of testing a theory is a conjunction of categorical and conditional sub-statements, such as

(for all x, fx) & ... & (for all y, if gy, then hy) & ... &
(for all z and u, if Rzu, then Szu) & ...   (7)

for properties f, g, h, ... and relations R, S, ...[12] For ease of reading, different letters are used for quantified variables, although there are no clashes of scope. In a given case, there may be no categoricals.

The first move is to take each conditional sub-statement in (7) in isolation and use *9·21 (or *10·27) from the predicate calculus to convert into a form that for the second clause in (7) reads

if (for all y, gy), then (for all y, hy);   (8)

the next sub-statement of (7), with two variables, is delivered by *11·32. Each original sub-statement is asserted, and statements such as (8) are deduced via the *modus ponens* rule of inference *3·35 (namely, A; if A, then B; therefore B). Then all the converted sub-statements are conjoined to form a new jumbo conjunction of implications:

[if (for all y gy), then (for all y hy)] & ... & [if (for all z and u Rzu), then (for all z and u Szu)] & ...   (9)

Next we use Leibniz's 'beautiful theorem' *3·47 from the statement (or propositional) calculus, which states that a conjunction of two implications implies the companion implication of conjunctions:

if [(if A, then B) & (if C, then D)], then
[if (A & C), then (B & D)].   (10)

The antecedent of (10) is then asserted, and the consequent deduced by *modus ponens*. If required, then (10) may be iterated to any finite number of sub-statements by conjoining the next conditional to each previous antecedent, invoking (10) to obtain the new consequent, and then applying *modus ponens* each time. (9) now becomes an implication of jumbo conjunctions

if [(for all y, gy) & ... & (for all z and u, Rzu) & ...] then
[(for all y, hy) & ... & (for all z and u, Szu) & ...].   (11)

---

[12] I follow the usual assumption in the classical predicate calculus that the quantifiers do not operate over non-empty sets; it can be avoided by the usual techniques of free logic.

An especially attractive alternative form is also available, in which all the quantifiers are gathered together as the front:

(for all y, z, u,...) [if gy & Rzu & ...), then (hy & Szu & ...)].[13] (12)

Finally, any categorical sub-statements are adjoined to each jumbo clause of (11) or (12). This is done via the special case of (10) given by C=B, which is known as 'Peano's factor theorem' (*3·45). It states that

if (if A, then B), then [if (C & A), then (C & B)]; (13)

again it may be iterated as many times as necessary. From (11) we obtain one version of Imp:

if [(for all x, fx) & ... & (for all y, gy) & ... & (for all z and u, Rzu) & ...], then [(for all x, fx) & ... & (for all y, hy) & ... & (for all z and u, Szu) & ...]. (14)

while (12) becomes another version of Imp:

(for all x, y, z, u,...) [(if fx & gy & Rzu & ...), then (fx & hy & Szu & ...)]. (15)

If inconsistencies among the premises are present, then, as mentioned earlier, their exposure is most likely to emerge when (14) or (15) is formed.

Should all be safe, then the antecedents in (14) or (15) are affirmed as the collective premise (C & A) of Imp; their consequent, (C & B), is deduced via the ever-ready *modus ponens* rule of inference. The specific details of the test and the required test statement(s) are then deduced as explained above. Should falsification be the outcome, it is driven back through the negated (C & B) and (C & A) to the negated C, A and B.

This presentation is very schematic concerning the formulation of theories. Usually they contain several basic assumptions, each of which supplies a sub-statement to the jumbo conjunction (7). Relations may well have more than the two variables exhibited for R and S in (7); and some sub-statements may draw upon higher–order quantification. Much more complicated relationships may well follow the initial quantifiers than has been shown here, possibly including further interior quantifications of other variables; subsidiary assertions using *modus ponens* may be needed to accommodate their requirements. All these complications should be

---

[13] As this deduction is not a routine result, I sketch the proof here in a generic form. Given

(for all x) (if fx, then gx) & (for all x) (if hx, then kx) (i)

we wish to deduce

(for all x)[if (fx & hx), then (gx & kx)]. (ii)

First, eliminate the quantifiers from (i) to obtain

(if F, then G) & (if H, then K), (iii)

where the capital letters denote the corresponding propositions which result.

Assume (F & H) (iv)   and deduce from it   F. (v)
From (iii) deduce   if F, then G,   (vi)
and from (v) and (vi) deduce   G.   (vii)
Similarly, from (iv) the second clause of (iii) deduce K. (viii)
From (vii) and (viii) form the conjunction   (G & K). (ix)
Remove the status of assumption from (iv) by using (ix) to form the conditional

if (F & H), then (G & K). (x)

Finally, put the quantifiers back into (x) and obtain (ii) as desired.

accommodable, in both Conj and both versions of Imp: the appropriate versions of (8) are available, so that the *forms* (7), (9), (11), (14) and (15) should be secured. It might be worth converting all predicate statements, especially (15), into prenex normal form, where all quantifiers are placed in order at the front and negations have been removed; this would yield the form 'for all x, there exists a y and z such that for all u and v ...'. For non-logicians, all these manipulations are just kit, which can be taken for granted.

Finally, I have followed Popper in using only the classical two-valued logic, which is certainly applicable to a wide remit of testing situations. However, the *choice* of logic to use in a scientific context is, in my view, a legitimate question; and if another logic were used, the availability of logical moves corresponding to the ones used above would have to be examined (and also in the traditional construal of testing in terms of a conjunction (7) of implications). Popper admitted only classical logic in testing, in order to increase its strength; but sometimes such strength might be brute. His logical monism here was rather dogmatic, and thereby perhaps a little irrational (§5). But of course classical logic applies to a class of scientific phenomena sufficiently vast to make the problems of induction central to philosophy.

Acknowledgement

I am most indebted to the referees for picking up some cute points, including suggesting the version (12) of Imp.

Bibliography

Babbage, C. 1837. *The ninth Bridgewater treatise. A fragment,* 2nd ed., London: Pickering. [Repr. as *Works,* vol. 9, London: Pickering, 1989.]

Franklin, A. 1990. *Experiment, right or wrong,* Cambridge: Cambridge University Press.

Grattan-Guinness, I. 2004. 'The place of the notion of corroboration in Karl Popper's philosophy of science', in F. Stadler (ed.), *Induction and deduction in the sciences,* Dordrecht: Kluwer, 251-262. [§3]

Kuhn, T. S. 1972. *The structure of scientific revolutions,* 2nd ed., Chicago: University of Chicago Press.

Maxwell, N. 1998. *The comprehensibility of the universe,* Oxford: Oxford University Press.

Medawar, P. B. 1969. *Induction and intuition in scientific thought,* London: Methuen.

Miller, D. W. 1982. 'Conjectural knowledge: Popper's solution of the principle of induction', in A. Levinson (ed.), *In pursuit of truth,* Newark: Humanities Press, 17-49.

Miller, D. W. 2002. 'Induction: a problem solved', in J. M. Bohm, H. Holweg and C. Hoock (eds.), *Karl Poppers kritischer Rationalismus heute,* Tübingen: Mohr/Siebeck, 81-106. [Repr. in Miller, *Out of error,* Aldershot: Ashgate, 2006, 111-132.]

Nicod, J. 1924. *Le problème logique d'induction,* Paris: Alcan. [English transs. 1930 and 1969.]

Popper, K. R. 1930. *Die beiden Grundprobleme der Erkenntnistheorie,* manuscript; publ. Tübingen: Mohr (Siebeck), 1979.

Popper, K. R. 1959. *The logic of scientific discovery,* London: Hutchinson. [German original dated 1934.]

Popper, K. R. 1963. *Conjectures and refutations,* London: Routledge and Kegan Paul.

Popper, K. R. 1972. *Objective knowledge,* Oxford: at the Clarendon Press.

Popper, K. R. 1974. 'Replies to my critics', in P. A. Schilpp (ed.), *The philosophy of Karl Popper,* 2 vols., La Salle, Illinois: Open Court, 961-1197.

Popper, K. R. 1977. 'Normal science and its dangers', in I. Lakatos and A. Musgrave (eds.), *Criticism and the growth of knowledge,* Cambridge: Cambridge University Press, 51-58.

Popper, K. R. 1982a. *Realism and the aim of science,* London: Hutchinson.

Popper, K. R. 1982b. *The open universe. An argument for indeterminism*, London: Hutchinson.

Swann, A. J. 1988. 'Popper on induction', *British journal for the philosophy of science, 39,* 367-373.

Whitehead, A. N. and Russell, B. A. W. 1910. *Principia mathematica*, 1st ed., vol. 1, Cambridge: Cambridge University Press. [2nd ed. 1925. Partial paperback reprint 1962, including the sections used here.]

Worrall, J. 1989. 'Why both Popper and Watkins fail to solve the problem of induction', in F. d'Agostino and I. C. Jarvie (eds.), *Freedom and rationality,* Dordrecht: Kluwer, 257-296.

5 Levels of Criticism:
Handling Popperian Problems in a Popperian Way

Popper emphasised both the problem-solving nature of human knowledge, and the need to criticise a scientific theory as strongly as possible. These aims seem to contradict each other, in that the former stresses the problems that motivate scientific theories while the one ignores the character of the problems that led to the formation of the theories against which the criticism is directed. A resolution is proposed in which problems *as such* are taken as prime in the search for knowledge, and subject to discussion. This approach is then applied to the problem of induction. Popper set great stake to his solution of it, but others doubted its legitimacy, in ways that are clarified by changing the form of the induction problem itself. That change draws upon logic, which is the subject of another application: namely, in contrast to Popper's adhesion to classical logic as the only welcome form (because of the maximal strength of criticism that it dispenses), can other logics be used without abandoning his philosophy of criticism?

> It doesn't matter if we are over-critical: if we are, we shall be answered by counter-criticism.
> Popper on logic in science [1972, 305]

> In mathematics we do not need the maximum generality: we need the *right* generality.
> Attributed to the algebraist Saunders Mac Lane

1. Introduction

1.1 Across the range of Karl Popper's philosophy the following four features are prominent:

{ThMeta} *The importance of distinguishing theory from metatheory.* In the early years of Popper's philosophical career from 1930 onwards, a development of major significance in philosophy was the recognition that logic of any kind was distinct from metalogic, and more generally that a theory of any kind was to be distinguished from its metatheory. One especially important aspect of this distinction is that the principles of a metatheory need not be the same as those of the parent

---

This article was first published in *Axiomathes, 18* (2008), 37-48; it is reprinted by kind permission of Springer Science+Business Media. Much of ¶2 has been omitted, since the topic was discussed in greater detail in §5; a few additions have been made to the text, and some sub-sections have been titled.

theory; for example, especially from the 1920s L. E. J. Brouwer developed a constructivist approach to mathematics in which the law of excluded middle and existence theorems were banned, but his metamathematics was classical.

Popper's principal initiation into the world of metatheory occurred early in 1935 when Alfred Tarski explained to him the semantic theory of truth, where statements about the truth (or falsity) of statements S as corresponding (or not) to the facts were located in the meta-language of the (formalised) language of S [Popper 1972, ch. 9].[1] His innovation was already known in Vienna, not only through Tarski himself but also because of other figures such as Brouwer and especially the young Kurt Gödel. A notable part of Popper's later philosophy was explicitly metatheoretic, where he argued for preferring one philosophical standpoint over another; for example, his book *The open universe,* where he marshalled arguments of various kinds against determinism and in favour of indeterminism [Popper 1982a].

{TestTh} *An emphasis on the testing of theories.* Popper represented the process of testing by the schemum

problem $\Rightarrow$ one or several theories as tentative solutions

$\Rightarrow$ error elimination $\Rightarrow$ new problem   (PS)

[1972, 241-244]; for convenience I shall refer below to the case of just one theory. He stressed both the provisional status of the theory itself, and advocated examining its consequences rather than trying to justify it or seeking to show that its foundations are secure. Initially formulated in terms of falsification or refutation of scientific theories, later he generalised testing to criticism of all types, intellectual (including logical) as well as experimental and experiential, of theories of all kinds, as well as philosophical ones [1963, ch. 8].

{MaxCrit} *The requirement that the testing of a scientific theory should be maximally severe.* I pick on one aspect of this demand; that the logic underlying both the theory and its tests be the classical bivalent logic [1972, 305]:

> If we want to use logic in a critical context, then we should use a very strong logic, the strongest logic, so to speak, which is at our disposal; for we want the criticism to be *severe*. [...] Thus we should (in the empirical sciences) use the full classical or two-valued logic.

That is, classical logic 'is *the organon of criticism*' [1972, 55; see also esp. pp. 304-308].

{Probl} *An emphasis on problems as such.* Prominent also in (PS) is the primacy of problems *as* problems. However, like many philosophers of science Popper did not discuss this feature very much (there is some

---

[1] Popper stated his position on these four features, and others, in various places; I cite either the main source, or an easily available representative one.

material in [1972, 344-347]), nor the processes involved in *creating* the theory in the first place. But this is a significant lacuna in his philosophy, worth filling: the formation of the problem, means of making it more specific, comparing it metatheoretically with related problems, and so on (§12).

1.2 The issue addressed in this essay is that there is a mismatch between, on the one hand, {TestTh} and {Probl} where problems and theories of all kinds are tentative and subject to criticism and error elimination, and on the other hand the rather dogmatic appeal {MaxCrit} to maximal criticism. The Popper motto at the head of this essay is an invitation made immediately after the passage quoted in {MaxCrit}. The stance adopted in this essay is inspired by the other motto there, which was made by Mac Lane in the context of abstract algebras, where generality is a major concern.[2] My version of his motto reads thus:

> In science we do not need the maximum criticism: we need the *right* criticism.

That is to say, the level or type of criticism is to be assessed relative to the problem at hand, be that scientific, philosophical or from any department of knowledge. In particular, the choice of a logic, or of a preferred logical form within a logic, can be debated, after the manner of (PS).

Further support for the adoption of this position comes also from the claim made by W. V. O. Quine that the distinction between analytic propositions based upon meanings and synthetic ones reliant upon experience is 'ill-founded' [Quine 1951]. This position contrasts between Popper's two claims in that (classical) logic is linked to analyticity whereas testing is intrinsically experiential. He does not seem to have discussed Quine's point, beyond a rather oblique remark in [1963, 74] where he associated it especially with formalised languages — of which logic provides examples.

2. The logic of testing:
the place of induction in the appraisal of scientific theories

The first application to be made of this approach is important in Popper's philosophy, due to the controversies that it excited from early on. 'I think that I have solved a major philosophical problem: the problem of induction' [Popper 1972, 1]. However, plenty of doubts were raised, especially that his requirements amounted to assuming induction. Advocates from each side are well represented by [Miller 2002] in favour and [Worrall 1989] in opposition. My own position differs from both in the spirit of {Probl}, namely, in my answer to the question: which philosophical problem are we trying to solve here?

The process of testing, as understood by both sides, was described in detail in §5. There I called it 'the inventory problem': compiling the *list*

---

[2] Unfortunately I have not been able to identify a source for the quotation in Mac Lane, and I never asked him about it. But it fits in with his contributions, especially to category theory; and in any case it stands on its own.

of theories, sub-theories and equipment and circumstances required for the test to take place. The disputants did *not* discuss the test itself; specifying that 'testing problem' (as I shall call it) is a *different* task. As we saw, the difference between the two problems rests on their logical forms. Popper's claim to have rejected induction can be affirmed if the testing problem is in mind, with its implicational form. But both he and his supporters and critics debated the absence or presence of induction within the inventory problem, in conjunctive form. The critics had a strong case, *for in debating the wrong problem he and his supporters adopted a logically stronger form than they needed.* The antecedent of each theorem is a conjunction of implications, appropriate for the inventory problem, while the consequent is an implication of conjunctions, as required for the testing problem. As usual the consequent is the logically weaker of the two sub-statements; hence the level of criticism is also weaker, but it is *right* (in Mac Lane's sense) for the testing problem.

This discussion is guided by Popper's conception of classical logic in which universal statements and implication play major roles. It is legitimate to wonder if theories can always be formulated in this logical form in the first place; but if not, then discussions of *both* the inventory and the testing problems fail. There are various other philosophical conceptions of scientific theories, in which other categories of knowledge are favoured [Dilworth 1994]; but if the problem of induction is addressed, then consideration of the refutation of a scientific theory would require the above formulation, or something similar to it. Thus some form of the distinction between the two types of problem should pertain also.

3. Criticising logics

The use of classical logic in the discussion above follows the usual practise of scientists themselves. However, while classical logic is quite satisfactory for many scientific contexts, is it *automatically* the logic to prefer when forming and testing a scientific theory? Normally scientists do not have any particular familiarity with the subject of logic and just take the classical version for granted;[3] but the choice of a logic, in scientific contexts (and also elsewhere), is an issue worth addressing. It is the second application of our approach, and is considered in this and the next section. While I shall clearly exhibit a preference for logical pluralism, the main purpose is not to advocate or reject any particular logic(s) but instead to address this meta-question: can a discussion of different logics be conducted in a spirit of Popperian fallibility?

I shall use the singular forms 'metalogic' and 'metatheory' below, without ignoring the fact that a theory has more than one metatheory available: for example, one involved with logical properties such as (in)consistency, and another one treating its history. Further, some theories

---

[3] Neither do scientists seem to have much skill in classical logic. In particular because of its special relevance to Popper's philosophy of refutation, their ability to spot the *modus tollens* rule of inference is very limited [Kern and others 1983]!

in mathematics and physics have symbolic formalised metamathematics as well as these more informal metatheories.

Popper worked on various aspects of logic [Schroeder-Heister 1984],[4] but he remained loyal to the classical version, at least for science; even in quantum mechanics he tried to retain it with his propensity interpretation of probability and statistical reading of the uncertainty principle ([Popper 1982b], which does not have an index entry under 'logic'). He also took this stance when advocating Tarski's semantic theory of truth; however, he extended it to natural languages within which scientific theories are very often expressed, and associated it with objectivity. While Tarski also formulated his theory within classical logic [Tarski 1944], he did not take those further steps with the same enthusiasm [Grattan-Guinness 1984 = §6]. Let us take a similarly cautious path.

The status and even fallibility of logic in general was posed by Popper's student W. W. Bartley, III, who carried the notion of criticism to the self-referring form of 'comprehensively critical rationalism'. Here an adherent 'holds *all* his beliefs, including his most fundamental standards *and his basic philosophical position itself*, open to criticism' [1964, 146, second occurrence of italics mine]; the focus of discussion falls upon the criticism of theories, instead of upon their justification. The range of criticism includes 'The check of *logic*: Is the theory in question consistent?' [p. 158], and even the possibility of revising it. On the whole he concerned himself with possibly changing specific features of classical logic, and preferred that a metalogically *stronger* substitute be made for the logic in place in a given situation [Bartley 1980, esp. pp. 71-72]; but he mentioned other logics, and they will form the main concern below. Once again Mac Lane's advice will help us; the (or a) right logic may be weaker, or stronger, than its competitors.

{The pluralist position adopted here is somewhat akin to satisficing, which was launched in the 1950s in connection with economics by Herbert Simon [Husseini 1990] and enjoys some currency in various disciplines [Byron 2004]. The aim is to choose one or maybe several satisfactory solutions to a problem rather seek only the allegedly optimal one. For example, instead of looking for the worst-case scenario, assessing several pretty bad ones may be more realistic.}

4. Choosing between logics

4.1 *A plethora*. Especially since the Second World War, a very wide range of logics has been developed: many-valued, modal, temporal, relevance, epistemic, deontic, fuzzy, paraconsistent, and so on [Gabbay and Woods 2006]. Often applications were in mind and even furnished the motivations, sometimes the physical and life sciences but more often fields such as logic itself, computer science, aspects of philosophy including

---

[4] Popper often used the word 'logic' in its other sense, concerning the logic of a situation, most prominently in his book 'The logic of scientific discovery' [Popper 1959]; I do not consider this sense in this essay.

language(s) and linguistics, and some social and psychological sciences (§21.5.1).

Can these theories really be called 'logics'? Should we support logical pluralism, or stick to a monistic dependence on one logic (traditionally, classical logic)? The questions are more than a matter of contention over the use of 'logic' as a name. Quine defended classical logic and resolutely opposed the growing tide of new logics ([1986, ch. 6] and many other places); but the vote seems to have gone the other way, not only in the sheer quantity of systems produced but also as a philosophical support for logical pluralism. Authors such as [Rescher 1969, ch. 3], [Haack 1996], [Beall and Restall 2005] and [van Bentham and others 2006] claim that genuine deduction is taking place, so that these systems *are* logics, both those that extend the classical version in some way and those that seek to rival it and those to replace it (to draw upon Haack's valuable distinction). I adopt this position (without feeling obliged to support any given case), both with respect to logics as sources of proofs and as sources of models [Restall 2006].

{An important guiding factor is the view that one takes about logical knowledge itself. Following the tradition of mathematical logic and of Tarski himself, Popper regarded logic as the theory of the transmission of truth; pluralists may prefer to go back to the tradition upheld by some algebraic logicians and others, that logic deals with the processing of information [Saguillo 2009].}

4.2 *Strategies*. We may interpret the search for logic(s) in scientific theories pragmatically: find a logic that works. Or we may adopt the principle of tolerance of Rudolf Carnap for logics [1934, art. 17]. But Bartley's considerations suggest that choosing a logic (with some metalogic) falls *within* Popperian fallibility, whether using refutation or criticism, by testing them as parts of scientific theories. We tentatively choose a logic for the scientific theory, formulate it in the implicational form (if axioms and rules of inference, then theorems) appropriate for the testing problem, and include it as a component at the testing stage (or earlier). If this logic has been suitable for the scientific theory and its tests, then it is corroborated; but if found inappropriate or too approximate to serve the theory, then it stands refuted. The notions of corroboration and of verisimilitude will need to be modified for the chosen logic.

Given a problem and a candidate scientific theory to solve it, criteria for choosing a logic would include the main properties of the theory and their associated predicates and relations, and the "values" (not necessarily numerical ones) that the variables take: maybe classical 0-or-1 but also others such as uncertainty, or even a continuum of values such as in [0,1] as used in fuzzy and some other logics. A (or the) preferred logic will have to be able to accommodate all the principal features of the theory, and perhaps different logics would be suitable in different parts of it. Classical logic is not the only candidate logic: indeed, Tarski's favourite example to illustrate his theory of truth was 'Snow is white', which, in its concern with the recognition of colours, is a singularly unfortunate choice when classical logic is the ground. For the strength of that logic is *brute* when applied to any theory that draws upon a continuum of gradations. Further, adherence to logical pluralism does not require abandoning the view that (a) logic is normative or prescriptive, if one holds such a stance in the first place [McNamara 2006]; just as at the pharmacist's, a prescription may be essential for a particular affliction while not suiting all of them.

Among the limitations of classical logic, its imposition of timelessness upon statements is overly strong in many contexts. Temporal logic is a particularly obvious example of a genuine logic that deserves serious attention [Øhrstrøm and Hasle 2006]. For example, a testable prediction drawn from a scientific theory often may assert that if one does A and then B and then C ... *in that order*, then some X will happen; and the apparent refutation of the theory may be due to the fact that the order was not correctly followed. In very many circumstances, not necessarily involving science, we frequently use temporal logic, especially with 'and' meaning 'and then', without noticing. Elsewhere, quantum theorists of a non-Popperian persuasion have long had their own non-classical logics; some kind of probability logic may suit the assessment of information in a questionnaire or in a study of beliefs; deontic logics have been found useful in some business and security situations, which can bear upon science; and some logics have been driven by applications of many kinds, including scientific ones. Here is an interesting example.

4.3 *Fuzzy logic*, an offshoot of fuzzy set theory, has gained much attention among various communities but not much from logicians. The logic is formulated from these sets as a valuation function onto [0,1], deploying structure similarity such as associating minimisation of truth-values with conjunction and universal quantification and maximisation with disjunction and existential quantification, after the manner of a de Morgan algebra. Classical logic comes out as the 'crisp' case where the only values are 0 and 1 — or [0,0] and [1,1] in my preferred version of the theory, where the range of the function takes sub-intervals of [0, 1] rather than numbers within it, and where the metalogic is also fuzzy [Grattan-Guinness 1979].

Whether one grants the valuation function the status of a logic is, as mentioned, a matter of convention. But I do not find Haack's reservations over fuzzy logic [1996, 232-242] very convincing. The purpose of fuzzy logic is not so much to study vague arguments [pp. 229, 233] as to develop arguments using vague *predicates and relations* (such as 'healthy' and 'the prettiest among') that are handled in fuzzy set theory. Dubois and Prade [1980] contains an early survey of the range of applications.

The theory has complications, but they show that even other many-valued logics are too simple, or else not sufficiently relevant to the applications. So the claims of simplicity made by some fuzzy logicians are indeed not happy, especially (for example) regarding linguistics; nor are some of the permitted qualifiers of truth. In general, the literature in fuzzy set theory and logic exhibits an unwelcome quantity of philosophical fuzziness!

4.4 *The places of logics in the psychological sciences* are particularly interesting, since the data can involve cognition, perception and/or mentation, and so involve logics in several of its branches. I set aside important questions about the actual processes of reasoning, which yield logics among other "products",[5] and focus upon the possibility of logical pluralism.

One context in which mental processes are among the objects of study is the formation and creation of problems and theories. We recall from {Probl} in ¶1.1 that Popper did not discuss them much; but these aspects of science are far too interesting and important to be left to sociologists! The formation of a problem is more than just an intuitive art, and erotetic logic may suggest some procedures [Wisniewski 1995]. The development of both problem and scientific theory involves levels of knowledge, belief and ignorance, which also fall under the remit of epistemic logic [Gochet and Gribomont 2006]. Popper rightly criticised the emphasis on belief made in some philosophies of science [1972, 23-29, 140-142]; but one person's subjective belief can be the focus of objective scientific study by others, as a type of (testable) metatheory.

The formation of a scientific theory about, say, psychotic behaviour is likely to profit from using some non-classical logic in handling the data, and maybe also the formation and testing of pertaining theories. However, the metatheory could be classical, in contexts such as the debate about whether, or not, schizophrenia should be treated as an illness at all [Read and others 2004, esp. chs. 2, 6 and 20].

4.5 *Metatheorising.* Another example concerning metalogics occurred in ¶2 where there was an over-simplification in stating, classical style, that the scientific theory fails its test or not. For in many scientific contexts the phenomena are so complicated, and/or little accessible to *direct* experience, that it can be hard to say whether success or failure has occurred for the theory, or to what extent (if at all) which of its various components have actually been tested; for example, fluid, aero- and gas dynamics show plenty of instances.[6] One should use here a three-valued metalogic, with true, false and uncertain as values.

We may also have competition between forms *within* a logic. For example, the discussion of ¶2 was an analysis, within classical logic, of two rival philosophical metatheories about testing scientific theories: the inventory version failed the test of induction, but the testing version passed it.

---

[5] For a review of the main positions on these topics, and on the psychology of reasoning see, for example, [Hanna 2006, chs. 4-5]. Part of my motive for focussing upon the psychological sciences is a reservation about the practise of most philosophies of science, Popper's included: an excessive emphasis on the physical sciences.

[6] [Darrigol 2005] surveys several parts of their histories, though without focussing much on testing. There must be many comparably complicated testing situations in technology, the life sciences and medicine, but they lie outside my specialist knowledge.

5. A reflection: the importance of hierarchies of theories in knowledge

This essay has been driven by the merit of distinguishing a theory from its metatheory {ThMeta}, which starts off a hierarchy of theories; a logic and its metalogic(s) were an important example. This kind of distinction is an important feature of several aspects of Popper's philosophy, as was mentioned in ¶1.1. But apart from logic in general, and in some related subjects such as mathematics and computer science, the centrality of hierarchies is not much emphasised, not even in some philosophical traditions.[7] Here I outline a general feature, using ¶2 as an exemplifying context.

Part of a scientific theory T under test is the logic that it uses, including its metalogic. This is an example of a *local hierarchy* relative to T; other examples include the various theories and metatheories involved with the instruments used in the test. The discussion of the inventory and the testing problems, itself metatheoretic relative to them, is an example of a *global hierarchy* because it involves the *whole context,* within which local hierarchies are sited. Further, should T be considered in some still wider context, then its previously global hierarchy becomes a local one relative to the new globality — and so on, with hierarchies of hierarchies. *No* contradictions or beggings of questions arise; nor need there be any worry about discussing logic within logic, or the circularity that threatens to attend the need to justify deduction.[8] Indeed, the usual Popperian cautions against justification in general [1963, 21-25] seem to apply here.

A hierarchy starts from some theory (maybe a logic) and its metatheory, or from a comparison of several theories and their metatheories; and their assessment takes place in *metametatheory*, which can have principles different from those of its two parents. This is the third level of knowledge, which has hardly begun to be developed in an explicit way [Tarski 1944, 350], just as metalogic only started overtly to make its mark around 1930 with Tarski and others, to be joined by young Popper soon afterwards. Its future should be fruitful.

6. Concluding remark

The millennia-long human faith in the reliability of logic in general, and especially the classical version, is very strange; for its "objects" are among the most elusive in all human knowledge, whether one seeks to ground them Kant-style in an innate logical category [Hanna 2006, chs. 1-3], or attempt Mill-like to regard logic as involving forms of thought generalised from experience [Scarre 1989, ch. 5], or use some other means. Maybe this faith formed part of the core of the human hope that some kinds

---

[7] For example, as a rule (as it were) Wittgensteinians explicitly reject or avoid hierarchies and prefer the showing-saying distinction.

[8] The intricate arguments surrounding this topic are well rehearsed in [Haack 1996, 183-213]; but I demur over the rejection of solution by metatheory [p. 187], as she stops the hierarchy at that level. For a general survey of the issues, see also [Hanna 2006, ch. 3].

of knowledge could be established as *certain*.⁹ Popper has been an important philosopher in challenging this hope, but ironically he adopted it where logic was concerned. (I tried twice to discuss with him the possibility of non-classical logics in testing, but failed utterly!) Nevertheless, it seems quite proper to extend fallibilism to logics and their applicability without abandoning features {ThMeta}, {TestTh} and {Probl} — and to do so in an extension of the normal conception of fallibilism, not as a rival to it.

Bibliography
Bartley, W.W. 1964. *The retreat to commitment,* London: Chatto and Windus.
Bartley, W.W. 1980. 'On the criticizability of logic', *Philosophy of the social sciences,* 10, 67-77.
Beall, J. L. and Restall, G. 2005. *Logical pluralism,* Oxford: Oxford University Press.
Byron, M. 2004. (Ed.), *Satisficing and maximizing: moral theorists on practical reason,* Cambridge: Cambridge University Press.
Carnap, R. 1934. *Logische Syntax der Sprache,* Vienna: Springer. [English trans. 1937.]
Darrigol, O. 2005. *Worlds of flow. A history of hydrodynamics from the Bernoullis to Prandtl,* Oxford: Oxford University Press.
Dilworth, C. 1994. *Scientific progress,* Dordrecht: Kluwer.
Dubois, D. and Prade, H. 1980. *Fuzzy sets and systems: theory and applications,* New York: Academic Press.
Gabbay, D.M. and Woods, J. 2006. (Eds.), *Logic and the modalities in the twentieth century,* Amsterdam: Elsevier.
Gochet, P. and Gribomont, P. 2006. 'Epistemic logic', in [Gabbay and Woods 2006], 197-298.
Grattan-Guinness, I. 1979. 'Forays into the meta-theory of fuzzy set theory', *Logique et analyse,* 22, 321-337.
Grattan-Guinness, I. 1984. 'On Popper's use of Tarski's theory of truth', *Philosophia,* 14, 129-135. [§6]
Grattan-Guinness, I. 2000. 'Mathematics and symbolic logics: some notes on an uneasy relationship', *History and philosophy of logic,* 20 (1999), 159-167.
Grattan-Guinness, I. 2004. 'Karl Popper and the "problem of induction": a fresh look at the logic of testing theories', *Erkenntnis,* 60, 109-122. [§4]
Haack, S. 1996. *Deviant logic, fuzzy logic. Beyond the formalism,* Chicago and London: The University of Chicago Press. [1st ed. 1974.]
Hanna, R. 2006. *Rationality and logic,* Cambridge, Mass.: The MIT Press.
Husseini, H. 1990. 'The archaic, the obsolete and the mythical in neoclassical economics: problems with the rationality and

---

⁹ Mathematics has long formed another part of this hope for certainty (§21.6), although it has always stood a long way apart from logic, and still usually does: mathematicians' logic ≠ logicians' logic [Grattan-Guinness 2000]! I discussed this curious situation several times with Popper, who was very intrigued by it.

optimizing assumptions of the Jevons-Marshallian system', *American journal of economics and sociology, 49*, 81-92.

Kern, L. H., Mirels, H. L. and Hinshaw, V. G. 1983. 'Scientists' understanding of propositional logic; an experimental investigation', *Social studies of science, 13*, 131-146.

McNamara, P. 2006. 'Deontic logic', in [Gabbay and Woods 2006], 198-288.

Miller, D. W. 2002. 'Induction: a problem solved', in J. M. Bohm, H. Holweg and C. Hoock (eds.), *Karl Poppers kritischer Rationalismus heute*, Tübingen: Mohr/Siebeck, 81-106. [Repr. in Miller, *Out of error*, Aldershot: Ashgate, 2006, 111-132.]

Øhrstrøm, P. and Hasle, P. F. V. 2006. 'A. N. Prior's logic' and 'Modern temporal logic: the philosophical background', in [Gabbay and Woods 2006], 399-446, 447-498.

Popper, K. R. 1959. *The logic of scientific discovery*, London: Hutchinson. [German original dated 1935.]

Popper, K. R. 1963. *Conjectures and refutations*, London: Routledge and Kegan Paul.

Popper, K. R. 1972. *Objective knowledge. An evolutionary approach*, Oxford: at the Clarendon Press.

Popper, K. R. 1982a. *The open universe. An argument for indeterminism* (ed. W. W. Bartley, III), London: Hutchinson.

Popper, K. R. 1982b. *Quantum theory and the schism in physics* (ed. W. W. Bartley, III), London: Hutchinson.

Quine, W. V. O. 1951. 'Two dogmas of empiricism', *Philosophical review, 20*, 20-43. [Various reprs.]

Quine, W. V. O. 1986. *Philosophy of logic*, 2nd ed., Englewood Cliffs: Prentice Hall.

Read, J. and others 2004. (Eds.), *Models of madness*, London: Brunner-Routledge.

Rescher, N. 1969. *Many-valued logic*, New York: McGraw Hill.

Restall, G. 2006. 'Relevant and substructural logics', in [Gabbay and Woods 2006], 289-398.

Saguillo, J. M. 2009. 'Methodological practice and complementary concepts of logical consequence: Tarski's model-theoretic consequence and Corcoran's information-theoretic consequence', *History and philosophy of logic, 30*, 49-68.

Scarre, G. 1989. *Logic and reality in the philosophy of John Stuart Mill*, Dordrecht: Kluwer.

Schroeder-Heister, P. 1984. 'Popper's theory of deductive inference and the concept of a logical constant', *History and philosophy of logic, 5*, 79-110.

Tarski, A. 1944. 'The semantic conception of truth and the foundations of semantics', *Philosophy and phenomenological research. 4*, 341-375 [cited here: repr. in *Collected papers*, vol. 2, Basel: Birkhäuser, 1986, 661-699].

van Bentham, J. and others. 2006. (Eds.), *The age of alternative logics. Assessing philosophy of logic and mathematics today,* Dordrecht: Springer.
Wisniewski, A. 1995. *The posing of questions. Logical foundations of erotetic influences,* Dordrecht: Kluwer.
Worrall, J. 1989. 'Why both Popper and Watkins fail to solve the problem of induction', in F. d'Agostino and I. C. Jarvie (eds.), *Freedom and rationality,* Dordrecht: Kluwer, 257-296.

6 On Popper's Use of Tarski's Theory of Truth

Susan Haack claimed in 1976 that Tarski does not present his theory of truth as a correspondence theory. Popper pointed out in a reply that Tarski does regard his theory as a correspondence theory. Since each author gives references to Tarski's writings to back up his position, there seems to be a contradiction in the air —- an ironic situation for a theory of truth. Perhaps the contexts of the passages will resolve the difficulty — or expose new ones.

1. Tarski's own declarations

In his reply [1979] Popper first cites this passage from the beginning of Tarski's paper on truth: 'I would only mention that throughout this work I shall be concerned exclusively with grasping the intentions that are contained in the so-called classical conception ("true — corresponding with reality") in contrast, for example, with the *utilitarian* conception ("true — in a certain respect useful")' [Tarski 1935, 153]. Two paragraphs later, Tarski summarises the first section of his paper: 'In §1 colloquial language is the object of our investigations. The results are entirely negative. With respect to this language not only does the definition of truth seem to be impossible, but even the consistent use of this concept in conformity with the laws of logic'.

Popper's second passage comes from the paper [Tarski 1936, 404]: 'We regard the truth of a sentence as its "correspondence with reality"'. Two paragraphs earlier Tarski writes:

> The language for which such a [structural] description can be given are called *formalized languages*. Now since the degree of exactitude of all further investigations depends essentially on the clarity and precision of this description, *it is only the semantics of formalized languages which can be constructed by exact methods.*

The impression provided by these contexts is that Tarski is restricting his theory of truth to formalised languages. The impression is confirmed in the passage from [Tarski 1944], cited by Haack in her paper [1976]. After describing various definitions of truth, including the correspondence theory, he writes: 'none of them is sufficiently precise and clear [...] at any rate, none of them can be considered a satisfactory definition of truth' [Tarski 1935, 343].

Thus Tarski affirms his own theory of truth as a correspondence theory for formalized languages, but is chary of defining truth (by any means, apparently) for natural languages. I think that Haack should have

---

This article was first published in *Philosophia*, 14 (1984), 129-135; it is reprinted by kind permission of Springer Science+Business Media. The section titles have been added.

made this point clearer by citing the passages that Popper has quoted. However, her paper already contains a comment on Popper's reply: 'Popper takes it — despite what Tarski says to the contrary — that Tarski's theory can be applied to natural as well as formal languages' [Haack 1976, 325]. In that reply Popper himself does not point out the distinction between natural and formal languages, and seems to convey the impression that Tarski asserts correspondence as a theory of truth for natural languages also, 'despite what Tarski says to the contrary'.

In his more extended writings on Tarski's theory, Popper points out that it is not possible to give a general criterion of truth for natural languages.[1] Nevertheless, he thinks that the view that Tarski's definition 'is applicable only for formalized languages is, I think, mistaken. It is applicable to any consistent, and even to a "natural" language, if only we learn from Tarski's analysis how to dodge its inconsistencies'. Thus it applies to 'a whole range of more or less artificial though not formalized languages' [1969, 223, 398-399]. The artificialities include the avoidance of semantic closure, and presumably the introduction of some logical machinery to define satisfaction.

Popper does not discuss one aspect of Haack's position. She points out that, since in Tarski's theory true propositions are satisfied by all sequences, it does not rely on any particular sequence of objects and so admits analytic cases — where correspondence to the facts is not entailed — as well as the synthetic cases that are Popper's concern [Haack 1976, 325]. Popper might reply that it is truth *with content* that is of interest, which analytic cases therefore necessarily lack.

2. Remarks on the correspondence theory

Theories of truth remain a controversial issue between philosophers. Correspondence theorists such as Popper, and semantics theorists such as Tarski, are both challenged by advocates of other theories, such as coherence, pragmatism and redundancy.[2] My purpose here is to consider various aspects of the correspondence theory *as it is sited in Popper's philosophy*. Seven issues seem to stand out.

2.1 Popper writes that Tarski solved the 'apparently hopeless problem' of elucidating the notion of correspondence by reducing it to the 'simpler idea' of satisfaction.[3] He emphasises that Tarski's definition of satisfaction is executed with respect to formalized languages, but presses ahead with its use in more general contexts, where natural languages presumably

---

[1] See, for example, [Popper 1962, 369-374], or [1972, 46, 317-318]; on p. 331 of the 1979 edition of the latter work he repeats his criticism of Haack.
[2] Haack provides an excellent survey in [1978, ch. 7].
[3] [Popper 1968, 274]. I do not understand the use of 'simpler' here. Tarski's ten-line definition of satisfaction [1935, 193] is not simpler than the two-line definition of truth [p. 194] from a formal point of view. And as Popper is not an instrumentalist or conventionalist, I cannot see why simplicity is worth claiming anyway.

obtain. But there even the 'simpler' notion of satisfaction presents difficulties for the establishment of an absolute objective theory of truth. For the choice of objects of the (physical) world as the referents of the arguments of the sentential functions could itself be theoretical, even subjective. If someone points to the sleeping cat and says: 'Look, of course snow is white, can't you see it for yourself?', who is to say that he is not a correspondence theorist? In a passage cited by Haack [1976, 330], Tarski points out that his theory is 'completely neutral' towards epistemological positions — even one, therefore, based on mis-identifications, or poor knowledge of the language involved [1936, 362].

2.2 'Snow is white' is a favourite example, with both Tarski and Popper, of truth defined as correspondence with facts. Assuming that snow is actually the referent of the statement, this example strikes one as excellent example of the difficulty of associating an objective theory of truth with facts. Is the snow white, or rather off-white by now? How is it being lit, anyway? Facts are of course highly theory-impregnated, as Popper has emphasised often; but then they must be unsuitable candidates for the site of the objectivity and absoluteness that he seeks.

2.3 As an example of the last point, consider this judgement by Popper: 'Tarski's theory allows us [...] to define reality as that to which true statements correspond' [Popper 1972, 329]. This definition is in fact consistent with some positivist and even solipsist epistemologies; indeed, it may be more in tune with them for the following reason, unremarked and presumably unnoticed by Popper. The set of true statements is denumerable, since it is lexicographically orderable. Hence reality is composed of a denumerable set of (classes of) referents of true propositions. I doubt whether one could actually prove or refute this claim in a formal way, but it seems to fit ill with the conception of reality that Popper expounds at length elsewhere (its independence of experiencers, and so on).

2.4 I would much prefer to see Popper's epistemology oriented not towards factual truth but towards what elsewhere I have called 'ontological correctness' [Grattan-Guinness 1979, 324-326 = §2.2, §9.6]. Theories are ontologically correct if they describe how the particular phenomena actually — that is, ontologically — occur. (I assume that we could invent languages sufficiently rich to express such theories.) In a physical world like ours, where its substances seem to be composed of 'microsized' elements and its causal connections are at least not experientially obvious, the distinction between ontological correctness and factual truth is important. For example, a theory can be both ontologically incorrect but factually true, if it describes phenomena in a way that actually does not apply but that corresponds to the facts as well as experiment and experience currently allow [Grattan-Guinness 1979, 326-330 = §9.7]. Scientific theories are often concerned with 'micro' rather than surface effects; quantum mechanics, one of Popper's favourite scientific areas, is full of examples.

2.5 'Snow is white' also clearly exemplifies another difficulty in applying Tarski's theory to general contexts: namely, its use of the classical logic as the underlying logic. Snow is white or it is not white, *tertium non datur*; but of course, as I indicated in ¶2.2 above, snow can go through various shades. Many scientific theories also use inexact concepts, and to base a theory of truth as correspondence with the facts on classical logic is to restrict that theory of truth to a limited class of phenomena. On the other hand, I suspect that a classical underlying logic could be used when truth is defined relative to ontological correctness; for a scientific theory is ontologically correct or not. (Unfortunately, we shall never be able to prove that a correct theory is correct, for the next test might refute it; but that is another matter.) If truth is to be defined relative to facts, whether by correspondence or some other way, then I think that some non-classical logic is needed, to try to capture the component of vagueness in our theories of which 'snow is white' is such a good example.[4]

2.6 The concept of vagueness is sometimes associated, and even confused, with that of belief. Popper places beliefs in his World 2 of subjective knowledge, but of course they can be the referents of objective World-3 theories of belief: among psychologists studying personality, for example, or neurophysiologists wondering if beliefs are explicable in terms of neural states. Such scientists will need a theory of the relationship between a belief and its referent; and thus arises the well-known criticism of the applicability of the correspondence theory to belief predicates, not out of the way in World 2, but present in Popper's World 3 of objective knowledge.

Popper might assert that, since 'World 3 is the world of the products of the human mind' [Popper and Eccles 1977, 449], there exists an object to which the belief corresponds. But, apart from the question-begging character of the assertion (how would one know that the object is there if the belief had not been uttered?), it surely cannot have' the same status as the facts to which his theory of truth has statements correspond.

2.7 Popper affirms the epistemological primacy of statements, and seems to regard analyses of concepts as essentialism or ordinary language analysis.[5] Sadly, he is very often right; but in addition there are many cases in scientific theorising where non-essentialist conceptual analysis is required. Foundational theories in a science contain many examples, especially if the theory has a mathematical component. Indeed, Tarski's theory of truth itself is, to one admirer, a *'successful description of a method for*

---

[4] My own favourite candidate for this approach is fuzzy set theory (§5.4.3), though I cannot recognise the validity of fuzzifying truth concepts themselves, as is carried out in [Bellman and Zadeh 1977].

[5] I have in mind Popper's table of opposites, where important use of propositions is contrasted with (allegedly) essentialist analysis of concepts (see, for example, [1969, 19] or [1972, 124]). The point made in this sub-section amounts to a request for a middle column to be added to the table.

defining "true"' [Popper 1972, 328]; thus it is a theory of the concept of truth. Hence an exposition of Tarski's theory as a correspondence theory for natural languages (that Popper asserts) surely requires some conceptual analysis of statements (that Popper disparages).

As I have mentioned points of discrepancy between Popper and Tarski, here is another quotation from Tarski: 'In fact, I am rather inclined to agree with those who maintain that the moments of greatest creative advancement in science frequently coincide with the introduction of new notions by means of definition' [1944, 359].

3. Different contexts, different truth theories?

I draw two conclusions from this discussion. Firstly, Popper's advocacy of truth as correspondence to the facts seems to me to be a problematic part of his epistemology, whether or not Tarski's semantic theory is held to explicate the machinery of correspondence; the emphasis on facts raises doubts, and even seems out of tune with other aspects of his philosophy. Secondly, and more broadly, the competition between different theories of truth may not be one for the possession of the concept of truth but a demarcation dispute between contexts in which these theories apply. For example, in ¶2.6 a scenario was sketched in which it would be possible to advocate one theory of truth for beliefs and another for the scientific investigation of beliefs. I do not see why any theory of truth should be expected to cover, say, social justice, analytic number theory, personality testing, quantum mechanics, paths of projectiles, musical temperament, religious conviction and viral infection; yet these are all topics in which truth can have a bearing. It may well be that the problem of defining truth is undecidable, and that we need different theories of truth for different contexts. In other words, we must seek a philosophy of truths.

Bibliography

Bellman, R. E. and Zadeh, L. A. 1977. 'Local and fuzzy logics', in J. M. Dunn and G. Epstein (eds.) *Modern uses of multiple-valued logic*, Dordrecht: Reidel, 105-165.

Grattan-Guinness, I. 1979. 'On Popper's philosophy and its prospects', *British journal for the history of science, 12*, 317-337. [§9]

Haack, S. 1976. '"Is it true what they say about Tarski?"', *Philosophy, 51*, 323-336.

Haack, S. 1978. *Philosophy of logics*, Cambridge: Cambridge University Press.

Popper, K. R. 1962. *The open society and its enemies*, vol. 2 (rev. ed.), London: Routledge and Kegan Paul.

Popper, K. R. 1968. *The logic of scientific discovery* (rev. ed.), London: Hutchinson.

Popper, K. R. 1969. *Conjectures and refutations* (rev. ed.), London: Routledge and Kegan Paul.

Popper, K. R. 1972. *Objective knowledge. An evolutionary approach*, Oxford: at the Clarendon Press.

Popper, K. R. 1979. 'Is it true what she says about Tarski?" *Philosophy*, *54*, 98.

Popper, K. R. and Eccles, J. C. 1977. *The self and its brain*, Berlin: Springer.

Tarski, A. 1935. 'The concept of truth in formalised languages', in [1956], 152-278.

Tarski, A. 1936. 'The establishment of scientific semantics', in [1956], 401-409.

Tarski, A. 1944. 'The semantic conception of truth and the foundations of semantics', *Philosophy and phenomenological research*, 4 (1943-44), 341-375 [cited here: repr. in *Collected papers*, vol. 2, Basel: Birkhäuser, 1986, 661-699].

Tarski, A. 1956. *Logic, semantics, metamathematics* (ed. J. H. Woodger): Oxford: Oxford University Press. [2nd ed. with intro. and index by J. Corcoran, Indianapolis: Hackett, 1983.]

PART 2

COMPETITIONS

> Man, we may say, appears to be
> not so much a rational animal as an ideological animal.
>
> Karl Popper, *The myth of the framework.*
> *In defence of science and rationality,*
> London: Routledge, 1994, 82

7 Russell and Karl Popper: their Personal Contacts

A handsel for Sir Karl Popper on his 90th birthday, 28 July 1992

Russell and Popper count as among the most influential philosophers of this century. Although they followed substantially different lines and traditions, especially concerning a priori knowledge and abstract objects and the role of induction in science, they held each other in high esteem; and they corresponded on a number of occasions, largely after the Second World War. The recent organisation of Popper's papers by the Hoover Foundation at Stanford University, California, USA, where they are now kept, complements the files at the Russell Archives, and enables both sides of the correspondence to be available. This article presents the contexts and texts of the most significant documents, divided around three groups of events, two just after the War (¶1-¶2) and the last one in 1959 (¶3). The texts are transcribed in ¶4.

1. Exchanges over books, 1946-1947

At the time of the War's end both Russell and Popper came to Britain and published important books. Russell arrived from the USA in the June of 1944, and was reinstated at Trinity College Cambridge as a lecturer.[1] Among other work he completed his *History of Western philosophy,* which first appeared in 1945 in the USA from the New York house of Simon and Schuster.

Popper left Vienna in 1937, when he secured a lectureship at Canterbury University College. Russell already knew him, and had supported him for the post in an open testimony dated 12 October 1936 from Tele-

---

This article was first published in *Russell, new ser.,* 12 (1992-1993), 3-18; it is reprinted with the kind agreement of the editor, Dr. Kenneth Blackwell. Three changes have been made. 1) Russell's testimony of 1936, quoted here in the body of ¶1, only came to light when the paper was on proof, and so it had to be shoved into a footnote there. 2) ¶4.3 contains a review by Popper, written in German, of Russell's *History of Western philosophy*, with my translation in parallel; originally the latter appeared in *Russell* immediately after the article. 3) Since the article was published, a letter written by Popper in 1957 has ben found; it is included as ¶4.7, and the later documents have been suitably renumbered.

There also exist two undated typescript pages entitled 'Note for Lord Russell' listing some notations for the propositional calculus; perhaps they relate to Popper's own work of the late 1940s on logic, on which see [Schroeder-Heister 1984].

[1] On the role of G. H. Hardy in Russell's appointment, see [Grattan-Guinness 1991, 176]. Mishandling of my word-processing facilities caused a footnote to disappear; the account of Hardy's 1929 lecture should have been supplemented by a reference to [Hardy 1929]. Since that article was prepared, R. Kanigel has published an excellent study [1991] of Ramanujan that contains much information on Hardy.

graph House, Harting, Petersfield: 'Dr. Karl Popper is a man of great ability, whom any university would be fortunate in having on its staff. I learn that he is a candidate for a post at Canterbury University College, Christchurch, New Zealand, & I have no hesitation in warmly recommending him'.

Popper left this post in 1945, and arrived in England in January 1946 to take a Readership at the London School of Economics. His *The open society and its enemies* had appeared the previous year in London from Routledge, but the American publication was not yet secured.

On 22 July Russell replied from Penrhyndeudraeth, Wales,[2] agreeing to a request (now lost) from Popper to recommend *Society* to Simon and Schuster but asking to be sent a copy (document ¶4.1). On 15 August he sent to Popper the promised 'appreciation'; he praised the book very highly, agreeing with Popper's evaluation of Plato as a totalitarian and finding 'deadly' his treatment of Marx (¶4.2.2). In a covering note he informed Popper that an unbound copy of *History* would be coming from Simon and Schuster (¶4.2.1). The 'Inner sanctum' of Simon and Schuster sent an encouraging letter to Russell on 30 September 1946, which Russell forwarded to Popper on 5 October; however, nothing came of this initiative, although Routledge used extracts of Russell's appraisal on the dust jacket of their 1947 reprint of the book. *Society* appeared in the USA only in 1950, from Princeton University Press, who included some of the extracts from Russell's appraisal on their dust jacket.

Russell's *History* appeared in Britain from Allen and Unwin in 1946, and Popper prepared a review of it for broadcasting on 19 January 1947 with the Austrian Broadcasting Service (¶4.3). The circumstances of a radio broadcast with limited allotted time prevented him from analyzing the book in detail; for example, while he placed Russell on a level with Kant, he did not discuss the deeply unsympathetic chapter on Kant contained in the book. He praised the author to the skies as the only great philosopher of our time, the most important contributor to logic after Aristotle when he published *The principles of mathematics* (1903) the kind of thinker who saw philosophy as 'argumentation' 'before Fichte and Hegel ruined it' and kept it free from the 'charlatanry and windbaggery' of 'our time', where the aim was 'to beguile' rather than 'to instruct'. Popper took these terms from Schopenhauer, but as an example he may have had in mind a more recent experience.

2. A game of poker: Popper, Russell and Wittgenstein at the Cambridge University Moral Sciences Club, 1946

An event notorious in the annals of modern British philosophy occurred on the evening of Saturday 26 October 1946. Popper had accepted an invitation from the Cambridge University Moral Sciences Club to speak on some 'philosophical puzzle', but he actually addressed it on the topic 'Are there philosophical problems?'. The lecture was delivered in the rooms of R. B. Braithwaite at King's College. Popper began it by stressing the difference between the invitation and its fulfilment, whereupon Witt-

---

[2] Russell wrote this letter from a hotel in Penrhyndeudraeth, not from the house there that he was to rent and occupy from August 1955 until his death in 1970.

genstein indicated that he had directed the form in which the invitation was to be sent. During the evening he interrupted, and at one point even resorted to the behavioural extreme of brandishing a poker at Popper; he then left the room after Popper claimed that this action had infringed the moral rule advocating the peaceful use of pokers, thereby exhibiting to the members of the Cambridge University Moral Sciences Club the existence of a philosophical category that Wittgenstein prohibited.

Popper included his own account of this event in his autobiography, which was published in 1974;[3] but versions of it were in circulation soon after its occurrence, and some have appeared in print.[4] As Popper indicates in his account, the truth-contents of the versions are extremely variable; in particular, a well-known variant states that Russell was not present. Letters between Popper and Russell written at the time allow us to avoid such mistakes.

The Cambridge lecture arose first in the correspondence in a letter from Russell of 16 October, now based at Trinity College. He promised to be present at the lecture (though he misdated it to Friday 25 October), and also offered to meet Popper. They did meet on the Saturday afternoon, and then Popper spoke and defended himself from various forms of attack in the evening, returning to his home in Barnet that night. The next day he wrote a long letter to Russell (¶4.4), thanking him both for their meeting and for his support during the lecture. He recalled Russell's mention of Locke's philosophy during the discussion after the lecture, and criticized Wittgenstein's conception of philosophy and attitude to other philosophies. The sentiments expressed, and even in places the wording, are similar to the broadcast review of Russell's *History* that he was to prepare three months later.

From the end of Popper's letter it emerges that on some occasion (of which there is no surviving record) Russell *himself* had suggested the anti-Wittgensteinian subject for Popper's lecture; and in his reply from Trinity College on 18 November Russell indicated his cool attitude to Wittgenstein, deploring the 'failure of good manners' during the meeting but not being surprised at its manifestation in Wittgenstein himself (¶4.5: he did not refer to any particular incident). At the end he invited a further meeting, and in a note of 30 November suggested a date early in December to rendevous at his London flat in Dorset House.

---

[3] [Popper 1974, 97-99]; also in [1976, 122-124]. Russell did not mention the event in his own autobiography.

[4] Different accounts have appeared in [Clark 1975, 494-495], [Blackwell 1981, 3 and n. 11], [Munz 1985, 1-2] and [Monk 1990, 494-495]. An admittedly fictionalized account appeared in a novel by Bruce Duffy [1987, 12-15]. {[Edmonds and Eidinow 2001] is a book centered upon this event and written after this article first appeared; they do not cite it, but they quote from some letters and query the role that I assign to Russell [pp. 39-42, 223-226].} On Wittgenstein's manner of conduct that year, see [Malcolm 1984, 39-40, 46-47, 52-55].

3. Exchanges over books, 1959

After 1947 contacts were much more occasional. {On 27 August 1957 Popper wrote, but maybe did not send, a letter appreciating Russell's defence [1957] of his theory of definite descriptions against an alternative view recently put forward by P. F. Strawson (¶4.6).} Some contacts were oblique, such as Russell to a sixth-former on 10 February 1960:

> I do not think that, if philosophy is what interests you, you need regret not going to Cambridge, as philosophy there is at low ebb. I should have thought that the London School of Economics would be the place for you. There is a good deal of philosophy there and it has the merit of being vigorous.[5]

Another coincidence of publication occurred in 1959. Popper's own translation/edition *The logic of scientific discovery* of his *Logik der Forschung* (1935) came out, and he wrote on 19 January from his home at Penn to say that he had asked his publisher to send Russell a copy. He also reported that a book-length *Postscript* was in preparation, of which the first chapter 'was written in the spirit of addressing you — of an almost personal discussion with you' (¶4.7).[6] In a reply of 22 January from his home at Penrhyndeudraeth, Russell thanked him for the copy of *Logic*,[7] and looked forward to the appearance of the *Postcript* volume. When his own *My philosophical development* came out in May, he sent Popper a copy, which excited on 29 May both an admiring response from Popper together with criticisms of his treatment of induction, and also a request to dedicate the *Postcript* to him (¶4.8). On 9 June Russell reported himself too concerned with nuclear war to deal with induction, but he warmly accepted the dedication (¶4.9).

However, Popper has always been more deliberate a publisher than Russell; in the end the *Postscript* did not appear until 1982 and 1983 (under the editorship of W. W. Bartley, III), and by then Popper had forgotten the planned dedication. And now in 1992 another connection comes into operation; for most books of both men will now be published by Routledge.

---

[5] Russell to Barry Sturt-Penrose (RAI 720). Compare the sentiments (and also the comments on Whitehead's philosophy) expressed to Popper on the following 6 May (¶4.10).

[6] In the book as published this material was presumably (a forerunner of) Popper's *Realism and the aim of science* [1982, ch. 1], which begins with the occasion of 1935 recalled in the next footnote.

[7] Russell also stated here that he had read Popper's *Logik der Forschung* upon its appearance in 1935, but he was being forgetful or over-polite: Popper sent him a copy that is in his library (now at the Russell Archives), but its pages are basically uncut. He may then have browsed in its first pages; and soon afterwards he will have heard some of Popper's views on the philosophy of science when Popper spoke in the discussion following Russell's lecture to the Aristotelian Society on 6 April 1936 on 'the limits of empiricism' [Popper 1974, 87-88; 1976, 109-110]. He spoke after L. S. Stebbing (Chairman), C. A. Mace, W. Kneale and A. J. Ayer [*Proceedings of the Aristotelian Society, new ser., 36*(1936), 274: Russell's paper is on pp. 131-150].

4. Documents

The ten documents, five for each, are transcribed here. Original documents are held in the Bertrand Russell Archives, McMaster University, Hamilton, Canada, and/or at the Karl Popper Archive, Hoover Foundation, Stanford University, USA (microfilms at the London School of Economics, and at the Karl Popper Library, University of Klagenfurt, Austria). The archives involved, designated by the letters 'RA' and 'PA', possess cover typed materials, top copies or carbons as appropriate. To save space the addresses at the heads of the letters are laid out in one line, and other printed letter-head information such as telephone numbers is omitted. Editorial interpolations or corrections are enclosed within square brackets, although the first pair is by Russell.

¶4.1 *Letter from Russell to Popper, 22 July 1946, holograph (PA).*
[Address till Aug. 15] The Hotel Portmeirion Penrhyndeudraeth North Wales   22.7.46
Dear Dr. Popper

Thank you for your letter. I should be glad to commend your book to my American publishers, Simon & Schuster. But owing to the fact that at the moment I have no house, my books (including yours) are inaccessible. Could you lend me a copy, as I could write a better recommendation after refreshing my recollection of the book?

My "History of Western Philosophy" was published last year in America, & will be published here in September. I think my point of view is very similar to yours, but it would have been very regrettable if, on this account, you had refrained from publishing your book, the more so as, in mine, the kind of thesis that interests you occurs only in relation to some of the men discussed.

I shall be back in Cambridge early in October, & shall be glad to see you there if you ever visit that place.

Yours sincerely
    Bertrand Russell

¶4.2 *Russell's letter covering a testimony for Popper's* The open society and its enemies, *30 August 1946, holograph (PA)*. First complete and textually faithful publication of the testimony.
[¶4.2.1] The Hotel Portmeirion Penrhyndeudraeth North Wales
August 30, 1946
Dear Dr. Popper

I have just finished re-reading your book, & think even better of it than at first reading. My only serious disagreements are: (1) I do not think there is adequate historical evidence for your favourable view of Socrates (or for its opposite); (2) I think Plato's view of Justice is more complex than appears in your account, & more connected with traditional Greek ideas; (3) I take a more unfavourable view of Marx's character than you do. Otherwise I agree closely with your hostility to Plato & Hegel, with your view that history has no "meaning", & with your positive opinions.

I have asked my publisher to send you an unbound copy of my History of Philosophy — published in U.S. a year ago, but still unpublished here. I have asked them also to send you my "Freedom & Organization, 1814-1914", where there is a long examination of Marx.

I am writing to Simon & Schuster on your behalf, & I enclose an appreciation, of which you may make whatever use you choose. (I have already sent it to Simon & Schuster.)

Yours sincerely
Bertrand Russell
Address after Sep. 20, Trinity College, Cambridge.

[¶4.2.2] "The Open Society & its Enemies", by Dr. K. R. Popper (of the London School of Economics & Political Science) is, in my opinion, a work of first class importance, & one which ought to be widely read for its masterly criticism of theoretical enemies of democracy, ancient & modern. His attack on Plato, while unorthodox, is to my mind thoroughly justified, & is in line with my own view of that arch-totalitarian. Uncritical admiration of has vitiated political thought ever since the Renaissance.

He points out that Fichte, the philosophic progenitor of German nationalism, was willing to throw in his lot with either the French or the Russians until he was made Professor in Berlin. His analysis of Hegel is deadly, & very able. Marx, whom he discusses at length, is treated more leniently than Plato or Hegel, but is dissected with equal acumen, & given his due share of responsibility for modern misfortunes.

The book as a whole is a vigorous & profound defence of democracy & of a philosophic outlook likely to promote belief in democracy. It is timely, & calculated to have an important beneficent influence. It is also very interesting & very well written. I cannot doubt that it will appeal to a large circle of readers.

Bertrand Russell.
August 1946

¶4.3 *Popper's unpublished review for radio of Russell's* History of Western philosophy, *19 January 1947 together with my translation (PA)*. The original was typed, but with handwritten alterations by Popper. Only those underlinings which seem to indicate scribal rather than vocal stress have been adopted. There is also an extra folio containing three lines of text, which appear in the definitive version virtually unaltered. Popper checked my English translation.

AUSTRIAN TALKS (408) for 19th January 1947 18[.]30
BOOK REVIEW: THE HISTORY OF WESTERN PHILOSOPHY
by BERTRAND Russell: reviewed by K. R. Popper.

Bertrand Russell hat ein neues Buch geschrieben. Es ist ein grosses Werk, gross in seinen Ideen, groß in seiner Anlage und groß in seiner Bedeutung. Der Titel ist: A History of Western Philosophy, auf Deutsch, Geschichte der Abendlaendischen Philosophie. Das Buch kann wohl einzigartig genannt werden. Jedenfalls ist es das erste	Bertrand Russell has written a new book. It is a great work, great in its ideas, great in its inspiration and great in its significance. The title is: A History of Western Philosophy, in German, Geschichte der Abendlaendischen Philosophie. The book can well be called unique. In any case, it is the first of its kind.

seiner Art. Es gibt viele Philosophie[-]Geschichten, vielbaendige und einbaendige, gut[e] und schlechte. Aber bisher gab es noch keine, die von einem wirklich grossen und originellen Denker beschrieben wurde. Die meisten wurden von gutunterrichteten Gelehrten geschrieben, aber zwischen einem gelehrten Professor der Philosophie, der Philosophiegeschichte schreibt[,] und einem Mann wie Russell, der selbst Philosophiegeschichte macht, ist ein grosser Unterschied. Das erklaert vielleicht teilweise das einzigartige an diesem Buch. Es ist ein Buch, das mit Klarheit und gleichzeitig mit liebenswuerdiger Leichtigkeit geschrieben ist; ein Buch das[s] es wagt, und das[s] es wagen kann, die Geschichte der Philosophie mit Humor und Grazie zu behandeln.

Was ist das Besondere und das Grosse in Russel[l]'s Buch? Das Inhaltsverzichnis ist nicht wesentlich verschieden von anderen Philosophiegeschichten. Russell selbst beschreibt das besondere Ziel, dass er sich gesetzt hat, folgendermaßen; er will jeden Philosophen aus seinem sozialen Milieu heraus verstehen, und seine besondere Philosophie (soweit so etwas moeglich) durch die sozialen Umstaende und die politischen Institutionen und Probleme seiner Zeit erklaeren. Aber das Ziel ist, meiner Ansicht nach, kaum das Besondere an Russell's Buch. Denn diese Methode kann heutzutage kaum mehr als originell bezeichnet werden. Es ist wahr, dass Russell der erste ist, der diese sozial-geschichtliche Methode auf die ganze Geschichte der Philosophie ausdehnt; und daß es ihm gelingt, neues Licht auf viele Probleme zu werfen — insbesondere auf Probleme und Gestalten der Philosophie des fruehen Mittelalters, zum Beispiel Augustinus und Boethius. Aber trotz des Reizes dieser

There are many histories of philosophy, multi-volumed and single-volumed, good and bad. But up to now there has not been one written by a really great and original thinker. Most were written by well trained scholars, but there is a great difference between a scholarly professor of philosophy who writes the history of philosophy, and a man like Russell who is himself a maker of the history of philosophy. Perhaps this explains in part the uniqueness of this book. It is a book that is written with clarity and at the same time with a cheerful ease; a book that ventures, and that can venture, to handle the history of philosophy with humour and with grace.

What is special and great in Russell's book? The table of contents is not essentially different from other histories of philosophy. Russell himself describes the special purpose upon which he has settled as follows: he wants to understand each philosopher from his social context, and to explain the philosopher's special philosophy (as far as possible) through its social circumstances and the political institutions and problems of his time. But in my view the aim is hardly the special thing in Russell's book. For nowadays this method can hardly be claimed to be original any more. It is true that Russell is the first to extend this socio-historical method to the entire history of philosophy; and that this allows him to throw new light on many problems — in particular, on problems and forms of philosophy of the early middle ages, for example Augustine and Boethius. But despite the attraction of these problems, I do not believe that the greatness of Russell's book lies in its socio-historical method.

What makes the book great is the man who has written it. The book is the man. By that I do not want to say that the book is less objective that other histories of philosophy. On the contrary, other books seek earnestly to be objective, but they never achieve it. What they achieve is only that they seem to be objective, and with that to give a false impression to the reader. Russell does not attempt to be objective. He permits himself to state his opinion simply and openly, and he makes it quite clear that this is his personal opinion — his well considered opinion — but not more; certainly not the judgement of history.

In my eyes, Bertrand Russell is without doubt the only man in our time of whom one can say that he is a great philosopher — a philosopher who can be named in the one breath with men like Descartes, John Locke, David Hume, or Immanuel Kant. He is the man whom we thank that philosophy has not entirely lapsed into one of the intolerable fashions of our time, and into charlatanry and wind-baggery. The expressions 'charlatanry' and 'windbaggery' were deployed by Schopenhauer, who saw these things and fought against them, as did Kant.

Until Fichte and Hegel ruined it, philosophy was argumentation. Arguments counted — otherwise nothing. Since Fichte and Hegel philosophy has moved towards spell-weaving. It has given up instructing us, and instead seeks to beguile us, as Schopenhauer said. The trendy

hoerten auf, Argumente fuer ihre Meinungen beizubringen. Sie posieren als Propheten, als Menschen, die durch tiefes Denken zu tiefer Weisheit gelangt sind, und die uns in der Fuelle ihrere Weisheit aus ihrem Ueberfluss ein paar Brocken zukommen lassen.

Diese Philosophie der grossen philosophischen Fuehrer und Verfuehrer, der grossen Propheten, Pedanten und Schwindler, dieser philosophische Faschismus, ist immer noch maechtig. Diese Philosophie ist ein maechtiger und ein verderblicher Einfluss. Aber sie ist nicht all maechtig. Dass sie in unserer Zeit nicht wirklich allmaechtig wurde, dass die Tradition der Vernunft den Angriff der Unvernunft bisher ueberlebt hat, das verdanken wir niemandem mehr als Bertrand Russell.

Russell hat viele wichtige Beitra[e]ge zu einer wissenschaftlichen Philosophie geliefert, insbesondere zur Logik. Seine [']Prinzipien der Mathematik[', 1903] waren wohl zur Zeit ihrer Veroeffentlichung der wichtigste Beitrag zur Logik der seit dem Tode des Begruenders der Logik, Aristoteles, gemacht worden war. Der Einfluss dieses Werkes auf die weitere Entwicklung der Logik und der Philosophie der Mathematik war ungeheuer. Aber das alles ist nicht wirklich das Grosse an Russell. Was macht ihn gross? Ich traue es mich kaum zu sagen: Er war der erste Philosoph seit Kant, der es wagte, seine Meinung offen und ohne weitere Umschweife zu aendern. Der einzige Philosoph, der nicht als allwissend posierte, sondern der offen zugab, dass er sich irren konnte; der durch die Tat bewies, dass ihm nur eines wichtig war: zu lernen und die Wahrheit zu suchen. Ich weiss nicht,

philosophers, who beguile us instead of instructing us, found an uncommonly simple means. They stopped putting forward arguments for their opinions. They pose as prophets, as men who have come to deep wisdom through deep thought, and in the richness of their wisdom give us a few lumps out of their surplus.

This philosophy of the great philosophical leaders and tempters, of the great prophets, pedants and swindlers, this philosophical Fascism, is still strong. This philosophy is a strong and a pernicious influence. But it is not all powerful. That it was not really all powerful in our time, that the tradition of reason in the attack upon unreason has survived until now, for that we thank no-one more than Bertrand Russell.

Russell has made several important contributions to an intellectual philosophy, especially to logic. His The principles of mathematics [1903] was the most important contribution to logic that had been made at the time of its publication since the death of the founder of logic, Aristotle. The influence of this work on the later development of logic and the philosophy of mathematics was enormous. But all that is not really the greatness in Russell. What makes him great? I hardly dare say: he was the first philosopher since Kant who ventured to alter his opinion, openly and without beating further about the bush. The only philosopher who did not pose as infallible, but who openly admitted that he could err; who through this act proved that to him only one thing was important: to learn, and to seek, the truth. I

wie oft Russell seine Meinung geaendert hat, [a]ber ich weiss, dass jedesmal wenn er es tat, es einen Fortschritt in der Philosophie bedeutete. Er aenderte nie seine Meinung ohne gute, sehr gute Gruende fuer die Aenderung vorzubringen. Und immer gabe er diese Gruende mit grosser Offenheit und Schlichtheit an.[8] Diese Aufrichtigkeit und intellektuelle Unbestechlichkeit, diese selbstvergessene Hingabe an die Sache der Vernunft, diese schlichte Menschlichkeit, das ist der Mann. Und das ist sein Buch — eine Philosophiegeschichte voll von glaenzenden Ideen; geschrieben von dem klarsten, dem schlichtesten und dem menschlichsten Denker unserer Zeit.

END

do not know how often Russell has altered his opinion, but I know: every time when he does it, it signifies progress in philosophy. He never altered his opinion without bringing forward good, very good reasons for the modification. And he would always give his reasons with great openness and simplicity. This sincerity and intellectual incorruptibility, this selfless devotion to truth and to reason, the simple humanity, that is the man. And that is his book — a history of philosophy full of enlightening ideas; written by the clearest, simplest and the most human thinker of our time.

END

¶4.4 *Letter from Popper to Russell, 27 October 1946, typed (PA).*

The London School of Economics and Political Science Houghton Street, Aldwych, London, W.C.2   October 27th, 1946.

Dear Lord Russell,

I am typing this letter — it won't be very short, and I do not wish you to have to decipher my handwriting.

I returned from Cambridge fairly late yesterday night, and the first thing I am doing this morning is to write this letter. I wish to tell you how much I enjoyed the afternoon with you, and the opportunity of co-operating with you, at night, in the battle against Wittgenstein.

As to the battle itself (and the victory we won) I did not quite enjoy it much as I had hoped. I must admit that it dispelled my doubts in the reality of Wittgenstein which I had mentioned to you in the afternoon. He is real. But he is not quite the Wittgenstein I had expected to meet. My Wittgenstein, as it were, is unfortunately unreal — as I had expected — although in somewhat different way! In other words, I was very disappointed by the debate. Heated as it was, it did hardly produce anything new — no new light was thrown on the situation. My own paper contained only little (as, you will remember, I had warned you); it was, for this reason, that I had considered discussing something else.

Your bringing in Locke helped a great deal. Indeed the situation is now, feel, as clear as it can be.

(a) Locke says something about ideas. What he says is neither very clear nor very enlightening, but obviously he has in mind something relevant to philosophy.

(b) Wittgenstein says "Why does Locke say such queer things?" thereby expressing (not very clearly) that he wants to give various

---

[8] At this point Popper annotated his text with the instruction '(langsam!)'.

interpretations of what Locke might have meant. Obviously, to give such interpretations Wittgenstein has in mind is also relevant to philosophy.

(c) Wittgenstein's assertion that nothing but what he is doing under (b) is relevant to philosophy is, quite obviously, dogmatic and narrow.

Besides, it is itself a philosophical assertion not falling under (b), that is to say, it presupposes a philosophy (of philosophy and of language) which, if we discuss it, leads us beyond the "why does he say this?" of (b).

(d) it is only the fact that (a) was a philosophical assertion of sorts, which makes (b) philosophically relevant. That is to say, the primary problem is contained in (a), and (b) is only a preliminary for the improved re-formulation of (a). Thus Wittgenstein's assertion (c) is an attempt to confine philosophy to its preliminaries.

(e) Besides, the discussion of (b), i.e. the philosophical activity in Wittgenstein's sense, is not exoterically arguable. It cannot, and does not, consist of more than clever guesses about various intended meanings. It leads to a series of "He may have meant ...", but it does not lead to any assertion which can be open to argument. This fact completely destroys any link with the rationalist tradition in philosophy and must lead to esotericity.

(f) Thus, the only assertion available and capable of being discussed is the methodological assertion (c). It offers the only possibility to break through the magic circle with the help of rational argument. This is why I had choose (and why, advised by you, ultimately did choose) this topic.

Please excuse this long letter. You will understand that I had to write it to somebody who was present (and who is sympathetic to the line I took).

I should very much like to see you again and to give you, if that is agreeable to you, an oral report of my solution (I believe it is a complete solution) of the problem of induction. I think I could do it in about 20 minutes.

Yours very sincerely,
[K. R. Popper]

¶4.5 *Letter from Russell to Popper, 18 November 1946, typed (PA).*
Trinity College, Cambridge   18th November, 1946.
Dear Dr. Popper,
Thank you for your letter of October 27th which I meant to have answered sooner. I agree with you in what you say about the debate at the Moral Science[s] Club. For my part I was much shocked by the failure of good manners which seemed to me to pervade the discussion on the side of Cambridge. In Wittgenstein this was to be expected, but I was sorry that some of the others followed suit. I was entirely on your side throughout, but I did not take a larger part in the debate because you were so fully competent to fight your own battle.

I should very much like to see you again at any time when it is possible. After December 6th I shall be in London where my wife and I have a flat; the address is:-
27 Dorset House, Gloucester Place, N.W.I.
Yours very sincerely,
Bertrand Russell

¶4.6 *Letter from Popper to Russell, 27 August 1957 (PA, photocopy at RA)*.

Fallowfield, Manor Road, Penn, Buckinghamshire
August 27th, 1957
My dear Lord Russell,

This is only to tell you how much I, and all my best students, have enjoyed your little note on Strawson. There are still quite a few people left, also among the young ones, who have some judgement, and some feeling of intellectual responsibility; but you may be assured that they are all on your side.

Yours sincerely,
K. R. Popper

¶4.7 *Letter from Popper to Russell, 19 January 1959, typed (PA, RA)*.

Fallowfield, Manor Road, Penn, Buckinghamshire
January 19th, 1959.
Dear Lord Russell,

I feel very diffident about writing to you, and I can only ask you not to answer this letter if you do not feel like it. I know how worrying and time-consuming the answering of letters can be.

It is a long time since I saw you last, and I have all this time been working on a book entitled Postscript: After Twenty Years. The book has been in galley proofs for the last two years, but owing to various difficulties, (among them trouble with my eyes), I have not finished the correction of these galley proofs.

One of the reasons for the delay is my trouble with my English. I have always considered your way of writing as my model, and as a consequence, I am always terribly dissatisfied with what I have written.

The first chapter of the Postscript is very long (129 galleys — about 260 pages, I should say), and it was written in the spirit of addressing you — of an almost personal discussion with you.

I had always hoped, for these last two years, to finish my corrections very soon, and to send you this chapter for your comments. But since I have still not completed my corrections, I have meanwhile asked my publishers to send you a copy of my Logic of Scientific Discovery. I do not expect that you will find time to read it, but you might find time to read the two Prefaces. The second contains my criticism of the present language analysts (both English and American). I feel fairly certain that you will not disagree. I wished I could feel as certain that you will be amused.

Yours sincerely,
K. R. Popper

¶4.8 *Letter from Popper to Russell, 29 May 1959, typed.* Popper misplaced the top copy and sent the carbon; the originals are held at PA and RA respectively.

Fallowfield, Manor Road, Penn, Buckinghamshire May 29th, 1959.
My dear Lord Russell,

Thank you so much for sending me My Philosophical Development. It is a wonderful book — to me it is more fascinating than any book I have read since your Portraits from Memory. I think that it is much better than the Portraits or your Autobiography in The Philosophy of Bertrand Russell. There are very, very few books which can compare with it.

I have, of course, some criticisms, and I should like to mention one of them.

As you explain your theory of scientific inference (pp. 200ff.), you make it clear that your credibility is to satisfy Keynes' formal system of probability, and that your five postulates have the purpose of conferring a finite prior (or a priori) probability upon certain generalizations and not upon others.

But the interesting laws of science can be shown to have a probability which is $\leq$ that of the less interesting laws. Loosely speaking, one can measure the interest of a general theory by its improbability: the scientist tries to pack as much information or content as possible into his theories; and it is obvious that a theory that tells us more than another must have a smaller prior probability.

Thus what we may call, intuitively, the credibility or acceptability of a law cannot satisfy the laws of the (Keynesian) probability calculus.

It is a prejudice that science aims at high probability. I do not know whether you remember the following inciden[t]: In Amsterdam, in 1948, I said the same things in a discussion (after a lecture by von Wright).[9] You talked to me afterwards, saying that my argument showing that science simply does not aim at high probability for its laws was new to you, and that you regretted not to have known it when you wrote Human Knowledge. I replied that it was already in my Logik der Forschung; see for example p. 273 of the English translation.

But I do not want to give you the impression that my attitude towards your book is largely critical. Although I am a very critical reader, I am in almost all of the really important points in agreement with you, and always, even where I do not agree, full of admiration.

The fact that you have sent me your book gives me the courage to ask you whether you would give me permission to dedicate to you my next book, still in proofs, Postscript: After Twenty Years. I do not dare to hope that you will agree with its contents, and it contains much criticism of you (and some that does no longer apply to your latest formulations). But for this very reason I wish to make it quite clear that I do not belong to those of your critics who think that another "style" of philosophising is required. In fact, I should like to put my dedication roughly as follows:

TO BERTRAND RUSSELL
whose lucidity
sense of proportion
and devotion to truth
have set us an unattainable standard of
philosophical writing.

---

[9] Popper was referring here to the Tenth International Congress of Philosophy. G. H. von Wright gave a paper 'On confirmation'.

P.S. Please excuse the delay in sending off this letter, and the fact that I am sending you only a carbon copy; I lost the original letter, typed by my wife, somewhere among my papers, and although I searched for days, I could not find it again.

Yours sincerely

K. R. Popper

¶4.9 *Letter from Russell to Popper, 9 June 1959, typed (PA, RA).*
Plas Penrhyn, Penrhyndeudraeth, Merioneth 9 June 1959
Dear Professor Popper,

Your letter of [M]ay 29 has pleased and gratified me in a high degree. I am very glad you think well of My Philosophical Development.

The things you say about probability and the laws of science require serious consideration which I ought to have given sooner and would have if I had not been so preoccupied with nuclear warfare.

I feel much honoured by your intention of dedicating your next book to me with such a very flattering inscription.

Yours sincerely,

Bertrand Russell

¶4.10 *Letter from Russell to Popper, 6 May 1960, typed (RA).* Reply to an invitation from Popper of 3 May (PA, RA), to address the British Society for the Philosophy of Science on the philosophy of A. N. Whitehead.

Plas Penrhyn   6 May, 1960.

Dear Popper,

(I think formality between you and me is unnecessary)

Thank you for your letter of May 3, and for the enclosed lecture on "The Sources of Knowledge and Ignorance".[10] I have not yet had time to read the whole of your lecture, but I can see that it is interesting and important.

I am sorry that I cannot agree to give such a lecture as you suggest. I never studied Whitehead's philosophical work at all thoroughly and I made a point of not saying anything publicly in criticism of it. What I did know of his philosophical work displeased me, partly because of what I thought unnecessary obscurity, and partly because of the trail [trait?] of Bergson. I could not give such a lecture as you suggest without first reading a lot of Whitehead's work, which, on the whole, I do not wish to do as my time is very fully occupied in trying to induce the human race to let itself survive.

I am very glad that you are so vigorously conducting the fight against the "Oxford" philosophers, some of the worst of whom are at Cambridge.

Yours sincerely,

Russell

---

[10] Popper's paper was published as his [1960], and is most easily available now as the opening chapter of his [1963].

Acknowledgements

For permission to publish these documents I express thanks to Sir Karl Popper (copyright © Karl R. Popper, 1992) and to McMaster University. Advice on details was received from Popper and from Dr. K. Blackwell. Mr. D. Reed, Deputy Archivist of the Hoover Institution, kindly referred to Sir Karl's files there on my behalf.

Bibliography

Blackwell, K. 1981. 'The early Wittgenstein and the middle Russell', in I. Block (ed.), *Perspectives on the philosophy of Wittgenstein,* Oxford: Blackwell, 1-30.

Clark, R. W. 1975. *The life of Bertrand Russell,* London: Cape; Weidenfeld and Nicolson.

Duffy, B. 1987. *The world as I found it,* New York: Ticknor & Fields.

Edmonds, D. and Eidinow, J. 2001. *Wittgenstein's poker: the story of a ten-minute argument between two great philosophers,* London: Faber.

Grattan-Guinness, I. 1991. 'Russell and G. H. Hardy: a study of their relationship', *Russell, new ser., 11*, 165-179.

Hardy, G. H. 1929. 'Mathematical proof', *Mind, new ser., 38,* 1-25. [Repr. in *Collected papers,* vol. 7, Oxford: Clarendon Press, 1979, 581-606.]

Kanigel, R. 1991. *The man who knew infinity: a biography of the genius Ramanujan,* New York: Scribner's.

Malcolm, N. 1984. *Ludwig Wittgenstein: a memoir,* 2nd ed., Oxford: Blackwell.

Monk, R. 1990. *Ludwig Wittgenstein: the duty of genius,* New York: Free Press.

Munz, P. 1985. *Our knowledge of the growth of knowledge: Popper or Wittgenstein?,* London: Routledge.

Popper, K. R. 1960. 'On the sources of knowledge and of ignorance', *Proceedings of the British Academy, 46,* 39-71.

Popper, K. R. 1963. *Conjectures and refutations,* London: Routledge and Kegan Paul.

Popper, K. R. 1974. 'Intellectual autobiography', in P. A. Schilpp (ed.), *The philosophy of Karl Popper,* La Salle Ill.: Open Court, 3-181.

Popper, K. R. 1976. *Unended quest. An intellectual autobiography,* London: Fontana.

Popper, K. R. 1982. *Realism and the aim of science,* London: Hutchinson.

Russell, B. A. W. 1957. 'Mr. Strawson on referring', *Mind, new ser., 66,* 385-389. [Repr. in *My philosophical development,* London: Allen & Unwin, 1959, 238-245.]

Schroeder-Heister, P. 1984. 'Popper's theory of deductive inference and the concept of a logical constant', *History and philosophy of logic, 5,* 79-110.

8 Karl Popper for and against Bertrand Russell

Although Popper and Russell prosecuted very different kinds of philosophy, they held each other in high respect, as their correspondence shows. Their similarities and differences are explored here; among the latter, influence from Kant is especially marked.

1. Impacts

The day that Popper died early in September 1994, I was in Basel, Switzerland; next day the local Sunday newspaper carried an article about him as the first page of the arts section. The day that he was cremated I was in Zaragoza, Spain; the local newspaper carried a substantial piece, probably syndicated, written by a Spaniard. Since Popper had had no special links with either city, then presumably I had randomly sampled the (non-stupid end of the) international press, and I concluded that it had given him worldwide acknowledgement. Such tribute to a philosopher had probably not happened since the death of Bertrand Russell in February 1970: then the reaction was still greater, though driven more by his general social concerns, especially the political activities of his final decade, than by his philosophical achievements.

Russell's activities crossed my path in the mid 1960s, when I was taking a Master's degree at the London School of Economics in Popper's department. In addition to the compulsory courses on philosophy, I specialised in mathematical logic and the philosophy of mathematics (as did David Miller and my colleague Allan Findlay). During the first year of some of the instruction in logic was given by Popper himself, in the form of a late afternoon discourse. On one day — to be precise, Monday 15 February 1965 — our discussion was cut short, so that we three could go downstairs to hear a speech in the Old Theatre by Russell. The hall was packed and the audience expectant; Russell's status among the young was then very high, especially concerning the Vietnam War and the great fear of nuclear war after the Cuba crisis, and his topic was 'The Labour Party's foreign policy' in the time of Harold Wilson's premiership.

But the atmosphere sunk ever lower as the speech proceeded. One reason was Russell's delivery: suffering problems with his throat (in his last decade he lived largely on fluids due to an inability to swallow), he spoke so quietly as to be hard to hear anyway. However, the main drawback was the *manner* of delivery; as usual with him, he read out a manuscript, but in so robotic a fashion as to show little awareness of its content — straight from eyes to lips, as it were. The text is published in the third volume of his autobiography [Russell 1969, 205-215]; its, shall we say, unusual features

---

This article was first published in *Russell, new ser.*, 18 (1998-1999), 25-42; it is reprinted with the kind agreement of the editor, Dr. Kenneth Blackwell.

for a piece by Russell include lengthy quotations from newspapers, one being *The news of the world*.[1]

At that time Russell needed to raise considerable sums of money to support his various ventures, and the publication of his autobiography was one consequence. The manuscript of that volume is entirely typescript, so that the authorship(s) of the second half of it, including this speech, cannot be determined. The previous two volumes *were* written by him, at various times from 1931 onwards; intended to appear posthumously as a single book, they came out as [Russell 1967, 1968], expanded into two volumes by the insertion of rather ill-chosen and -explained chunks of correspondence at the end of each chapter. I have asked various people who worked with Russell in the 1960s, and they all assert that he was not senile; however, questions of his judgement hang over decisions taken during those years. Soon after his death there was published a lengthy repudiation of actions taken by his former secretary, Ralph Schoenman;[2] but by then much damage had been done, years after worried letters had been sent to Russell by friends.[3]

These letters can be seen in the Russell Archives, another result of his need for money. Putting his manuscripts on the market [Feinberg 1967], they were purchased by McMaster University in Hamilton, Canada, where the Archive was established [Blackwell 1968]: at Russell's insistence, it was renamed 'The Russell Archives'. I had begun to work in the history and philosophy of mathematics and logic; and by coincidence at that very time I came across a large and very important collection of letters written by Russell to his former student Philip Jourdain during the 1900s, when he had concentrated upon logic and mathematics. This finding oriented my researches specifically towards Russell;[4] it also started my connection with the Archives that has continued ever since, in connection with its development and its journal *Russell,* and also the multi-volume edition of Russell's *Collected papers*. The current publisher of the edition is Routledge, which also became Popper's house in his final years.

In the rest of this article I shall 'compare and contrast' Russell and Popper in various ways, especially as philosophers. The effect will be enhanced by further comparisons with some other figures, principally Ludwig Wittgenstein (1889-1951) and Rudolf Carnap (1891-1970). On

---

[1] My university colleague Philip Maher was also present at the lecture (we did not then know each other, and the fact emerged only during the preparation of this article); he corroborates my impression of Russell's performance. {On this lecture and its place in Russell's anti-war activities see also [Grattan-Guinness 2009].}

[2] [Russell 1970]; on p. 645 Russell alluded to some 'folly' committed by Schoenman after the lecture at the London School of Economics, which I did not witness.

[3] Disapproving of Russell's attitude to the USA, Popper did not join his students downstairs that Monday evening; he told me later that he did not know about Russell's circumstances during those years. He had been in the USA during the Cuba crisis of 1962, and maintained an enormous interest in it thereafter, reading many of the books written about it.

[4] This correspondence served as the base for my book [Grattan-Guinness 1977].

occasion I quote Popper from our conversations; the texts are based on notes that I made at the time, and not on my decidedly fallible memory.

2. From the correspondence between Russell and Popper

One early duty that I undertook on behalf of the Archives was to ask Popper in 1973 if they might have copies of his side of his correspondence with Russell. He replied regretting his inability to locate them; but when the Hoover Foundation at Stanford University sorted out his manuscripts, he was able in 1991 to invite me down to his home at Kenley in Surrey to give them to me, and also to talk about various things Russellian. On putting these documents together with those held at the Archives I found a small but fascinating exchange, which I edited as a paper (§7) in *Russell*, 'the best English-language journal in philosophy' (K. R. Popper, on two occasions in my hearing). Two episodes are worth noting here.

The first one concerns a lecture on 'philosophical problems' that Popper delivered to Wittgenstein and his coterie on 25 October 1946. Apparently Wittgenstein waved a red-hot poker, in response to the speaker's irrational insistence that there really are philosophical problems; was told by Russell to put it down; did so but shortly afterwards departed the company in anger (§7.3). Two days later Popper wrote to Russell at length about the occasion, including Russell's own advocacy, in Popper's favour, of John Locke as real philosopher. He also thanked Russell for having advised him to defend philosophical problems; that is, *Russell himself* had proposed the anti-Wittgensteinian topic, or at least encouraged it (§7.4.4).

Russell's opinion of Wittgenstein seems to have decreased monotonically over time. In their first encounters in the early 1910s Russell was deeply impressed by Wittgenstein's incisive criticisms of his logic and logicism, and his developing empiricist epistemology, and he supported the publication of the *Tractatus* in 1921 and 1922. However, doubts were soon to develop. 'He *was* very good', he wrote of Wittgenstein to the logician H. M. Sheffer perhaps in 1923,

> but the War turned him into a mystic, and he is now quite stupid. I suspect that good food would revive his brain, but he gave away all his money, and won't accept charity. So he is an elementary schoolmaster and starves. I do not believe his main thesis; I escape from it by a hierarchy of languages. He wrote his book during the War, while he was at the front; hence perhaps his dogmatism, which had to compete with the dogmatism of bullets.[5]

The second phase of Wittgenstein's thought, where philosophy was denied a proper place, earned Russell's contempt (for example, [Russell 1959, 215-223]), and he must have seen Popper's lecture as an occasion for a confrontation; but the consequences were greater than he expected. In reply to Popper's letter on 18 November 1946, Russell informed Popper that he 'was

---

[5] The original of this letter is lost; Sheffer included this passage in a letter of 27 October 1923 to the philosopher R. F. A. Hoernlé (Sheffer Papers, Houghton Library, Harvard University, Correspondence Box; copy in the Russell Archives). See also footnote 19.

much shocked by the failure of good manners that seemed to me to pervade the discussion on the side of Cambridge. In Wittgenstein this was to be expected, but I was sorry that some of the others followed suit' (§7.4.5).

The other episode straddles the first one chronologically. It concerns the British publication of Russell's *History of Western philosophy* and possibilities for the American appearance of Popper's *The open society and its enemies*. In August 1946 Russell recommended Popper's book (§7.4.2.2), though finally unsuccessfully, to an American house (it had already appeared from Routledge in Britain). When Russell's book came out in Britain (from Allen and Unwin), Popper praised it to the heights in a broadcast on Austrian Radio in November 1947 (§7.4.3). Possibly with certain recent philosophical experiences in mind, he contrasted Russell with

> The trendy philosophers, who beguile us instead of instructing us, [and have] found an uncommonly simple means. They stopped putting forward arguments for their opinions. They pose as prophets, as men who have come to deep wisdom through deep thought, and in the richness of their wisdom give us a few lumps out of their surplus.
>
> This philosophy of the great philosophical leaders and tempters, of the great prophets, pedants and swindlers, this philosophical Fascism, is still strong. This philosophy is a strong and a pernicious influence. But it is not all powerful. That it was actually not all powerful in our time, that the tradition of reason in the attack upon unreason has survived up till now, for that we thank nobody more than Bertrand Russell.

Popper's performance is indeed rather fawning; one silence will be picked up in ¶7 below. But the defence of philosophy remained a theme with him at that time; for example, he placed it at the head of a (rather insipid) lecture [1948] to the Aristotelian Society on 'What can logic do for philosophy?'.

In 1959 another coincidence of publication loomed. Russell published *My philosophical development,* and sent a copy to Popper. In his reply Popper reported that he had on proof a *Postscript to the logic of scientific discovery* and sought permission to dedicate it to Russell. The request was readily accepted (§7.4.7-§7.4.8); however, no Russellian in publishing, Popper withdrew the book, and it appeared only as [Popper 1982a, 1982b, 1983], under the editorship of W. W. Bartley, III and with different dedicatees for its three volumes.

By the time of the request Russell had largely abandoned philosophical work for political and related activities mentioned in ¶1. When not answering philosophical enquiries himself, he would refer correspondents to Popper (and to A. J. Ayer), and logical ones to W. V. O. Quine.

3. Socialism

One feature of *The open society* that Russell would have liked is its implicit, even at places explicit, advocacy of socialist values. He maintained such views throughout his life; his speech in 1965 clearly manifested his disappointment over the current Labour government, whoever crafted the text. This side of Popper's book was unexpectedly confirmed when I asked him (on a later occasion) about influences upon him of his parents when he was young; for he chose the following episode.

In Vienna at that time, at least in the bourgeoisie to which Popper's family belonged, there operated the 'Dienstmädchen' system of 'slave labour', as he described it to me. A woman worked as servant to a family for 13 days per fortnight, from a Sunday to the following Saturday week; then her employment would continue unless the head of the household decided that it be terminated at the end of the next fortnight. When Popper was about nine years old (around 1911, therefore) his father accused their servant of stealing an amount equivalent to £15, and dismissed her under this rule. Upon asking his father about the woman's prospects, 'I did not receive a satisfactory reply'. Thus the influence of his father was negative — in his own later terms, a falsification. In response to my query, he confirmed that *The open society* had been written to oppose that sort of system as well as the ones that the Nazis and the communists[6] were trying to impose in the 1940s.

4. Philosophies of science

Popper was of course well aware that the *kind* of philosophy espoused by Russell differed fundamentally from his own. Russell's kind of position(s) is characterised as 'analytic philosophy';[7] but the adjective is very unhelpful. Apart from the overuse of the word 'analytic' in philosophy anyway, this kind is associated with the philosophy of language (or sadly, to be more accurate, the philosophy of English), which never became a principal concern for Russell although his theory of definite descriptions has been a key technique in it. His own name for his position, 'logical positivism', is far better: 'logical' reflects the major influence on his epistemology of the logical enterprise culminating in *Principia mathematica* (1910-1913); 'positivism' captures the aims of his epistemological period beginning with *The problems of philosophy* (1912) and especially *Our knowledge of the external world as a field for scientific method in philosophy* (1914) — to quote for once its highly instructive title in full (as does not happen even in some reprintings of the book). Both stances were importantly inspired by his adoption around 1899 of G. E. Moore's anti-idealist position, which involved a desire to avoid deploying abstract objects [Russell 1959, ch. 4]

Russell did not often consider the philosophy of science; but one case is a popular book of 1931 on *The scientific outlook*, which began by stating inductivist epistemology as the scientific method: observations first, build-up of theories from 'a careful choice of significant facts' afterwards.[8] Popper owned a copy of this book, and showed me this passage as an

---

[6] 'Communism will always revive', Popper said to me on another occasion.
[7] On this question see, for example, [Monk and Palmer 1996, esp. chs. 1-4]. On Russell's own uses of 'analysis' see [Hager 1994, pt. 1].
[8] [Russell 1931, 15-16]. Compare Popper's discussion of Russell's affirmation of induction in *Postscript* [1982a, 52-92 *passim*].

example of how not to philosophise about science. For example, the 'careful choice' transcends observation, and also empiricism.

Russell's philosophy greatly influenced Carnap, especially from the mid 1920s when he greatly (though not entirely) reduced the role of the neo-Kantian philosophy that he taken largely from Hugo Dingler (1881-1954).[9] Indeed, Carnap's epistemological programme from then to the Second World War, outlined initially in a book with the Russell-like title *Der logische Aufbau der Welt* [1928], very much fused the techniques of Russell's symbolic logicism (especially *Principia mathematica*) with the aims of Russell's prosodic epistemological writings (especially *Our knowledge*) to produce a formal quasi-axiomatic epistemology with a strong preference for notions from physics.[10] This kind of philosophy became predominant in the group around Moritz Schlick that (to the annoyance of most members) became known around 1929 as the 'Vienna Circle'.[11] Carnap and Schlick were perhaps its two leading thinkers; they shared a distaste for metaphysics, Carnap even looking forward in his [1932] to its 'overcoming' by 'the logical analysis of language'. Russell may be another influence here; his acceptance of Moore's conversion engendered much reluctance over metaphysics, although he always granted it a place in philosophy, for example to accommodate the unavoidable need for universals [Russell 1940, 149].

Onto this threadbare philosophical scene came the young Popper, with his ideas about science being falsifiable guesswork. His book *Logik der Forschung* [1935] was published in a book series associated with the Circle of which Carnap was a co-editor, and it was praised by several members;[12] but they seem not to have realised that the clash between falsification and

---

[9] In the 1910s and 1920s Dingler was an influential figure in the (neo-Kantian) philosophy of mathematics and logic. While not of the calibre of, say, Ernst Cassirer of that ilk, his work needs revival for historical purposes. Its general lines are disclosed in [Weiss 1991].

[10] For the central role of Russell in Carnap's philosophy at this time, see [Grattan-Guinness 1997b]. This view conflicts in balance with the opinion in much writing on Carnap, but accords with Carnap's own recollection [1964, 13]. The word 'logicism' is due to Carnap, proposed in a book on Russell's (and Whitehead's) logic that is almost completely ignored by Carnap specialists: [1929, 2-3]. See also footnote 19.

Among retained neo-Kantian elements in Carnap, note, in particular, the 'autopsychological basis' of the single person rather than the 'heteropsychological' basis of a community in the *Aufbau*. For commentary, see [Friedman 1992].

[11] Outstanding among the writings on the Circle is the descriptive survey and bibliographies in [Stadler 1997]; it includes a conversation between the author and Popper [pp. 525-545].

[12] Felix Kaufmann was especially admiring (Carnap Papers, University of Pittsburgh Archives, file 28-20), while Ernst Nagel even thought Popper to be similar to the positivists (29-05). By contrast, Otto Neurath was critical, because the incompleteness of scientific theories apparently made falsification into a philosophical error (29-09).

verification was a fundamental issue, both serving as a special case of fallibilism against certainty and involving the status of metaphysics.[13]

At that time and until about 1950 Carnap and Popper maintained a cordial connection;[14] but after that contact seems to have fallen off considerably, as the differences between their philosophical directions became clearer. One divide concerned the status of metaphysics, as Popper showed in his contribution [1964] to the Schilpp volume for Carnap. Another cleft involved the role and epistemological interpretation of probability theory: here Carnap filled a considerable gap in the philosophy of Russell, who for some reason never attended properly to the subject.[15]

When a conference on the philosophy of science was planned to take place in London in July 1965, Carnap was hesitant to attend in case relations with Popper became difficult.[16] However, no disaccord arose, and an extra gathering took place on the Saturday morning when the two men debated induction and probability. While each man stood up and spoke, the other sat at the front; and by chance during Popper's contributions Carnap sat next to me on the second row. At one point Popper tried to present on the blackboard one of his criticisms, namely the failure of the transitivity law for conditional probabilities when zero probabilities were involved; but surprisingly he messed up the derivation, and Carnap did not dispute the property anyway.[17] Old irritations now surfaced, for my neighbour muttered a question that maybe only Popper and I heard: 'why don't we discuss whether 21 is a prime number or not?'. After that the atmosphere became detectably frosty: Popper had intended to slay the inductivist dragon, but he had muffed it, and moreover not on an essential matter.[18] Apart from

---

[13] Did Popper, or his editor uncle [Popper 1974, ch. 16], design the published version of this book to try to accentuate these differences? In particular, the chapter on corroboration is badly placed as the last one (ch. 10 instead of ch. 5), and the notion is often not granted the importance needed for a proper appreciation of fallibilism (§3). Also, Popper's choice of 'the logic of scientific discovery' as the title of the English edition (1959) is unfortunate: 'investigation' would have been both more accurate and appropriate a rendering of 'Forschung'.

[14] Carnap Papers, file 102-59; Popper Papers, file 282.24.

[15] Russell's most substantial account of probability theory occurs in [1948, pt. 5], a section of around 80 pages; it treats axioms, frequentist and logical interpretations, probabilistic inference, and scientific induction. The Russell Archives contains about 65 typed pages of preparatory notes for this Part and others. A shorter presentation had occurred in [1932, ch. 25]. Neither passage seems to be greatly significant or innovative. Popper's contributions to probability theory are well encapsulated in [Miller 1997].

[16] Carnap Papers, file 27-31-63. Russell was too busy with his various activities to take part.

[17] Compare [Popper 1964, art. 6]. In his classes on logic mentioned in ¶1 he stated that, since Carnap took the logical probability of a scientific theory to be zero, his enterprise seemed to be pointless.

[18] For Quine's recollection of this session, see his [1985, 337].

personal factors, the differences between their philosophies were central, with Russell's position a principal source of cleavage.[19]

5. Theory of truth and philosophy of logic

Russell adopted the correspondence theory of truth between propositions and facts, although he found it difficult to accommodate the 'objective falsehoods' corresponding to false propositions.[20] The status of truth was especially hard to locate in his logicism because he did not envision metalogic as distinct from logic; thus, to take another example, he was notoriously unclear on the relationships between implication, inference, entailment and consequence. Curiously, as he mentioned in the quotation in §2 above, he did alight upon the idea of hierarchy of languages in 1922 when writing his preface to the English translation of Wittgenstein's *Tractatus* [Russell 1922, 111-112]; but he never recognised its importance, especially when ignoring it completely on revising *Principia Mathematica* very soon afterwards! The general recognition of metatheory from theory is due principally to two other logicians: Kurt Gödel (1906-1978) (partly under the influence of David Hilbert's programme of metamathematics), whence it bore heavily upon Carnap; and Alfred Tarski (1902-1983), who duly imparted it to Popper in 1935 [Popper 1974, ch. 17]. The distinction has major consequences for logic and philosophy, and underlies many differences between Russell's and Popper's philosophies (and also of both from that of the later Wittgenstein, who rejected it entirely).

One consequence of the distinction is Tarski's assignment of the truth of a proposition in a formal language to its metalanguage, by means of a semantic device called 'satisfaction' of a proposition under correspondence with facts [Tarski 1935, ch. 8]. Like Tarski and Russell, Popper was a correspondence theorist, and claimed that Tarski's theory buttressed his own philosophy of science by vindicating his objective view of (fallible) knowledge [1972, ch. 1]. However, Tarski [1944] had asserted that his theory was epistemologically neutral; hence, for example, both Carnap and Popper gained great benefit from it. In this and other ways Popper went beyond the bounds that Tarski had correctly set for his theory of truth (§6). Russell appreciated Tarski's theory in his *An enquiry into meaning and truth* [1940, 62-65],[21] but he did not see it as requiring any major revision of

---

[19] One should add, however, that by the late 1930s Russell himself had become sceptical about Carnap's enterprises, especially over the degree of formalism deployed and the physicalistic reductionism: see, for example, [Russell 1940, esp. pp. 93, 267, 275, 310-311]. Later he judged Carnap's conception of language to be somewhat detached from reality [1945]. In later life Sheffer came utterly to deplore the activities of 'Carnap and Co.' [Berlin 1978, vii-viii].

[20] See, for example, [Russell 1906]. He omitted the last section, on objective falsehoods, from the reprint in [1910, ch. 6].

[21] Russell had been aware of Tarski's work but not of its details since at least since 1929; on 23 December 1929 he preferred Leon Chwistek for a chair for Lvóv University [Jadicki 1986, 243]. He became better acquainted with Tarski's work during their common sojourn in the USA: see his letter of 1939 to Quine in [1968, 225-226].

his epistemology; but by the time of *Human knowledge* eight years later, truth was once again 'a property of beliefs' and of its attendant sentences, and Tarski was out of sight [1948, 164-170].

One main source of the distinction was Gödel's incompletability theorem of 1931 and its corollary about not being able to prove the consistency of first-order arithmetic. Another consequence was that a logicist reduction of mathematics to mathematical logic could not be achieved. Afterwards logico-mathematical programmes became relatively more modest in scope and pragmatic in practice. In particular, when Popper became interested in logic in the late 1940s, he tried to develop an approach based upon taking logical consequence as a primitive notion.[22]

However, on one aspect Russell and Popper were agreed: adhesion to bivalent logic. Russell saw the law of excluded middle (LEM) and its equivalents as essential for logic and did not embrace the modal systems proposed in his time; he also argued for the law in his epistemology, preferring the 'logical' theory of truth (LEM always valid, but the truth-value of a proposition not always known) over its 'epistemological' competitor (knowledge tied to experience, so that LEM not always tenable).[23] Popper grounded his preference for LEM in his fallibilism; that logic provides the strongest criticism. An important figure for comparison is Quine, also an adherent who protects the honour of LEM with the 'maxim of minimum mutilation' [1986, 7, 86], and at his start the principal follower of Russell's type of logico-mathematical construction [Ferreiros 1997]. Logical pluralists such as myself find the criticism rather too brute for this often continuous and vague world (and thereby its science), and see delicacy rather than mutilation in the broader view (§5). The favourite example, 'snow is white', is indeed a good source for unclarity!

6. The formation of language

Popper's main address to the conference of 1965 dealt with 'Rationality and the search for invariants'{; it was eventually published as [1998]}. While acknowledging that scientific theories often had to follow the 'Parmenidean' tradition and propose invariants or constants of some kind as key notions, he recommended that rationality could permit 'swimming against the tide' in contexts where effect may not equal cause, or life is not always zero sum. He gave as examples thermodynamics, quantum mechanics and economics.

Another target for anti-invariance was Noam Chomsky's theory of deep structure to explain the formation and development of natural languages in children. On several of the occasions when we met Popper assailed me on this topic, clearly hinting that I should take it up; the

---

[22] As Popper recalled [1974, ch. 32], his efforts were not very successful. For modifications see [Lejewski 1974]; and for appraisal see [Schröder-Heister 1984]. Popper also considered intuitionistic negation at that time.

[23] I hope that this summary captures the purpose of Russell's long and difficult discussion in [1940, chs. 20-21].

outcome was a short essay [Grattan-Guinness 1995 = §20], of which Popper read the draft late in 1993.

Popper's position grew out of the evolutionary epistemology that he advocated from the 1970s onwards, and specifically from the four different functions of a natural language that he had learnt from his teacher Karl Bühler (for example, [Popper and Eccles 1977, 57-59]). Popper thought that one-word sentences, along with body actions, gestures, and tones of voice, are the basic building-blocks not only of the learning of language but also of its formation. Further development happens (if at all, for a given person or community) in response to problems (such as ambiguities over the words themselves), and takes the form of the emergence of two-, three-, ... word sentences or phrases; in due course syntax and grammar gradually supplant actions and tones. As parents and child psychologists know well, the initial speech utterances by children normally take the form of single words or even noises. They are usually expressed in varied and varying tones and levels of voices, and accompanied by body signals and gestures; all these qualify as sub-languages of their own, often of a (too) general character. Further, these processes are integral to the human sciences; they cannot be explained by evolution from animal languages. They are also enough; no supplementary structure is needed, deep or otherwise. On the contrary, structures then develop in the language in order to distinguish the various functions mentioned above.

Unlike Popper, Russell was a parent, and even an educator, running a school for children in the 1920s. But when he wrote a book *On education, especially in early childhood* at that time, he omitted the question of language formation; indeed, then in his behaviourist phase, he emphasised gestures and even doubted the merits of teaching words at all.[24] In a later book he considered single words and sentences in children, but he did not allow for anything in between [1932, 54-57]. As Popper was aware, Chomsky had put forward his theory in opposition to behaviourism; but he had gone too far for Popper, who saw this neo-evolutionary approach as a kind of middle way.

7. Contrasts

A main point of contrast between the two philosophers concerns *the certainty for knowledge.* Russell began a manuscript note for his *History* thus:

> R's Philosophy
> (1) Quest for certainty   (2) [Derivative] Analysis of data and premisses.
> [...]

---

[24] [Russell 1926, 77]; compare [1940, 65-68]. In a curious coincidence Quine has recently been considering this problem, in a manuscript of 1997 on 'The growth of mind and language'. Much more empiricist than Popper but less behaviouristic than Russell, he tries to build up knowledge inductively from 'observation sentences' of (nearly) similar perceptions. {It was published posthumously as [Quine 2008].}

<u>Method</u> Occam's Razor: assume science true, what is minimum assumption involved.

Minimum vocabulary and minimum premises.[25]

By contrast, for Popper philosophy was an 'unended quest', as he entitled his autobiography on its (slightly revised) separate publication as [Popper 1976]. Many philosophies, especially those greatly concerned with epistemology (of some kind), have hoped for certainties;[26] but Popper's falsificationism and its maturity into fallibilism brought him to *un*certainty as the key,[27] and to emphasise 'the *growth* of scientific knowledge' (the subtitle of *Conjectures and refutations* of 1963) from state to state rather than resting in any particular state. In later years he came to espouse evolutionary epistemology, although its kinship to the scientific theory of evolution is not clear;[28] by contrast, Russell had told Jourdain in 1910 that 'anything evolutionary always rouses me to fury'.[29]

Allied to this difference is the role of the cognitive agent. The passively empiricist side of Russell's logical empiricism is an example of 'the bucket theory of the mind', as Popper called it [1972, ch. 2 and appendix]. By contrast, his own activist line sees the source of conjecture in the fallibility of our human enterprise. Further, while empiricist Russell wished to avoid any third world of abstract objects, Popper was content to admit such a 'World 3',[30] and indeed to over-populate it to such an extent that anything and everything could be real there. If Russell's threadbare thought-world may seem to be a philosophical stupidity, then Popper's World 3 is the opposite one, in which all questions concerning ontology are overly trivialized [Cohen (J.) 1980; §2.12.4].

The divide between the two philosophies was cast principally by Immanuel Kant — or rather, by positive and negative reactions to him.

---

[25] Russell Archives, ms. 210.006746; compare the rather perfunctory treatment in [1946, book 3, ch. 21].

[26] Russell's quest faltered in 1901-1902 with the discovery of his paradox of set theory, together with his new awareness of suffering and the loss of love for his wife Alys. In his recollection of the period in his autobiography he described this triple setback in an intermingled manner that seems to be unintentional; moreover, they refute isomorphically the three aims of his life set out in the opening prologue [Russell 1967, 144-147, 13]. I made these points in my [1977, 160], and then found that Popper had partly anticipated me: see his comments in [Magee 1971, 144].

[27] And is that for sure? On the meta-epistemological issues surrounding 'comprehensive critical rationalism' see, for example, the papers [CCR 1971].

[28] See, for example, [Ruse 1977]. In a lengthy and positive appraisal of evolutionary epistemology Munz does not seem to address this issue fully [1985, ch. 6].

[29] Quoted in [Grattan-Guinness 1977, 126]. A related point of difference concerns C. S. Peirce. Russell almost entirely ignored him, concerning both logic and philosophy; for example, nothing of significance is stated in the *History*. Popper was quite praising, especially in his [1974, 1072]; see also his [1972, 212-216]. However, he did not much use Peirce's philosophy, and gave to Miller his copy of the 1930s edition of Peirce's works.

[30] See, for example, [Popper 1974, ch. 38]; or [Popper and Eccles 1977, esp. ch. P4].

When Popper broadcast his eulogy to Russell's *History* he must have realised how unsympathetic was its chapter on Kant, regarding both content and influence.[31] By contrast, Popper developed his philosophical position very much under the influence (largely negative) of Kant's treatment of induction and demarcation,[32] and in his maturity he wrote admiringly on Kant, especially the activism (for example, [1963, chs. 7-8]).

Another consequent difference is the role of history. Popper's fallibilism is intrinsically historical, in that the sequence of conjectures and refutations (or criticisms) that a theory undergoes is built in to the form that it takes (even the first theory in a given context probably uses antecedents from somewhere); such background should be recognised and understood, at least in its main lines. By contrast, the positivists see themselves on their own each time, and regard the past (including their own) as largely nostalgia at most. Carnap regarded himself 'as unhistorically minded a person as one could imagine', according to the testimony of the historian of scienec Bernard Cohen [1974, 310].

There were also differences of personality that may have accentuated the intellectual ones, at least for Popper. One marks in particular the contrast between a worldly member of the Empire-owning British Victorian aristocracy who led one of the main philosophical traditions of his time and country, and an emotional loner from a persecuted race living in foreign lands to whom such traditions were alien.[33]

In 1962 Popper hoped to write 'on what I consider to be Russell's greatest contribution to philosophy'; sadly, he never did, and the source of praise was not identified.[34] A hint may come from a letter of 1980 to the Bertrand Russell Society, when he stated his favourite Russell books to be *The Problems of philosophy* (1912) and *Mysticism and logic* (1917)[35] — two of the shortest ones, albeit the first a prolegomena for Russell's logical empiricism. There may be here a residue of the great influence of both Russell's logic and his logicism upon the Vienna Circle (especially for Carnap and Schlick[36]), the environment within which the young Popper developed his first philosophical ideas. Personally he never belonged to the Circle; but his early philosophy and his first book did, and Russell was a

---

[31] [Russell 1946, book 3, ch. 20}. The following ch. 21 on Hegel is no better — but many people say the same of ch. 12 of Popper's *Open society*.

[32] See in particular, [Popper 1979], edited from a manuscript of the early 1930s; it seems to me superior to *Logik der Forschung* and this and some other respects.

[33] The isolation of Popper, and even more that of his wife, are finely captured in [Hacohen 1996]; see also §1.

[34] Correspondence between Popper and Schoenman, Popper Papers, file 276.19. The book was to be, eventually, [Schoenman 1967].

[35] Popper Papers *ibidem*.

[36] Although Schlick published little on mathematical logic, he took a great interest in it, using Russell's version for his student seminars in 1925-1926 (one student was a certain K. Gödel) and 1932 (Schlick Papers, State Archives of North Holland, Haarlem, The Netherlands, files 52/B32-2 and 58/B38). From his early days he was an enthusiastic anti-metaphysician (file 82/C1).

major father figure there. Popper judged Russell's logical enterprise as exceptionally heroic, even though its aim was not achieved.[37]

Popper appraised Russell to me one day as 'not a great philosopher, although a brilliant writer'. But he also had in his home a large photograph of Russell, which he encouraged me to photograph him holding it {see the frontispiece of this book}: 'I loved Russell. This is Russell as I remembered him'.

Acknowledgements

For information on various points I am indebted to J. W. N. Watkins and K. Blackwell. The text was improved by reflections upon the sharp questioning of the audience of my presentation of this material at the Annual Conference on the Philosophy of Sir Karl Popper, held at the London School of Economics on 14 March 1998.

Bibliography

Berlin, I. 1978. *Concepts and categories. Philosophical essays* (ed. H. Hardy), London: Hogarth Press.

Blackwell, K. 1968. 'The importance to philosophers of the Bertrand Russell Archive', *Dialogue* (Canada), 7, 608-615.

Carnap, R. 1928. *Der logische Aufbau der Welt,* Berlin: Welt-Kreis.

Carnap, R. 1929. *Abriss der Logistik, mit besondere Berücksichtigung der Relationstheorie und ihre Anwendungen,* Vienna: J. Springer.

Carnap, R. 1932. 'Überwindung der Metaphysik durch die logische Analyse der Sprache', *Erkenntnis, 2,* 219-241.

Carnap, R. 1964. 'Intellectual autobiography', in [Schilpp 1964], 1-84.

CCR 1971. Papers on 'comprehensive critical rationalism', *Philosophy, 46,* 43-61.

Clark, R. W. 1975. *The life of Bertrand Russell,* London: Cape; Weidenfeld and Nicolson.

Cohen, B. 1974. 'History and the philosopher of science', in F. Suppe (ed.), *The structure of scientific theories,* Urbana: University of Illinois Press, 308-373.

Cohen, J. 1980. 'Some comments on Third World epistemology', *British journal for the philosophy of science, 31,* 175-180.

Feinberg, B. 1967 (Ed.), *A detailed catalogue of the manuscripts of Bertrand Russell,* London: Continuum.

Ferreiros, J. 1997. 'Notes on types, sets and logicism', *Theoria, 12,* 91-124.

Friedman, M. 1992. 'Epistemology in the *Aufbau*', *Synthese, 93,* 15-59.

Grattan-Guinness, I. 1977. *Dear Russell - Dear Jourdain. A commentary on Russell's logic, based on his correspondence with Philip Jourdain,* London: Duckworth; New York: Columbia University Press.

---

[37] See Popper's appraisal in [Magee 1971, 142-144]. However, in an irony of history, he relied on Russell's own recollection of writing *The principles of mathematics* (1903) in [1959, 72-74], which is *very* inaccurate: the much more complicated but human story is reconstructed in [Grattan-Guinness 1997a].

Grattan-Guinness, I. 1995 'Experience or innateness? Sir Karl Popper on the origins and acquisition of natural languages', *Languages Origins Society forum*, no. 20 (Spring 1995), 16-25. [§20]

Grattan-Guinness, I. 1997a. 'How did Bertrand Russell write *The Principles of Mathematics* (1903)?', *Russell, new ser.*, 16, 101-127.

Grattan-Guinness, I. 1997b. 'A retreat from holisms: Carnap's logical course, 1921-1943', *Annals of science*, 54, 407-421.

Grattan-Guinness, I. 2009. 'Bertrand Russell (1872-1970), man of dissent', *Notes and records of the Royal Society of London*, 63, 365-379.

Hacohen, M. H. 1996. 'Karl Popper in exile: the Viennese progressive imagination and the making of the *Open society*', *Philosophy of the social sciences*, 26, 452-492.

Hager, P. 1994. *Continuity and change in the development of Russell's philosophy*, Dordrecht: Kluwer.

Jadicki, J. J. 1986. 'Leon Chwistek — Bertrand Russell's correspondence', *Dialectic and humanism*, 13, 240-263.

Lejewski, C. 1974. 'Popper's theory of formal or deductive inference', in [Schilpp 1974], 632-670.

Magee, B. 1971. *Modern British philosophy*, London: Secker and Warburg.

Miller, D. 1997. 'Sir Karl Raimund Popper, C. H., F. B. A.', *Biographical memoirs of the Royal Society of London*, 43, 367-409.

Monk, R. and Palmer, A. 1996. (Eds.), *Bertrand Russell and the origins of analytic philosophy*, Bristol: Thoemmes.

Munz, P. 1985. *Our knowledge of the growth of knowledge: Popper or Wittgenstein?*, London: Routledge.

Popper, K. R. 1935. *Logik der Forschung*, 1st ed., Vienna: J. Springer.

Popper, K. R. 1948. 'What can logic do for philosophy?', *Proceedings of the Aristotelian Society, suppl. vol. 22*, 141-154.

Popper, K. R. 1963. *Conjectures and refutations*, London: Routledge and Kegan Paul.

Popper, K. R. 1964. 'The demarcation between science and metaphysics', in [Schilpp 1964], 183-226. [First published as [Popper 1963, ch. 11].]

Popper, K. R. 1972. *Objective knowledge*, Oxford: Clarendon Press.

Popper, K. R. 1974. 'Replies to my critics', in [Schilpp 1974], 961-1197.

Popper, K. R. 1976. *Unended quest*, London: Fontana/Collins.

Popper, K. R. 1979. *Die beiden Grundprobleme der Erkenntnistheorie* (ed. T. E. Hansen), Tübingen: Seebeck/Mohr.

Popper, K. R. 1982a, 1982b, 1983. *Postscript to the logic of scientific discovery*, 3 vols. (ed. W. W. Bartley, III), London: Hutchinson.

Popper, K. R. 1998. 'Beyond the search for invariants' in *The world of Parmenides* (ed. A. F. Pedersen), London and New York: Routledge, 146-222.

Popper, K. R. and Eccles, J. C. 1977. *The self and its brain*, Berlin: Springer.

Quine, W. V. 1985. *The time of my life*, Cambridge, Mass.: MIT Press.

Quine, W. V. 1986. *Philosophy of logic*, 2nd ed., Cambridge, Mass.: Harvard University Press.

Quine, W. V. 2008. 'The growth of mind and language', in *Confessions of a confirmed extensionalist and other essays* (ed. D. Føllesdal and D. B. Quine), Cambridge, Mass.: Harvard University Press, ch. 12.

Ruse, M. 1977. 'Karl Popper's philosophy of biology', *Philosophy of science, 44,* 638-661.

Russell, B. A. W. *Papers. Collected papers* (ed. various), in progress, [now] London: Routledge.

Russell, B. A. W. 1906. 'On the nature of truth', *Proceedings of the Aristotelian Society, new ser., 7,* 28-49.

Russell, B. A. W. 1910. *Philosophical essays,* London: Longman's.

Russell, B. A. W. 1922. 'Introduction', in L. Wittgenstein, *Tractatus logico-philosophicus,* 2nd ed., London: Routledge and Kegan Paul, 1961, ix-xxii. [Repr. in *Papers,* vol. 8, 96-112: cited here.]

Russell, B. A. W. 1926. *On education, especially in early childhood,* London: Allen and Unwin.

Russell, B. A. W. 1931. *The scientific outlook,* London: Allen and Unwin.

Russell, B. A. W. 1932. *An outline of philosophy,* London: Allen and Unwin.

Russell, B. A. W. 1940. *An enquiry into meaning and truth,* London: Allen and Unwin.

Russell, B. A. W. 1945. 'Logical positivism', *Polemic,* no. 1, 6-13. [Repr. in *Papers,* vol. 11, 147-155.]

Russell, B. A. W. 1946. *History of Western philosophy,* London: Allen and Unwin.

Russell, B. A. W. 1948. *Human knowledge. Its scope and limits,* London: Allen and Unwin.

Russell, B. A. W. 1959. *My philosophical development,* London: Allen and Unwin.

Russell, B. A. W. 1967, 1968, 1969. *Autobiography,* 3 vols., London: Allen and Unwin.

Russell, B. A. W. 1970. 'Bertrand Russell's political testament', *Black dwarf, 14,* no. 37 (5 Sept. 1970), 7-10. [Repr. with attestations of authenticity in [Clark 1975, 640-651]; cited here.]

Schilpp, P. A. 1964. (Ed.), *The philosophy of Rudolf Carnap,* La Salle, Illinois: Open Court.

Schilpp, P. A. 1974. (Ed.), *The philosophy of Karl Popper,* La Salle, Illinois: Open Court.

Schoenman, R. 1967. (Ed.), *Bertrand Russell. Philosopher of the century,* London: Allen and Unwin.

Schröder-Heister, P. 1984. 'Popper's theory of deductive inference and the concept of a logical constant', *History and philosophy of logic, 5,* 79-110.

Stadler, F. 1997. *Studien zum Wiener Kreis. Ursprung, Entwicklung und Wirkung des logischen Empirismus im Kontext,* Frankfurt am Main: Suhrkamp. [English trans.: *The Vienna Circle. Studies in the origins, development, and influence of logical empiricism,* Vienna and New York: Springer, 2001.]

Tarski, A. 1935. 'Der Wahrheitsbegriff in den formalisierten Sprachen', *Studia philosophica*, *1*, 261-405. [Repr. in *Collected papers*, vol. 2, Basel: Birkhäuser, 1986, 51-198. English trans. in *Logic, semantics, metamathematics*, 1st ed. Oxford: Clarendon Press, 1956; 2nd ed. Indianapolis: Hackett, 1983, ch. 8.]

Tarski, A. 1944. 'The semantic conception of truth and the foundations of semantics', *Philosophy and phenomenological research*, *4*, 341-375. [Repr. in *Collected papers*, vol. 2, 661-699.]

Weiss, U. 1991. *Hugo Dinglers methodische Philosophie*, Mannheim: Wissenschaftliche Buchgesellschaft.

9 On Popper's Philosophy and its Prospects

This essay review treats various aspects of Popper's philosophy of science, with especial reference to the alternative positions put forward by Thomas Kuhn and Imre Lakatos, which Popper criticised strongly. Some attention is also paid to the history and philosophy of mathematics, which was Lakatos's other main concern.

Philosophical papers, Volume ii: *Mathematics, science and epistemology*. By Imre Lakatos. Edited by John Worrall and Gregory Currie. Cambridge: Cambridge University Press, 1978. Pp. x + 285. £10.50.

> [...] philosophy has to choose, between several equally logical alternatives, the most rational one. The use of this principle, which is frequent, seems to me not very happy, and indeed may be employed to cloak what is really intellectual capitulation.
> Bertrand Russell [1897, 114]

> Much of my work never will be published. If I can, before I die, get so much made accessible as others may have a difficulty in discovering, I shall feel that I can be excused from more. My aversion to publishing anything has not been due to want of interest in others but to the thought that after all a philosophy can only be passed from mouth to mouth, where there is opportunity to object and cross-question and that printing is not publishing unless the matter be pretty frivolous.
> C. S. Peirce, letter of 2 December 1904 to Lady Welby, in [Hardwick 1977, 44] but not in [Cust 1931]

1. Purpose

Some opening words of explanation are required concerning the form and content of this chapter, and its original appearance in the *British journal for the history of science*. It began as a review of the second volume of Imre Lakatos's *Philosophical papers* (1978), which I was invited to review because the book contains several articles on the history and philosophy of mathematics. I have covered these articles in ¶2, and certain aspects of his historiography of science in ¶3 and ¶4. But as I read the other non-mathematical articles in the book, with the comments on Popper's philosophy of science, and the constant reference to the broader *methodo-*

This essay review was first published in the *British journal for the history of science, 12*(1979), 317-337; it is reprinted by kind permission of the current editor. Some of the overlaps with previous chapters have been shortened and cross-references given.

logical issues raised in articles that are now published in the first volume of the edition, I felt the need to look at Popper's philosophy in general and its bearing in both Lakatos's methodology and the historiography of science, with reference to the following two problems.

Firstly, since Popper's philosophy of science is based on the idea of falsification of theories, it follows that a theory is to be assessed in part in terms of falsifications of its predecessors, which is a historical matter; hence the history of science is intertwined with Popper's philosophy of science. However, while some of his ideas have been advocated in the past (especially concerning the fallibility of knowledge), philosophies of science that were dominant earlier were not Popperian in character; indeed, they were often of a kind that Popper has criticized (inductive epistemology, for example). Hence, how do we use Popperian philosophy in a past, which was not Popperian in its philosophical environment?

Secondly, several features of Popper's philosophy have been criticized, including its bearing on the history of science; in particular, Lakatos's methodology was intended to be both an improved philosophy and a richer historiography of science. What are the matters for concern in Popper, and how does Lakatos's approach improve them? The discussion here is inevitably of a largely philosophical character, but its bearing upon the history of science should be clear: for the problems are perennial, and in fact first attracted my attention when trying to understand the thought of scientists in the past.

The results of these considerations take up ¶5-¶9 of the review. Particular attention is given to the epistemological status of falsification, the use made of Alfred Tarski's theory of truth, the relationship between ontology and facts, the distinction between science and pseudo-science, and the logic of discovery in non-scientific areas. In a concluding ¶10, I bind together these considerations in a brief survey of the place of rationality of science and in its history. {Page references to the Lakatos edition take the form '[L1, 213]', which specifies page 213 of the first volume.}

2. Philosophy and history of mathematics

2.1 *Lakatos on the philosophy of mathematics.* The best-known product is his 'Proofs and refutations', published as the paper [1963-1964] and posthumously in a revised book form in [1976]; five other works in this area constitute the first part of this book. His main contribution to the philosophy of mathematics was to accept Georg Polya's emphasis on the modification of proofs as a source of mathematical ideas, and build it up into a large-scale philosophy both for mathematics and its history. In 'What does a mathematical proof prove?', written around 1960 [L2, 61-69], he distinguishes between pre-formal, formal and post-formal proofs. Post-formal proofs seem to be informal versions of formal (that is, fully detailed) proofs, and he calls them 'formal proofs with gaps'. Pre-formal proofs are proofs for which there is not (yet) available a formal proof for comparison. The process of making a proof more rigorous is intimately linked with the study of the foundations of a mathematical theory; their status is discussed in 'Infinite regress and the foundations of mathematics' of 1962 [L2, 3-23].

I make first an historical remark about Lakatos's philosophy of mathematics. It is not true that his proof-oriented philosophy of mathematics was without precedent. In addition to Polya's writings, where this approach is in embryonic form, René Dugas and François Rostand both wrote extensively on incomprehension and unclarity in mathematics: indeed, Rostand's discussion of errors and omissions that can occur in proofs is more extensive than Lakatos's. It is most unfortunate that their fine work has been ignored; in particular, Lakatos appears to have been unaware of it.[1]

I comment now on Lakatos's own contribution. His philosophy of mathematics certainly works well for the polyhedron theorem, in the context of which it was partly developed; but often it is much less informative elsewhere. For example, a sloppy manuscript of 1966, entitled 'Cauchy and the continuum: the significance of non-standard analysis for the history and philosophy of mathematics' [L2, 43-60], edited for publication by John P. Cleave, considers the history of A.-L. Cauchy's 1821 "theorem" that the sum-function of a convergent series of continuous functions is itself continuous. There were apparent counter-examples to the theorem, and yet Cauchy had a proof.

Lakatos looks at the history of the theorem from Cauchy onwards in terms of Abraham Robinson's modern theory of infinitesimals. It seems to me that he grossly exaggerates the similarities between Robinson's infinitesimals and those of the classical theories [Bos 1974, 81-84]. With regard to Cauchy's 1821 proof, I think that he is right to believe that Cauchy worked with a system of infinitesimals; but Cauchy's use of 'infinitely large' is only the usual *façon de parler* of the time for 'finite but very large', and the parallel with non-standard analysis does not hold concerning actual infinities. With very few exceptions, the infinitely large was abhorred until Georg Cantor's work in the 1870s, and Cauchy is one who explicitly rejected it for its allegedly contradictory properties.[2] Again, care must be taken over the order in which multiple limits are handled in proofs of this kind, and Cauchy was not aware of this requirement at the time of his theorem.

In addition, Lakatos's command of the history of this theorem is inadequate. E. H. Dirksen's attempt in 1829 to reconcile theorem and counter-example is overlooked. Cauchy's word-for-word 1833 repetition of

---

[1] See [Dugas 1940]; parts of this rare book are published as the paper [1939]. Rostand [1952, 1960, 1962]. Among Polya's writings, I recommend especially [1954].
[2] See [Bunn 1977]. Cauchy denied the infinitely large numbers in [1833a, 422], in lectures that were first published in 1868 by Cauchy's follower the Abbot Moigno, who emphasized Cauchy's point by including his own essay against the actual infinite. In 1833 Cauchy also repeated his 1821 theorem and proof word-for-word in his *Résumés analytiques* at Turin [Cauchy 1833b, 55-56]. The same misunderstanding of Cauchy's 'infinite' occurs in [Fisher 1978]; and David Bloor accepts without criticism Lakatos's history of Cauchy's theorem in his [1978, 262-263]. {See now also [Grattan-Guinness 1987] on the role of the Swedish mathematician E. B. Björling in the reaction to Cauchy's theorem.}

both theorem and proof, after apparent counter-examples had been published, is not considered. Karl Weierstrass is said to have 'lectured' since 1841 on uniform convergence (which clears up the ambiguities of Cauchy's proof) 'with text book clarity' [L2, 46], whereas Weierstrass was then only using the idea in research (almost certainly he obtained it from his teacher Christoph Gudermann) and was not likely to teach it at the Münster Gymnasium or the school at Deutsch-Krone. He certainly did not publish anything on it at the time, and thus it was unavailable for Ludwig Seidel's alleged '"translation" of Cauchy's proof into Weierstrass's theory' [L2, 50] to take place in 1847.

Finally, Seidel's definition of not arbitrarily slow convergence, which was probably inspired by the contemporary interest in the rate of convergence of an infinite series, is basically different in form from Weierstrassian uniform convergence anyway (although it is logically equivalent to one of its modes), so that even if Seidel had known anything about schoolteacher Weierstrass's manuscripts, he did not use it when making his own modification to Cauchy's theorem [Grattan-Guinness 1970, ch. 6].

2.2 On Lakatos. I have been asked several times why I have not discussed Lakatos's interpretation of Cauchy's theorem in my own writings on the history of mathematical analysis. The reason is that I learnt of its existence only after seeing it mentioned, after my own work was published, in [Cleave 1971]. Cleave cited the version contained in Lakatos's Ph. D. thesis and now published in the book form [1976] of 'Proofs and refutations', not the version that he has edited for this volume. I had been told by Lakatos's colleagues at the London School of Economics — in error but doubtless in good faith — that the version of 'Proofs and refutations' published as [1963 -1964] formed the entirety of his thesis. My own interpretation of Cauchy's theorem was contained in my own thesis, which I applied to write under Lakatos's supervision at the School in 1966 (at exactly the time, I now discover, when he wrote the manuscript under review). My application was rejected, and passed over to the School's new Department of Mathematics, where I was supervised by Professor A. C. Offord and Dr J. R. Ravetz (of Leeds University). It was defended in 1969.

I had no contact with Lakatos after 1966. In particular, I received no reply to a letter of 26 May 1971, six months after the publication of my work, rejecting the charges of plagiarism that had by then percolated down to me through mutual acquaintances — and continued to do so for more than two years, although at no time were they addressed to me directly.

3. On the prominence of rivalry between scientific theories

3.1 Lakatos on the philosophy of science. Lakatos largely abandoned his promising studies in the history and philosophy of mathematics to write on the philosophy of science in the last decade of his life. Starting from Popper's position that scientific theories are characterized by their falsifiability (or refutability) and that critical analysis is the bouquet of science, he developed a more complicated version of this view known as 'sophisticated methodological falsificationism'. Apparently in science we work with various types of acceptance of a scientific theory: acceptance$_1$ as being worthy of testing, acceptance$_2$ (at a given time) if it has yielded 'novel facts', and acceptance$_3$ as reliable for making new predictions. There are also companion modes of rejection and ad hocness.

A prominent feature of this view is the structuring of a theory into a 'research programme' containing a 'hard core' surrounded by relatively soft material. The hard core 'develops slowly, by a long, preliminary process of trial and error' [L1, 48]. The distinction between hard and soft is a crucial stage in the construction; so it is surprising to find it achieved by a process that Lakatos describes elsewhere as 'long, pedestrian' [L1, 131] and to be associated with '"*immature science*"' [L1, i, 87; compare pp. 4, 21, 111, 150]. Another important feature is the constant rivalry between research programmes. They are to be appraised by 'meta-criteria', which display the 'progressive' or 'degenerating' character of their 'problem shifts' as they confront 'observable states of affairs'; the progressive research programmes succeed over the degenerating ones. Sophisticated methodological falsification 'offers new standards for intellectual honesty' [L1, 7], and provides a theory of rational procedures in science.

3.2 *On rivalries*. I question here the ubiquity of rivalries, and the apparently imperative need for them, that feature in both the methodology and its claimed bearing on the historiography of science. In the process I quibble not only with Lakatos but also with Popper and other philosophers of science, who (for example) focus so much attention on the 'revolutions' in science, where basic changes have taken place — and hence large-scale rivalries have been settled. These occasions are undoubtedly of great historical importance, but they constitute only a fragment of scientific activity. In the vast remainder rivalries are often not significant, and when they occur are of relatively medium or small scale: they concern theoretical and/or experimental details, even personal animosities. However, for Lakatos, '*The history of science has been and should be a history of competing research programmes (or, if you wish, "paradigms"), but it has not been and must not become a succession of periods of normal science: the sooner competition starts, the better for progress*' [L1, 69, his italics].

Lakatos's comment occurs in an attack on Thomas Kuhn. Kuhn's theory of normal science and paradigms was a most welcome attempt to attend philosophically to the majority of scientific activity, with his emphasis on one predominating theory during a period of development of a science. Some of Kuhn's critics, including Lakatos and Popper, have been too harsh in deprecating the allegedly routine nature of some of normal science's "puzzles": I, for one, would have been proud to have solved some of the puzzles presented by the normal sciences of which I have studied the history. However, several of the many criticisms of Kuhn were pertinent. His multiple use of the term 'paradigm' was very confusing; his choice of 'puzzle' to describe normal scientific problem-solving was unfortunate, and invited the excessive criticism to which I have just referred; and his account of paradigm change on the important occasions of large-scale rivalry between theories was unsatisfactory. In addition, he seemed to assume the epistemological thesis of 'historical relativism' — that scientific discussions require a commonly agreed framework before they can take place — and Popper brought strong arguments against it.[3] However, I am sure that Kuhn's approach can be preserved, with historical relativism itself treated as part of the historical study: instead of the Myth of the Framework, we have the history of frameworks.

Since it is now fashionable to denigrate Kuhn's work, I shall outline the development of theoretical astronomy in the period 1750-1850, as an historical example where a normal theory was predominant and rivalry in Lakatos's large-scale sense was seldom present[4] — and one, moreover, of which Lakatos used part to illustrate what he took to be the 'characteristic story' of scientific progress. Newtonian mechanics provided the hard core for theoretical astronomy, but much soft material could be added to it when falsification seemed at hand. To start with, planets were regarded as mass-points, which affected the Sun and each other according to Newton's law. Later modifications included replacing the mass-point with a homogeneous sphere; realizing that the spheres were in fact not quite spherical, and rather than being homogeneous were solid centres surrounded by shifting liquid shells; positing the existence of as yet unobserved planets in certain orbits (this possibility formed Lakatos's 'characteristic story'); allowing for error in observation due to the refraction of rays of light when passing through our atmosphere, and to the capillary action of fluids in barometers; and wondering if the astronomical instruments might suffer distortion of shape due to exposure to the Sun while in use. Meanwhile, the mathematical part of the theory was subject to criticism or caused difficulties: terms additional to Newton's inverse square term were tried; certain integrals were only approximately integrable and/or evaluable; and quantities were calculable only by an infinite series of terms of which the sum was unknown and the rate of convergence very slow. Even the question of whether one could hope for exact mathematical solutions at all, or of whether one should use techniques of interpolation directly on the observational data, was taken seriously.

---

[3] For these and other criticisms of Kuhn see the articles by M. Masterman and Popper in [Lakatos and Musgrave 1970], and [Stegmüller 1976, 170-180]. Kuhn responded to some criticisms in the second edition [1970] of his book, and further in 'Second thoughts on paradigms' [1974]. Kuhn's defence of his objectivity in [1977, 320-339] seems anaemic to me, since it is concerned only with the criteria under which scientists take decisions. Thus once again he talks methodology rather than epistemology.

For a brief but enlightening survey of the history of methodology in science, see [Whitrow 1970].

[4] Until the mid 18th century Newtonian mechanics faced substantial rivalry from vortex theories [Aiton 1972]. Although Lakatos claims various examples of the power of his historiography in the history of astronomy, he gives this one only a slight mention [L1, 217]. Another area of rivalry in which Newtonianism largely won by the mid 18th century was Newton's conception of matter as opposed to Leibniz's.

As was mentioned, Lakatos used one small part of this history — the positing of the existence of another planet in the discovery of Neptune — for his 'characteristic story' [L1, 16-18], and Popper rightly rejected it as such [1974b, 1007]. But the story as a whole does seem to me to be characteristic of an important aspect of articulation of scientific theories: I call it 'desimplification'. Here the central part of a theory is adorned with auxiliary hypotheses, and the effectiveness of each is studied in the context of the problem situation at hand. This is one reason why a falsification leads to a complicated situation: it is often not clear which part(s) of the theoretical structure may be at fault, whether or not a significant auxiliary hypothesis has been overlooked, and whether the revision need be radical or marginal.

3.3 *On paradigms.* This particular history also displays examples of another significant feature of the history of science. In analyzing the activity of a scientific community over a period of time, Kuhn now distinguishes between its 'paradigms' (the exemplifying problems) and its 'disciplinary matrices' (the group commitments). I would distinguish both of these from an individual scientist's paradigms and commitments, especially in his early years when he may not fully belong to his community. Here are two examples of this feature that are related to the history that I have just outlined.

Firstly, as Kuhn [1976] has described well, until the late 18th century there was a marked tendency in both terrestrial and celestial physics to demarcate between the mathematical and the experimental aspects of the work. However, after that there seem to have been hopes that mathematics could provide predictions for direct experimental test: the move away from classical solutions to interpolative techniques noted in ¶3.3 is an important example. {The success of the attempted merger as such was limited, but several major changes in the personal and even some group paradigms and commitments occurred in mathematical physics and mathematical analysis in France between 1800 and 1830, in several concurrent competing and indeed partly conflicting large-scale enterprises [Grattan-Guinness 1990]: an ambitious mathematicised programme of molecular physics led by P. S. Laplace (§2.7.5) and continued by S. D. Poisson (§11.5); a competing and eventually victorious alternative theory of light due to A. J. Fresnel, relying upon supposed wave motions in the ubiquitous aether (§12.6.5); new ontological questions about the relationship between electricity and magnetism posed for Fresnel's landlord and fellow aetherist A. M. Ampère by the discoveries of electromagnetism and for him especially 'electrodynamics'; a penchant for the empirical shared by several engineers such as G. G. Coriolis affirming the importance of 'work', and also by Joseph Fourier in connection with heat diffusion, in a common sentiment that August Comte will soon call 'positivism'; Fourier's own solutions of the diffusion equation, inaugurating Fourier analysis (series and integrals) in both pure and applied mathematics (§11.6); an improved new grounding for mathematical analysis initiated by Cauchy (¶2.2), replacing an ideology due to J. L. Lagrange that was mistakenly dependent upon the supposed generality of the Taylor-series expansion of a mathematical function; and Lagrange's own 'analytical mechanics', an algebraisation a world away from the work mechanics of the engineers.[5]}

The other example concerns Lord Kelvin, who was an important figure in the 19th-century problem of the age of the Earth, where physics and astronomy met up with geology and evolution. Kelvin used Fourier's analysis of heat diffusion in his predictions. Around 1840, in his 16th year, he read Fourier's 1822 book on heat, and transformed his own conception (that is, his paradigms and commitments) of mathematical physics (§11.6). But by 1840 Fourier's work was both paradigmatic mathematical physics and also a source of several disciplinary matrices, both mathematical and physical. Thus Fourier's mathematical physics can be interpreted historically as both revolutionary and normal science at the same time — and without contradiction, since different historical points of view are involved.[6]

I must say that I find Kuhn more enlightening than Lakatos in trying to understand this development — and many others, indeed — where rivalry was sometimes not prominent and never concerned with really fundamental questions. Rivalries centred on alternative strategies for desimplification, at both the theoretical and experimental level; or else they were concerned with related metaphysical factors, which both Lakatos [L1, 95-96] and Kuhn under-estimate in importance; for example, our modes of experience of the physical world, the bearing of science upon theology, or the relationship between observation and theory. (There was a noteworthy change in the early 1800s from the view that the foundations of a theory are securely grounded by induction to the view that a theory, including its foundations, is hypothetical.) The mathematics involved raised further metaphysical problems: the status of numerical methods for example, and the rivalry between algebraic, geometrical and analytical conceptions. I shall return to metaphysical factors in science in ¶5.

4. On the normative character of Lakatos's methodology

I take up now Lakatos's claims quoted in ¶3.1 that the history of science 'should be' a history of competing research programmes, and 'must not become' a study of normal science. He felt that his methodology explicated rationality in science and its history, and so he gave it the normative character that the words 'should' and 'must' express. Unfortunately this attitude leads to a historiography of science that extols verifications of its own principles. As he puts it, 'if we abandon naive falsification in method,

---

[5] In footnote 10 of his 'Second thoughts on paradigms' Kuhn [1974] uses the unfortunate expression 'law sketches' to describe Hamiltonian and Lagrangian principles in mechanics, since the particular function needs specification for a given problem. The phrase 'law schemata' more precisely conveys the idea, and also avoids the sense of incompleteness that 'sketches' conveys.

[6] Fourier himself exemplifies the hope of merging mathematics and experimentation in the context of heat diffusion, although the idea was not original with him (see, for example, his comments quoted in [Grattan-Guinness and Ravetz 1972, 35]). His experimental work shows the disappointing results of comparison; but he is a fine example of a desimplifier of a theory [chs. 20, 5-19; §12.6].

why stick to it in *meta-method*?' [L1, 151; compare p. 131]. Instead, 'the historian is highly selective: he will omit everything that is irrational in the light of his rationality theory' [L1, 119]. This is a process of 'normative selection', and has as a consequence that 'internal history is not just a selection of methodologically interpreted facts: it may be, on occasions, their radically *improved* version'. He then gives as an example Bohr's omission of electron spin in his 1913 programme. 'Nevertheless, the historian, describing with hindsight the Bohrian programme, should include electron spin in it, since electron spin fits naturally in the original outline of the programme' [L1, 119]. Lakatos's approach is thus the same as in his history of mathematical analysis that I discussed in ¶2; for he thought that 'a rational reconstruction in Robinson's spirit [of a modern theory of infinitesimals] can illuminate' problems in the history of the calculus [L2, 44].[7]

Indeed, Lakatos does not stick to naive falsificationism in his historiography. In fact, he is not even sticking to his theory of scientific research programmes; for whereas there we protect the hard core against anomalies by adding layers of protective soft material, in history we immunize the hard core against anomalies by omitting 'everything that is irrational', and putting in any feature that 'fits naturally' into the rational reconstruction. In the course of demolishing as history an essay by Lakatos on Newton [L1, ch. 5], Paul Wood [1979] has nicely characterized Lakatos as an 18th-century *philosophe*, interested in history in order to find out what "should be" in it.[8]

One of the most intriguing conclusions that Lakatos's *philosophie* entails is that internal history, which he conceives of as "rational" reconstructions based on his normative methodology — the quotation above about Bohr is an example — is *autonomous* of external history of science, which seems to be descriptive sociological and psychological history [L1, 120; L2, 202]. Further, 'external history is irrelevant for the understanding of science' [L1, 102; compare L2, 242]. Obviously, then, the distinction between internal and external history (in his senses of the terms) is crucial; but on which side of the line does the history of science education lie, for example? I cannot find the question answered in these volumes, although historians of science know how important the history of science education can be. The point to be made is surely that social history of science that takes science as only a social activity, bereft of individuating technical content, will be of intrinsically minor value, however well it is done. Lakatos over-criticizes such history in [L1, 120-121], when he describes it as 'worthless'.

---

[7] Robinson himself held the same mistaken view of the bearing on history of his new theory. On 15 November 1965, while his book [1966] containing his new theory of infinitesimals was in proof, he gave a lecture at Popper's seminar at the London School of Economics on the history of the foundations of the calculus and its 'correction' by his theory. {See also *ROL,* chs. 2, 13.}

[8] To be fair to Lakatos, according to his editors he regarded this essay 'as in need of substantial revision' [L1, 193]; but the editors published it just the same, and also several other essays that he felt to be unready. They also omitted from [L1, 8] an important question mark from the title of a section.

All this is very sad and unfortunate, for Lakatos was undoubtedly right to emphasize the importance of history, and to try to increase the historiographical component of Popper's philosophy. For history is culturally unavoidable; just as ignorance of the law is no defence against the law, so ignorance of history is no immunity against history. We should regard the study of history as fulfilment of our duty to memory, our obligations to culture. But the very process of cultural change over time affects the vision that we have of our inherited past. If we think of history as the re-cycling of past information, then we put that information to new uses as well as old ones.

Now if Popper's philosophy has a bearing on the history of science, as Lakatos and I agree, then how is it best used? I find it most useful as a source of *historiographical falsifications*. At least at the level of conscious intent, our historical figures often held epistemological and methodological views that Lakatos and Popper have criticized: they believed in induction, they thought that facts were theoretically neutral, and so on. Nevertheless, Popper's philosophical points still have force, and thus a *clash* between Popperian views and the historical record can be most profitably explored (§10). This is a special case of an approach to history that I call 'the method of mutual illumination', where the historian looks at each of several historical figures, scientific traditions, or whatever from the point of view of the others: the clashes can be very illuminating. This whole review is based on treating Popper's philosophy and the history of science in this way.

Sometimes there is an element of historiographical verification in Popper's historical writings, in that he tends to emphasise those elements in the work of historical figures that are most congenial to his philosophy.[9] Lakatos's methodology has gone on much further in the direction of verifications, and that is where many of the drawbacks of his historiography have their source.

I subscribe to the view that Lakatos describes as *'history "falsifies" falsificationism (and any other methodology)'* [L1, 123], as opposed to his own *'History — to varying degrees — corroborates its rational reconstructions'* [L1, 131]. However, there are various basic issues involved in this clash (for example, rationality itself) with regard to which Popper's philosophy, from which both Lakatos and I start, seems unsatisfactory. The rest of this chapter is devoted to such questions. In order both to reduce its length and keep the discussion general, I shall not use historical case studies much, although the reader can surely find them in his own experience: the historical outline that I gave in ¶3.2 contains many examples.

5. Unfalsifiability and the status of falsification

A point of friction between historians of science and Popperian philosophers of science involves a misunderstanding over the use of the term 'unfalsifiable propositions'. When Popper and Lakatos speak of them, they have in mind what I prefer to call 'intrinsically unfalsifiable proposi-

---

[9] For the example of Newton's conceptions in astronomy, contrast Popper's 'The aim of science' (1957) in [1972, ch. 5] with [Whiteside 1970].

tions', whose nature precludes any possibility of falsification. They include not only general existential propositions but also propositions that express metaphysical world-views such as realism, idealism and solipsism.

Now these propositions must be distinguished from *experimentally* unfalsifiable propositions, in which defects in instrumentation or experimental design prevent an effective test of a theory being carried out for the time being. This category of propositions is quite significant in the history of science; for example, at some period in the history of astronomy the comparison between theory and experiment may be prevented until instruments are developed that are capable of measuring to within 15 seconds of arc, say. Such problems have been an important motivation in the history of instruments and technology: they require 'the desimplification of instruments', one might say.

The importance of intrinsically unfalsifiable theories in science needs stressing.[10] For example, as I mentioned in ¶3.3, some of the disputes that occurred in the development of theoretical astronomy were centred on issues of that kind, concerning the ways that scientific theories should be formulated and how they should relate to the physical world. Popper has been aware of such issues, and has written on metaphysical research programmes (for example, [1974a, chs. 33, 37]). Lakatos accepted metaphysical theories into his methodology of research programmes [L1, 41-42], and indeed made no special point about them.

However, Popper and Lakatos clashed over the issue of falsification. From the start of his philosophical work Popper was aware that we can 'find some way of evading falsification, for example by introducing ad hoc an auxiliary hypothesis'; indeed, some singular statements when used this way 'to assist the theory [...] are quite harmless'. However, 'auxiliary hypotheses should be used as sparingly as possible', in order to maintain the high level of testability of a theory [1959, 42, 83, 273]. My own view of desimplification, mentioned in ¶3.1, is roughly a restatement of Popper's views on auxiliary hypotheses: I use 'desimplify' because it conveys the idea of *grades of addition* of hypotheses better than does Popper's (rather pejorative) use of 'ad hoc'.

By contrast with Popper's (over-)cautious attitude to auxiliary hypotheses, Lakatos encouraged their use, for they furnish the soft material that surrounds the hard core of a research programme. Their profusion entails that 'No single experiment can play a decisive, let alone "crucial", role in tilting the balance between rival research programmes' [L2, 212]. Indeed, since there must always be rivalry, '*There is no falsification* [of a theory] *before the emergence of a better theory*' [L1, 35]. However, Lakatos opposed W. V. O. Quine's version of the 'Duhem-Quine thesis', that there is no rational means of selecting from alternative theories after a

---

[10] The place in science of conceptual questions of this kind is welcomely stressed in [Laudan 1977]. See also his criticisms of Lakatos's methodology, especially on pp. 76-88, 168-170, 232-234.

theory has been falsified;[11] for his apparatus of acceptance and meta-criterion would perform that task [L1, 96-101].

The question of how Lakatos's system performs these tasks relates to the distinction between the falsification of a theory (an epistemological question) and its rejection (a methodological matter). Lakatos claimed that 'Popper clearly identifies "falsification" with "rejection" and "elimination"' [L1, 142; compare p. 155]. Yet even in Popper's early writings it is quite clear that we may *decide* to *accept* a theory even though it has been falsified (because, say, we have no better replacement: see, for example, [Popper 1959, 53-54, 108-111]. With justice he repudiated Lakatos's charge of the conflation of falsification and rejection, together with Lakatos's conclusion that in Popper's philosophy Newtonian mechanics and psychoanalysis are equally experimentally unfalsifiable. He thought that Lakatos provided 'an interpretation of my theory of falsifiability that makes nonsense of all my views'.[12]

6. Are there two types of objective truth?

Popper uses a correspondence theory of truth, where truth is the property of a proposition that it corresponds to the facts. To use a favourite example: 'snow is white' is true if and only if snow is white. But his use of this theory faces various philosophical difficulties, as we saw in §6.

Some of them also attend Lakatos's methodology, since if a research programme is progressive it predicts 'novel facts' (see, for example, [L1, 32-35]. I infer that he also uses a correspondence theory of truth, although he does not make himself clear: the closest approximation to a statement seems to occur at [L2, 108-109]. The issues bear on Popper's category of corroboration of a theory (§3), and the distinction between ontologically correct and factually true propositions was proposed in §2.7.1, where a proposition can be ontologically incorrect and factually true at the same time, given the state of the facts at the time. To apply it to the example used in the last section, a fact in theoretical astronomy is indeterminate to within 15 seconds of arc; the theoretical prediction falls within this band of error, and so corresponds to that fact; hence the prediction is corroborated, and so is factually true. This is the kind of context in which concepts such as 'a more accurate measurement' are to be understood: a measurement is more accurate if a previously experimentally unfalsifiable prediction now becomes experimentally falsifiable.

---

[11] For a survey of views on the Duhem-Quine thesis, see [Harding 1976]. In this connexion I may mention that Lakatos's claim that 'theories with a *ceteris paribus* clause, have no empirical basis' [L1, 18] must be wrong; the exploration of the referents of such a clause for sources of hitherto overlooked but possibly relevant auxiliary hypotheses is as empirical an activity as one could imagine.

[12] [Popper 1974b, 1004-1013]. On p. 1009 he says that the falsification of a theory is a 'matter of logic' while its rejection 'is a question of methodology'. However, on the next page he says that his theory of falsifiability is 'methodological or philosophical'. This puts it in the same category as the rejection of a theory, whereas a sharp distinction between falsification and rejection has just been made. I take Popper's use of 'methodological' on p. 1010 to be a slip of the pen.

7. Some consequences of the distinction between ontological correctness and factual truth

7.1 *The search for truth.* Popper's view of science as a search for truth is, I think, better understood as a search for ontology (§2). How, then, does ontology relate to facts? Lakatos's description noted in ¶3.1 of a scientific theory as a hard core surrounded by auxiliary hypotheses suggests a valuable analogy; for ontology can similarly be regarded as a theory about the hard core of "essences" surrounded by auxiliary factors such as our experience of the universe and the design of scientific experiments. Further, the search for ontology with theories is paralleled by the search for corroboration or falsification of its predictions in the experimental facts. This line of thought led to the schematic representation of the principal relationships between ontology, facts, theory, and prediction in science that was discussed in §2.7.6.

Popper advocated a 'World 3' of abstract objects that exist independently of anyone knowing them (§2.12.1). But the eclectic population of this World — unthought-of theorems, melodies, propositions, theories of all sorts, and so on — raises many difficulties. For example, it trivializes all questions of ontology, since if you think of X then for Popper there exists the X that you think of. Lakatos seems to accept Popper's World 3 without apparent criticism — his research programmes inhabit it [L2, 226] — although he does not explain how his methodology can gain us access to it.[13] For Lakatos's methodology is affected by the ontological issues.

7.2 *On the factors in between ontology and facts.* The essences of the universe are minute, and in some cases so small that they are not directly experientiable. Similarly, the bonds connecting them (forces, fields, or whatever they are) are detectable only by manifestation. Hence ontologically correct theories will have to include non-experiential concepts and relationships between them.

From these "micro" essences, and the bonds between them, are constructed the plenum of "macro" objects that we can experience, and the events (motion, and so on) in which they take part. However, although they are experientiable, they do not require a sentient being to experience them. And it is he who chooses collections (or types of collection) of object and event out of the plenum around him, and regards them as some sort of interlinked unities that form facts. His choice of collection is made on theoretical considerations, which may not be at all obvious empirically; they may even be controversial.

The theories about the universe that we invent will attempt to refer either to the ultimate essences and bonds, or to the experientiable objects and events, or both; and the mixture of these two kinds of referent can vary for different theories of even the same aspects of the universe. For example, in quantum mechanics we study the structure of physical matter; but in

---

[13] For Lakatos's acceptance of Popper's World 3, see especially [L1, 110, 119; and L2, 108-109, 226-235]. The curious nature of World 3, and also the difficulties attending Lakatos's methodology, are noted in [Skolimowski 1976].

statics we take the structure for granted and examine its "macro" properties. In fact, theories referring to the same aspects of the universe can sometimes be formed into a sequence, with a changing balance between the unexperientiable and the experientiable: for example, quantum mechanics — statics — mechanical engineering; or molecular biology — embryology — physiology. Of course, the boundaries between member theories of a sequence are not sharp; but nevertheless the balance between unexperientiable and experientiable referents changes much along a sequence. An important case of the change in balance is the boundary between micro and macro itself; for example, at what stage does a collection of basic physical particles, apparently moving in all directions, become an object that moves only vertically downwards when released in vacuo?

Further difficulties attend those branches of science that refer to sentient beings' own constitution and experiences. Are the secondary qualities involved — sweetness, aggression, cold, and so on — admissible as components of facts, or must they be replaced by non-sensory qualities? If I feel cold, is there a coldness that I feel? Such questions have considerable ontological import; for example, is the objectivity of such qualities to be distinguished from that of unexperienced objects and events? There are difficulties of a rather similar kind concerning space and time; and also the referability of statistical theories (for can statistical theories be in principle ontologically correct?).

7.3 *The place of language in scientific knowledge.* As part — but only a part — of our scientific theorizing, we use languages. Thus we must invent and develop them: not merely extend their vocabulary but especially enrich their syntax, semantics and logics. Broadly similar developments occur in (relatively) formal languages, especially mathematics. In order to allow for the changes that occur in languages, we use the term 'sentence-token' to describe a proposition uttered at a particular stage of its language's development.

Popper has always been dismissive of the piffling puzzles "studied" in current ordinary language philosophy. While his view is understandable, I am sure that it is rather over-stated in terms of his own epistemology. I remember that in his lectures at the London School of Economics he often used the slogan 'words do not matter' to state his position. But this proposition is self-falsifying, as I recall noting at the time: for example, the presence of 'not' in it matters a great deal. Lakatos was more aware of the role of language: for example, 'The choice of a language for science implies a conjecture as to what is relevant evidence or what, or what is connected, by natural necessity, with what' [L2, 161].

Popper's adoption of Tarski's definition of truth as a property of propositions makes the role of language in his epistemology obviously prominent (for example, for Popper it provides a precise specification of 'true' under certain circumstances); and some of his technical ideas, such as verisimilitude (to be discussed in ¶7·5 below), are also heavily language-dependent. Further, his 'World 3' of abstract objects, mentioned in ¶7.2, includes propositions among its eclectic membership. But then if we are trying to gain access to this 'artefactual world', as I may call it, then surely the linguistic worlds that we invent — their syntax, logic(s), semantics and ontological commitments — are of paramount significance. So also are relationships and translations between them; and the possibility, seemingly allowed for by Popper, that there is an ideal language up in World 3 awaiting discovery rather than invention.

It is against the background of factors such as these that Popper's idea of science as a search for truth, and truth as correspondence to the facts, has to be appraised. Our search for ontology can still be seen as objective; but the considerations here show how problematic is Popper's and Lakatos's handling of truth and of facts. Ontology can be thought of as facts in disguise, and the disguise has to be penetrated.

7.4 *Discovery in science*. We can only discover what is there to be discovered in the first place; but falsification shows us that a particular search has been unsuccessful, and absence of falsification to date does not preclude falsification by a later test. (Even if a theory is ontologically correct, we can never prove that it is.) Hence falsification generates a logic of non-discovery, or of not proven discovery, when 'discovery' is used in this sense. To know more science is to know more mistakes. The analogy with discovery in explorations is apt, for the history of explorations includes cases of mistaken identification of the geographical feature that has been discovered. I may think that I have discovered the Nile, but — setting aside the tautological case that I may have discovered that river and *named* it 'the Nile' — later tests may show that I did not do so.[14]

These considerations lead to a significant criticism of Popper; for the very title of his *Logic of scientific discovery* [1959] is misleading. Firstly, there is the problem of ontology and discovery discussed in ¶7.2. Secondly, Popper tells us nothing about the psychology of discovering scientific theories; the discussion is centred on the logic of testing theories already discovered. (Hence it ill behoves him (and Lakatos) to dismiss philosophers who try to develop theories of the psychology of discovery; it also reduces the relevance of their work to the history of science.) Thirdly, the book is concerned with methodological as well as epistemological questions, as was properly reflected in the title of the original 1935 German version ('Logik der Forschung'); 'investigation' or 'research' would have been a much better translation than 'discovery'. I am surprised that Lakatos did not challenge Popper on this matter; instead, he quoted Popper's own account of the choice of title with apparent approval, grossly extended a

---

[14] N. Rescher rejects the analogy with geographical exploration as 'fundamentally mistaken' because the analogy assumes that 'science progresses by cumulative accretion' and 'the magnitude of these additions is steadily decreasing' [1978, 29]. My use of the analogy makes no use of these assumptions, which I do not accept: so I do not find Rescher's grounds for rejecting it to be conclusive. See also footnote 20 below.

remark made by Popper that science is a game, and earned Popper's ire for this extension.[15]

7.5 *Verisimilitude*. Popper has a theory of the verisimilitude, or truth-likeness, of a scientific theory. Verisimilitude is an indication of the extent to which a theory has accumulated (to date) factually true consequences.[16] Unfortunately it has been shown recently that false theories are not comparable by verisimilitude (§2.7.4). I am not surprised, since verisimilitude uses the loose criterion for truth as correspondence to the facts together with the very strict logical principle known as 'the law of excluded middle': propositions are true or false, predictions are corroborated or not, and no other possibilities are allowed. Verisimilitude as an idea is certainly worth preserving; and it may be that, if correspondence to the facts is still to be its guiding criterion, some kind of non-classical logic, using values between, or other than, 'true' and 'false', is needed to replace the classical logic that the law of excluded middle requires. Popper is very conservative about logic (§5.2); Lakatos seemed also to accept the classical form, for he rejected the construction of inductive logics as part of his Popperian criticism of inductivist epistemologies [L2, ch. 8].[17]

Another difficulty for verisimilitude concerns the allegedly cumulative character of theories that refer to the same type of phenomena; if theory T* supersedes T, then its verisimilitude will be greater than that of T. Popper has insisted that T* should contain all the explanatory power of T; and Lakatos seems to have been of a similar view, since the progressiveness or degeneration of problem-shifts of rival research programmes relates to the idea of supercession of theories (see especially [L1, 33]). However, there have been many cases in the history of science where T* fails to encompass all the details of T while apparently surpassing it in its principal concerns. But then verisimilitude will be definable only relative to some area A where T* and T both offer explanations; and A may either be difficult to specify or uninterestingly minute. For example, Einstein's theory of relativity has a greater verisimilitude than Newton's laws of gravitation; but this does not imply that the physical world does actually — that is,

---

[15] See [L1, 141], quoting [Popper 1959, 53]. For strong criticisms of Popper's eschewing of cognitive aspects of scientific theorizing see [Rescher 1978, esp. pp. 51-57].

[16] The terminology here is exceptionally difficult to explain in German, and Popper has, as far as I know, discussed verisimilitude only in his English writings. 'Wahrscheinlichkeit' is normally used to translate both 'probability' and 'verisimilitude', but this must not be done here. Indeed, in view of the fact that 'verisimilitude' is being used as a technical term, it seems to me to be a doubtful term to use even in English. 'Truth-content' ('Wahrheitsgehalt'), used in a qualitative rather than a quantitative sense, may be sufficient in itself to express the idea of 'verisimilitude'; and then 'Wahrscheinlichkeit' can be confined, as normally, to 'probability'.

[17] N. Koertge offers in [1978] an interesting defence of probabilistic considerations against both Popper and Lakatos, and an appealing contrast between the two.

ontologically — resemble an Einsteinian field structure more closely than a Newtonian matter force system.

7.6 *Lakatos's avoidance of truth*. Lakatos had a solution to all the problems raised in this section. He regarded Popper's claim that we may invent (or discover) ontologically correct theories as 'a flaw in his fallibilism', and he replaced it 'in a true Marxist spirit by my doctrine of infinite sentences in God's Blueprint of the Universe' [L2, 126]. He reinforced this view by assuming without argument that 'the universe is infinitely varied' and inferring that 'it is very likely that only statements of infinite length can be true' [L2, 123], which does not follow even if the premise is ontologically correct.

I do not see why we should put God to such transfinite trouble, or deny ourselves a capacity to develop sufficiently rich languages. Further, I see no advantage in Lakatos's intrinsically unfalsifiable view over Popper's tri-distinction between ontologically correct propositions, possession of such propositions, and proof of their possession (§2.5.6). For this tri-distinction is one of Popper's most important contributions to both epistemology and methodology, and very much at the root of his realism, whereas Lakatos's view tends towards epistemological relativism. Without ontology there is no realist epistemology, for instead we can say: facts are theoretical, and so disputable; you follow your version of them and I will follow mine, and there is no ultimate criterion to adjudicate between us.

8. Can we demarcate science from pseudo-science?

Popper's original philosophical concern was to demarcate scientific from pseudo-scientific theories in terms of their falsifiability and unfalsifiability. Unfortunately, while pseudo-scientific theories will contain no falsifiable propositions, scientific theories may contain unfalsifiable ones. For example, when we say that Newtonian mechanics is falsified, it is only the law that explains the action of the inter-particulate forces that is so falsified. The assumption that there exist such forces is existential, and thus is intrinsically unfalsifiable: in fact, it may be ontologically correct.

Perhaps under the pressure of such considerations, Popper has broadened his position by saying that we criticize scientific theories, with falsification as an important special case; and Lakatos seems to have broadened this position still further. The criteria of demarcation are different for the two; Popper's hinges on the absence or presence of unfalsifiable components, while Lakatos's is based on the progression or degeneration of problem shifts (see especially [L2, 148-153]). But neither may be satisfactory. I take astrology as my case study, for both Lakatos and Popper pick on it as a favourite example of pseudo-science (or 'metaphysics', a term that Popper uses as a synonym).[18]

---

[18] See [Popper 1963], especially pp. 1-3 for the synonymy of 'pseudo-science' and 'metaphysics' and the status of astrology. For Lakatos's agreement with Popper on astrology see for example, [L2, 227].

I regard this appraisal of astrology as mistaken. Certainly the interpretation of the stellar bodies, the houses, and the configurations in an astrological chart are based on intrinsically unfalsifiable doctrines; but astrology includes 'prognostication', where events and features pertaining to the individual's life can be predicted into the future (or retrodicted onto the past), by conjoining the birth chart (or nativity) with its progressed position for the chosen time. Astrology also predicts that the death chart of an individual will display planets occupying positions that other planets took at important times of his life. Such predictions may not be as exact as (some) scientific ones, but they are certainly falsifiable (and capable of corroboration, too). Falsifications lead to desimplifications of the astrological theory involved, just as in science. Indeed, even Lakatos's theory is relevant, for there is rivalry between competing astrological research programmes in his sense (for example, between Hindu astrology and the methods of the Ebertin school).[19]

I am not interested here in building up a case for astrology. The point is that there seems to be no basic epistemological difference between astrology and Newtonian mechanics (say), although there are large differences in balance between the falsifiable and unfalsifiable components in each theory (compare §16).

These considerations suggest to me that in addition to Popper's 'classical' demarcation criterion it may be worth exploring 'non-classical' gradations of the balance between the falsifiable and unfalsifiable components of theories, like the non-classical theory of verisimilitude suggested in ¶7.5. Popper employs 'classical' theories of truth and falsifiability; that is, when asking the questions (that are independent of each other) of whether a theory is ontologically correct or incorrect, and whether it is falsifiable or unfalsifiable, he allows only those two possibilities in each case. The suggestion here is that we may need non-classical truth-value systems and falsification theories, as with the theory of verisimilitude.

9. Can there be logics of discovery in non-scientific areas?

9.1 *Discovery and decision.* Lakatos hoped that his methodology would extend to non-scientific areas. For example, his idea of proofs and refutations is a prolegomenon to a logic of mathematical discovery, and mathematics is an experimentally unfalsifiable and thus a non-scientific body of knowledge (if I place a drop of water on a drop of water and obtain a drop of water, I do not conclude that $1 + 1 = 1$). Indeed, had Lakatos seriously considered the questions of ontology that I raised in ¶7, he would have been able to pursue profitable analogies with mathematics, since the

---

[19] Kuhn makes a similar defence of astrology in his 'Logic of discovery or psychology of research?' (1974), in his [1977, 274]. A clear account of the basic rules of astrology is given in [Carter 1963]. Information on Hindu astrology is contained in, for example, [Oken 1972]. For Ebertin's approach, see his [1972]. Reports on the current state of research into the birth chart, the testing of astrological theories, and the gullibility of the public, form sections of [Dean and Mather 1977].

question of whether we discover or invent mathematical objects is one of the chief problems in the philosophy of mathematics.[20] But he did pursue analogies between science and its history, for he wanted historians of science to regard themselves as appraising rival historiographical research programmes, just as scientists are (allegedly) appraising rival scientific research programmes. For the reasons given in ¶3 and ¶4 I doubt if his approach will be successful, but the problem of logics of discovery in non-scientific areas (including the history of science) remains important. It may be phrased as follows.

Popper argues for a logic of scientific discovery, with falsification as the epistemological lynchpin. Can we also have discovery in a non-scientific area (with a logic of testing to go with it), or can we only decide about things there?[21]

To explore the question further, let us first take a famous example in science itself. For a long time it was thought that Newton had discovered that the physical world worked according to his inverse square law of attraction, but eventually the discrepancies between his predictions and experimental information showed that he had not done so. Had his theory been ontologically correct, then he would have discovered the law as well as decided in favour of it; but then according to Popper he would never have been able to show that the discovery had occurred, for the next test might have falsified his law.

The distinction between discovery and decision is an example of the distinction between epistemology and methodology, which I discussed in ¶5. For me, ontology has priority over epistemology, which in turn is prior to methodology: what there is affects what we know about what there is, and what we know affects what we do about what we know. (This point underlies my reservations in ¶4 concerning much social history of science.) The connexions can work the other way, but more weakly; God is the First Source, not the sociologist. However, the methodological problems of science possibly present still harder challenges than do the ontological and

---

[20] Bertrand Russell once even drew on the geographical analogy that I used in ¶7.4 with regard to mathematical objects: 'Arithmetic must be discovered in just the same sense in which Columbus discovered the West Indies, and we no more create numbers than he created the Indians' [1903, 451]. Recent discussions of the problem include [Smith 1975], [Thom 1975] and [Kitcher 1978]. Lakatos's 'A renaissance of empiricism in the recent philosophy of mathematics?' of 1967 [L2, ch. 2] is more concerned more with proof-methods in mathematics than with the discovery of objects.

[21] This question is raised for moral discovery in [Sharratt 1974], the best short introduction to Popper's philosophy that I know. It is discussed with insight in R. Pirsig's deservedly popular *Zen and the art of motorcycle maintenance* [1974]. It ought to be prominent also in [Ward 1972], a comparison of science with economics. But in taking Kuhn as his main guide to science, Ward unfortunately highlights puzzle-solving as the principal scientific activity, and thus largely sidesteps epistemological questions, such as the possibility of a logic of testing economic theories.

epistemological ones, especially because they involve non-scientific areas such as education, institutionalization, and sometimes ethics and politics.

9.2 *Non-self-reference.* This brings me back to the theme of this section. Do we discover in non-scientific areas, as shown by the logic of testing theories? Does non-scientific knowledge grow, or does it merely change? Is 'non-scientific discovery' a contradiction in terms?

As I pointed out in ¶7.2, discovery is related to ontology; so the burden of these questions is thrown on to the ontological situations in non-scientific areas. Popper's falsification criterion is valid for science because the world actually — that is, ontologically — works in a certain way, and not in any other way. Is there such an ontology outside science, especially, for areas such as ethics, law, aesthetics, sociology, history, religion and politics; where human action and response is central? In terms of my discussion of truth in ¶6, can there be ontologically correct theories in these areas, or must we replace ontology by axiology? Can we even appeal to the much weaker criterion of correspondence to the relevant facts?

In addition to these general points of comparison, here is a specific one. Can there be theories in non-scientific areas that are universal over space and time, as scientific theories are (or — and this may be an important qualifier — are claimed to be)? Indeed, must theories in medicine, biology and botany, areas that are normally regarded as scientific, possess this property?

I cannot answer these questions, but the following difference may be germane. The doctrine of criticism, advocated by Popper for both scientific and non-scientific areas, is self-referential: it criticizes itself (§2.9.4). Lakatos's theory of research programmes is, I think, also self-referring; the purported rationality is its hard core, and the apparatus of acceptances, ad-hoc-nesses, and so on, form the soft material. By contrast, falsification in Popper's sense, which so far is confined to science, is not self-referential: it is an epistemological theory, and as such it is neither falsifiable nor unfalsifiable. Thus if a non-scientific area A is to have a logic of discovery, it may need a non-self-referring criterion C(A) of demarcation between A and non-A. Criticism, being self-referring, cannot qualify on its own. The discovery of a C(A) for A would be an important result; for a noticeable feature in many areas of human thought, especially since the 17th century, has been attempts in non-scientific areas to imitate scientific method — whatever the non-scientific practitioners of the day may have thought scientific method to be. Conversely, if it is shown that A cannot have such a C(A) and thus no logic of discovery, then this would be an important piece of negative information.

10. The place of rationality in science and its history

So far, I have omitted any discussion of rationality, in order to use it as a common theme to unite the various topics discussed earlier. I shall begin with its place in science itself.

10.1 *Rationality in science* is a prime concern of philosophers of science. For example, Popper has frequently emphasized the close connexion between rationality and the critical appraisal of scientific theories, and Lakatos saw his methodology of scientific research programmes as specifying rationality in science. A principal aim of such philosophers, especially Lakatos, was not merely to emphasize rationality but also to provide a calculus of rationality — or at least, a means of appraising competing strategies in terms of the measure of their "rational" components.

Previous sections of this chapter have discussed topics that make a theory of rationality so hard to achieve. I have in mind the indirect or even non-experientiability of (many of) the world's essences (¶7.3), with the attendant difficulties of rationally assessing ontological theories (¶7.1); the psychology of inventing a scientific theory, which is not so much a rational as a creative or even an inspirational activity, and whose understanding depends partly on the state of development of psychological theory (¶7.4); the conceptual problems, often general and/or unfalsifiable in character, which can comprise substantial components of scientific theorizing (¶3, ¶5); the linguistic element of scientific theories, clearly substantial, and far from yielding its syntactic, semantic, or ontological secrets to linguisticians (¶7·3); the vexing questions of the various logics (¶7.5) and theories of truth (¶6) that relate both to language and to its modes of reference; the notion of verisimilitude, which might contribute to a calculus of rationality if it worked at all (¶7.5); and even the problems connected with mathematics (¶2, ¶3), which play a large role in some sciences. There is also the problem of self-reference mentioned in ¶9; for can a theory of rationality itself be rational, or do we need a non-self-referring criterion for rationality?

Since rationality is concerned also with behaviour, both individual and communal, then it has to consider also the social aspects of science, especially science policy, education, funding and professionalization. Since it is not clear whether there can even be a logic of discovery for such areas (¶9), or how individual rationality relates to communal rationality (it was in connexion with this issue that I suggested in ¶3 that an individual scientist's paradigms be regarded as additional to community paradigms), the articulation of theories of rationality for them is especially daunting.

A further difficulty for rationalizing these topics is that they contain a considerable element of convention, and it is not clear where the limits of convention lie. Indeed, convention plays a role in aspects of science where falsification may be effective: for example, in nominal definitions (Poincaré made great use of them in advocating his conventionalist philosophy of science), or in measures and standards. Popper has always been aware of the considerable place of conventions in science, although Lakatos's characterization of his philosophy as 'another brand of revolutionary conventionalism' [L1, 108] is inaccurate.

The considerations above suggest to me that at present it is not realistic — rational, one might say — to take the explication of rationality as a fundamental aim of the philosophy of science.[22] Obviously scientists can make rational choices over a certain range of contexts, especially those of a straightforwardly experiential character. But the topics mentioned above will soon make their appearance, and rationality theory is placed in serious difficulties. I am not proposing a theory of irrationality as an alternative, but I do suggest that there is a great deal of arationality in science: that is, aspects of science that are not yet within our rational purview, and perhaps never can be. This is a way of restating the message of ¶8, that the demarcation between science and pseudo-science is not clearly drawn.

10.2 *Rationality in the history of science* involves many or all of the above topics occurring for our historical figures. There are many other similarities with science: for example, the scientist tries to measure as accurately as possible while the historian gives references as precisely as he can. But there are also basic differences, principally because the rationality theory of the historical figure is itself subject to the historian's study. To take an important example, the historian can rationally appraise the conduct of his historical figure as irrational (even, perhaps, relative to that figure's own theory of rationality). This example also underlines my total disagreement with Lakatos's normative historiography of science and my alternative proposal of emphasising falsifications of historiographical expectations: we falsify our theory of rationality by testing it against the historical record with techniques such as the method of mutual illumination (¶4).

To unite the two topics of this section, I suggest that science contains less rationality than Lakatos hoped, but that the history of science can be more rational then he allowed it. In addition, his advocacy of rationality, being dogmatic, is irrational. If theories of rationality for both science and its history are to be developed, then the arational topics to which much of this chapter is devoted need our prime attention. For I agree with Lakatos that 'Philosophy of science without history of science is empty; history of science without philosophy of science is blind' [L1, 102].

Acknowledgements
I am greatly indebted for comments to Eugene Gadol, Susan Haack, Jerry Ravetz, the editor and his referees.

Bibliography
Aiton, E. J. 1972. *The vortex theory of planetary motion*, London: MacDonald; New York: American Elsevier.
Bloor, D. 1978. 'Polyhedra and the abominations of Leviticus', *British journal for the history of science*, 11, 245-272.

---

[22] For example, while impressed by Laudan's advocacy [1977] of a historiography based on research traditions, I am sceptical of his claims for its characterization of rationality in science.

Bos, H. J. M. 1974. 'Differentials, higher-order differentials and the derivative in the Leibnizian calculus', *Archive for history of exact sciences*, *14*, 1-90.

Bunn, R. 1977. 'Quantitative relations between infinite sets', *Annals of science*, *34*, 177-191.

Carter, C. E. O. 1963. *The principles of astrology*, 5th ed., London: Theosophical Publishing House.

Cauchy, A. L. 1833a. *Sept leçons de physique générale faites à Turin en 1833*, in *Oeuvres complètes*, ser. 2, vol. 15, Paris: Gauthier-Villars, 1974, 412-447 [cited here].

Cauchy, A. L. 1833b. *Résumés analytiques*, in *Oeuvres complètes*, ser. 2, vol. 10, Paris: Gauthier-Villars, 1895, 5-184 [cited here].

Cleave, J. R. 1971. 'Cauchy, convergence and continuity', *British journal for the philosophy of science*, *22*, 27-37.

Cust, Mrs H. 1931. (Ed.), *Other dimensions. A selection from the later correspondence of Victoria Lady Welby*, London: Cape.

Dean, G. and Mather, A. 1977. (Comps.), *Recent advances in natal astrology. A critical review 1900-1976*, Subiaco, Western Australia: Analogic.

Dugas, R. 1939. 'L'incompréhension mathématique', *Thalès*, *4* (1937-39), 168-183.

Dugas, R. 1940. *Essai sur l'incompréhension mathématique*, Paris: Vuibert.

Ebertin, R. 1972. *Applied cosmobiology*, Aalen: Ebertin Verlag.

Fisher, G. M. 1978. 'Cauchy and the infinitely small', *Historia mathematica*, *5*, 313-331.

Grattan-Guinness, I. 1970. *The development of the foundations of mathematical analysis from Euler to Riemann*, Cambridge, Mass.: The MIT Press.

Grattan-Guinness, I. 1981. 'Mathematical physics in France, 1800-1840: knowledge, activity and historiography', in J. W. Dauben (ed.), *Mathematical perspectives. Essays on mathematics in its historical development*, New York: Academic Press, 95-138.

Grattan-Guinness, I. 1987. 'The Cauchy-Stokes-Seidel story on uniform convergence again: was there a fourth man?', *Bulletin de la Société Mathématique de Belgique*, *(A)38* (1986), 225-235.

Grattan-Guinness, I. 1990. *Convolutions in French mathematics, 1800-1840. From the calculus and mechanics to mathematical analysis and mathematical physics*, 3 vols., Basel: Birkhäuser; Berlin: Deutscher Verlag der Wissenschaften.

Grattan-Guinness, I. and Ravetz, J. R. 1972. *Joseph Fourier 1768-1830*, Cambridge, Mass.: The MIT Press.

Harding, S. G. 1976. (Ed.), *Can theories be refuted?*, Dordrecht: Reidel.

Hardwick, C. S. 1977. (Ed.), *Semiotic and significs. The correspondence between Charles S. Peirce and Victoria Lady Welby*, Bloomington, Indiana: Indiana University Press.

Kitcher, P. 1978. 'The plight of the Platonist', *Noûs*, *12*, 119-136.

Koertge, N. 1978. 'Towards a new theory of scientific enquiry', in G. Radnitzky and G. Andersson (eds.), *Progress and rationality in science*, Dordrecht: Reidel, 253-278.

Kuhn, T. S. 1970. *The structure of scientific revolutions*, 2nd ed., Chicago: University of Chicago Press.

Kuhn, T. S. 1974. 'Second thoughts on paradigms', in [1977], 293-319.

Kuhn, T. S. 1976. 'Mathematical versus experimental traditions in the development of physical science', in [1977], 31-65.

Kuhn, T. S. 1977. *The essential tension*, Chicago: University of Chicago Press.

Lakatos, I. 1963-1964. 'Proofs and refutations', *British journal for the philosophy of science*, *14*, 1-25, 120-139, 21-243, 296-342.

Lakatos, I. 1976. *Proofs and refutations: the logic of mathematical discovery*, Cambridge: Cambridge University Press.

Lakatos, I. and Musgrave, A. 1970. (Eds.), *Criticism and the growth of know-ledge*, Cambridge: Cambridge University Press.

Laudan, L. 1977. *Progress and its problems*, Berkeley, Los Angeles and London: University of California Press.

Oken, A. 1976. *Astrology: evolution and revolution*, New York: Bantam.

Pirsig, R. 1974. *Zen and the art of motorcycle maintenance*, New York: Bantam.

Polya, G. 1954. *Mathematics and plausible reasoning*, 2 vols., Oxford: Oxford University Press.

Popper, K. R. 1959. *The logic of scientific discovery*, London: Hutchinson.

Popper, K. R. 1963. *Conjectures and refutations*, London: Routledge and Kegan Paul.

Popper, K. R. 1972. *Objective knowledge. An evolutionary approach*, Oxford: at the Clarendon Press.

Popper, K. R. 1974a. 'Autobiography of Karl Popper', in [Schilpp 1974], 1-181. [Repr. as *Unended quest*, London: Fontana, 1976.]

Popper, K. R. 1974b. 'Replies to my critics', in [Schilpp 1974], 961-1117.

Rescher, N. 1978. *Peirce's philosophy of science*, Notre Dame: University of Notre Dame Press.

Robinson, A. 1966. *Non-standard analysis*, Amsterdam: North-Holland.

Rostand, F. 1952. 'La notion de scrupule dans la psychologie des mathématiques', *Revue générale des sciences*, *59*, 325-336.

Rostand, F. 1960. *Souci d'exactitude et scrupules des mathématiciens*, Paris: Vrin.

Rostand, F. 1962. *Sur la clarté des démonstrations mathématiques*, Paris: Vrin.

Russell, B. A. W. 1897. Review of L. Couturat, *De l'infini mathématique* (1896), *Mind, new ser.*, *6*, 112-119.

Russell, B. A. W. 1903. *The principles of mathematics*, Cambridge: Cambridge University Press.

Schilpp, P. A. 1974. (Ed.), *The philosophy of Karl Popper*, La Salle, Illinois: Open Court.

Sharratt, M. 1974. 'Introducing Karl Popper', *The clergy review*, *59*, 390-400, 467-480, 543-555.

Skolimowski, H. 1976. 'Evolutionary epistemology', in R. S. Cohen and others (eds.), *PSA 1974. Proceedings of the 1974 Biennial Meeting of the Philosophy of Science Association*, Dordrecht: Reidel, 191-213.
Smith, B. 1975.'The ontogenesis of mathematical objects', *Journal of the British Society for Phenomenology*, 6, 91-101.
Stegmüller, W. 1976. *Structure and dynamics of theories*, Berlin: Springer.
Thom, R. 1975. 'Les mathématiques et l'intelligible', *Dialectica*, 29, 71-80.
Ward, B. N. 1972. *What's wrong with economics?*, London: Macmillan.
Whiteside, D. T. 1970. 'Before the Principia: the maturing of Newton's thoughts on dynamical astronomy, 1664-1684', *Journal of the history of astronomy*, 1, 5-19.
Whitrow, G. J. 1970. 'An analysis of the evolution of scientific method', *L'âge de la science*, 3, 255-280.
Wood, P. 1979. Review of Lakatos's *Philosophical papers*, *Annals of science*, 36, 289-298.

PART 3

APPLICATIONS

The Light of Lights looks always on the motive, not the deed,
The Shadow of Shadows on the deed alone.

W. B. Yeats, *The Countess Cathleen* (1892)
(*Collected plays*, 2nd edition, London: Macmillan, 1952, 50)

10 Notes on some Methodological Problems in the History of Science

One feature, and perhaps a consequence, of the growth of interest in the history of science in recent decades has been a developing awareness among some historians of methodological and philosophical problems underlying their work. What are we doing in the history of science? Can we discuss historical method at all, and if so how? Popper's philosophy is applied to these questions.

1. Introduction

The history of science would appear to be a subject highly appropriate for methodological study, since it contains, on the one hand, epistemological problems and the growth of knowledge, and on the other hand, problems of a purely historical kind — concerning science as a human and intellectual activity, scientists as people and their interaction with the times and societies in which they lived, and so on. In other words, the history of science is both history and science, and so contains a double measure of philosophical problems.

The notes offered here are based upon Karl Popper's philosophy. They do not constitute a comprehensive historiography; they are tentative suggestions concerning some methodological problems that seem to be significant for the subject *in general* and not just to particular branches of the history of science. So I refrain from giving detailed examples from those areas of the history of science, and the historical writings of others on them, with which I am familiar.

I make no particular claim of novelty for these notes; some have been made before in various forms, and in any case they probably only articulate what many historians have felt intuitively. Yet there is an advantage in making issues explicit rather than leave them implicit; for intuitively held views may be overlooked or forgotten on occasions. At the same time, I have no doubt that everyone will disagree with some of the ideas that I put forward; but even then, those who disagree with me may well also disagree with each other!

2. Some Popperian possibilities for the philosophy of the history of science

2.1 Popper's philosophy, developed mainly in the context of methodological and epistemological problems, require much supplementation and modification to be applicable to the history of science. To put this point

---

This chapter is based upon the parts of 'Notes on some methodological problems in the history of science', *Rete,* 1 (1971-1972), 1-14 that were not used in §0; they reappear here with the kind agreement of Gerstenberg Verlag GmbH & Co. KG.

in its bluntest form: if one reads the great works of science armed with ideas about tentative knowledge, critical discussion, bold tests of imaginative theories, then one will soon receive a nasty shock. Yes, the theories are there, the problems are stated, after a fashion; but what are all those paragraphs allegedly defining exact concepts, or indulging in lengthy investigations of apparently fruitless and unimportant features of a theory? The answer is simple but important: our predecessors were rarely fallibilists (and neither are we!). In fact, the history of thought, scientific and otherwise, is saturated throughout with justificationist methodologies of all kinds.

Thus we must try to determine our predecessors' own philosophical and methodological standpoint as well as we can, and expect it to differ widely from our own expectations. In fact, they probably worked in a mixed methodological situation — a mixture of some kind of justificationist methodology, which they though they were following; and simultaneously, fallibilistic problem-solving, which is what was actually taking place.

2.2 We may consider science from *three different aspects*.

The *creative aspect*: the knowledge or ignorance of past and contemporary work of individual (or intimate groups of) scientists; their (rational and irrational) methods of formulating problems and inventing theories to solve them. A scientific theory is viewed here as a personal creation.

The *social aspect*: science as a human and institutional activity; scientific education and textbook writing; fashionable and unfashionable problems and ways of solving them. A scientific theory is viewed here as a doctrine, influential (or uninfluential) for some period of time over an area of scientific development.

The *explanatory aspect*: science as sets of statements attempting to explain the world; knowledge and its growth. A scientific theory is viewed here strictly as a problem-solution, independent of the manner of its creation.

These aspects are of course mutually interacting; but at the same time the distinctions between them are fundamental enough to require historians to make clear which of them they are considering when making historical (or philosophical) points. Indeed, each aspect has its own methodological problems. In these terms, Popper's work has been largely (though not entirely) concerned with the explanatory aspect of scientific theories, and an important part of future work must be the further extension of explanatory methodology into social and creative methodology, and the metamethodological relationships between them.

2.3 A widely discussed work has been *Thomas Kuhn's theory of 'normal science'*, which attempts to explain the occurrence of revolutions as the overthrow of 'paradigms' that have arisen in the 'research programmes' of 'normal science'. There seems to be lasting value in this and the other ideas that he had proposed (§9.3), and the developments that are needed could be based on the more explicit utilisation of the tridistinction just made. I shall briefly describe three examples of the kind of point that may be worth examining.

Kuhn used the terms 'paradigm' and 'normal science' from both the personal and social points of view without distinguishing between them. But personal paradigms are of a character quite different from social paradigms; for example, an individual scientist's own paradigmatic or normal approach to problems may be highly unorthodox in the view of his contemporaries. Similarly, personal 'revolutions' are often quite different from social 'revolutions'. For example, a revolution in the thought of an individual scientist may take place when he reads a book or paper written by another, which itself may be quite 'normal' from the social point of view. Again, the revolutionary new ideas that he may himself create will have a significance and even form in his own work quite different from that which they take in the scientific community at large.

2.4 *Unveilism.* An historical method quite widely used nowadays has been recently characterised for me in conversation as 'unveilism' by Henk Bos, of the Mathematical Institute of Utrecht University, who, like me, criticises it (§2.10.2). It may be summarised as follows.

Today, as we work on the history of science, science itself is at a certain state of development. Among its mass of knowledge we hold certain results as important, certain techniques as influential, and so on, and our task as historians is to show how this situation came about — to 'unveil' it, emerging from the past. Thus we describe the past in terms of modern, or at least, later ideas, because by this means we shall clarify the form of the ideas that were actually developed at that time. In particular, we shall identify the first time that important ideas were introduced (or, at least, the early occurrences of these ideas: first-time judgements, in the sense of priorities of discovery, are not the type of historical problem under discussion here).

Unveilism can be criticised in several ways. For example, it is a form of historical verificationism; for the aim is to show how the knowledge that we know emerged in the form in which we know it, thus verifying our own version of it. Unveilism is also connected with historical determinism, for, since our knowledge emerged for us in the way that we wanted it to, then it could not have emerged in any other way. (Yet who can say today how science will advance in the future?) Thus the criticisms that apply to verificationism in general are relevant also to unveilism.

As opposed to unveilism, I suggest that the prime interpretative task of historians is not to resolve muddles by means of posterior knowledge, but *to reconstruct them in terms of then current ignorance.* {See §13 and *ROL,* §2 on history and heritage.} It may be thought that this standpoint regards ignorance as a virtue in the historian. However, this is not valid; one will have a certain amount of posterior knowledge anyway, and the better it is understood the more carefully can it be recognised as posterior.

2.5 An important part of historical work is the interpretation of the *new discoveries* of our predecessors. Unveilism may cause us to overlook the discovery of ideas that have been obvious to us for decades, since they will hardly be comprehensible as significant in their own time. But there is another and more important problem that unveilism is also likely to ignore namely, the changing priorities within knowledge.

Like the refutation of theories, the change of priorities is also a product of critical discussion, but it takes a rather different form. For it is not necessarily the case here that theories became modified through criticism or refutation, but that they are reinterpreted in the light of new knowledge and become assigned — in a basically intact form — to fresh positions in the general structure of knowledge, more, or less, important than before.

An unveilist approach to history is likely to transplant the modern priority structure onto the historical situation; but the reconstruction of that old priority structure is itself an important part of the historical task. In terms of that old structure the discoveries of the past could be surprising; for example, what we assess now as important may then have been only incidental to solve a problem that we have now forgotten. And by contrast, what was then held as important may seem to us quaint or uninteresting. Yet we must examine these old "blind alleys" carefully, however much we regard them as failures; if our predecessors found them significant, then we must also.

2.6 A common type of scientific discovery is the introduction of a *new distinction*, or criterion of demarcation, into a body of knowledge. The interpretation of the prehistory of a distinction is an extremely difficult problem, and unveilism is likely to lead to the worst interpretation possible; for the distinction will automatically be used, as part of the general aim of (allegedly) clarifying the past by means of later ideas. Yet this is highly dangerous; if we today know the difference between two types of idea A and B, then surely the prehistory of the difference will reveal examples of scientists happening to use A (or B), rather than deliberately using one and avoiding the other; or using both A and B without distinguishing between them; or even using A and B in different contexts without seeing the connection between them.

2.7 This last point is an example of the problem of which ideas are present *consciously and deliberately*, and which *unconsciously and accidentally*, in the scientist's work. It is extremely difficult to assess the significance of the absence of discussion of some idea; but at least the distinction should be made between explanatory interpretations, which assert that the theories, qua statements, have such and such a content, and social and personal interpretations, which try to assess how much of that content was understood and recognised at the time, and how much remained implicit and overlooked. Some extreme forms of unveilism do not even make this distinction; from the modern viewpoint, the content of a theory is so-and-so, and that is all that needs to be said.

2.8 Many historians and philosophers (including both Popper and Kuhn) stress the importance of *revolutions* in scientific development — mainly the social rather than the personal ones (in the sense of ¶2.2 above)

— and they obviously need especial attention. But again there are dangers. If a revolution is asserted to have taken place at a particular time, then there is a great danger that its prehistory will be written solely or chiefly as its prelude. Anticipations of the new ideas in the old science will be sought: they will be found, of course, and identified with those new ideas, even if the similarities are superficial. All pre-revolutionary activity will be seen in terms of some sort of decay; since the revolution came, then it must have been awaited. This is purely unveilist; its aim is to verify our historical theory that the revolution actually occurred. Yet new knowledge (especially if revolutionary) may not only be surprising but quite possibly unexpected.

2.9 All knowledge has *foundations*, which are themselves tentative. Thus we dig down to them through the mud of our knowledge, rather than lay them down in concrete and construct our knowledge up from them. Therefore it follows that we have a double growth of our knowledge: one growth that digs deeper into new (but tentative) foundations following the failure of their predecessors, and the other that pushes upwards into ever more detailed technicalities. Quantitatively speaking, most of our knowledge is of this latter type, but foundational questions always hold a prominent part in the development of the theories that they found. Fallibilistic methodology applies to both types of growth, but in foundational studies there are special difficulties. The process of criticism and replacement of theories is somehow much more subtle: unfalsifiable elements (in the sense of §9.5) are often present; foundational changes may lead to complicated reconstructions, both in the foundations and also in the technical knowledge "above"; even the knowledge that claims these new foundations will change, if several previously disparate theories, each with its own foundations, are unified under a new theory.

3. The history of science and its external relations

3.1 A problem of some current interest is *the relation between history of science and history of other subjects*. To ask if history of science can be called 'history' is a dangerous formulation, since as it stands it can be resolved only by deciding how to use the word 'history'; and nothing depends on the use of words. Rather, it is better to consider different types of historical method that are and have been used in the subject, and how they relate to each other and to methods used in the history of other subjects.

With regard to scientific and non-scientific knowledge, the following distinction may have some implications for all forms of historical work. In science our theories move towards the truth by the qualitative criterion of increasing corroboration, but the structure of that truth is not under our control — if the world is an atomistic place of some particular kind, for example, then that is that and we cannot do anything about it. (Nor, we recall, can we prove that it is like that, either.) But in non-scientific discussions — of politics, say, or of the creative arts — the truth is not invariant in this sense. To express this point in a phrase, scientific knowledge grows in an absolute sense in terms of which non-scientific knowledge only changes (compare §2.9 and §9.9).

3.2 The history of science has many implications for *the teaching of science*. Basically the situation is that justificationism has reigned supreme throughout the history of education (and still does); the students are presented with true (or at least reliable) knowledge secured on the foundations of exact concepts. In fact, they are usually served with the latest prejudices, plucked from the sky and put into books, clever nonsense that starts from nowhere and leads to nothing. I am sure that fallibilistically presented history can do much (though not all) to inject into teaching that element of motivation that it so often lacks, especially if corroboration is emphasised (§3).

But the balance is difficult to achieve; the teaching approach will be too historical if it never advances beyond historical questions. I do not know how to solve this difficult problem, but a mixed heuristic and historical approach may be the most successful solution, especially if it concentrates on important problem-areas, stresses that they transcend fossilised 'subjects', and takes them through several historical phases and up to modern or at least recent difficulties: some examples of this tactic in mathematics education are rehearsed in §13 and §14.

11 Structure-similarity as a Cornerstone of the Philosophy of Mathematics

How does a mathematico-empirical theory talk about the physical world? If scientific theories are guesses and may be wrong, then *what* do they talk about at all, and what is the mathematics doing in there? The best available answer to these questions is 'it depends'; here is an attempt to get further. After some explanations in Part One, a range of case-studies is presented in Part Two before proceeding to some general philosophy in Part Three. The notion of structure-similarity will be proposed as a fundamental component of this philosophy; set theory, logic and the normal "philosophy of mathematics" of today have a significant but restricted place.

PART ONE   INTRODUCTION

But I hope that I have helped to restart a discussion which for three centuries has been bogged down in preliminaries.
K. R. Popper [1972, vii]

1. Chains of reference

How does a mathematical statement mean in an empirical situation? Are the axioms of mechanics chosen for their epistemological role or for their empirical evidence, for example? Does the algebra of logic have to reflect the laws of thought?

Structure-similarity attempts a new and general answer to such questions. Its similarities to and differences from the more established philosophies need to be explained. The chief concern is with the *content* of a mathematical theory, especially the way in which its structure relates to that of other mathematical theories (which I call 'intra-mathematical similarity'), to that of a scientific theory to which it is on hire ('scientific similarity'), and on to empirical interpretations of that scientific theory in reality ('ontological similarity'). But some difficult questions arise. Scientific theories often imitate our universe in containing trans- or non-empirical components, so that it is not sufficient to claim similarity by pointing to simple correspondences between theoretical components and directly empirical notions.

Here are two very simple examples. The first concerns scientific similarity: if the integral is thought of as an area or a sum, and $\int f(x)\,dx$ says something about hydrodynamics, does it have to do so in areal or summatory terms? The second example uses empirical similarity: if $a = b + c$, and a is an intensity of sound, do b and c also have to be added intensities in

---

This article was first published in J. Echeverria, A. Ibarra and T. Mormann (eds.), *The space of mathematics. Philosophical, epistemological, and historical explorations*, Berlin and New York: de Gruyter, 1992, 91-111; it is reprinted by kind permission of the publisher.

order that the equations makes good acoustical theory? Both examples can be applied to intra-mathematical similarity also: if a and b are interpretable as lengths, say, or if the integral is so interpretable in a problem concerning conics.

The word 'applied' can itself be applied to all these examples, for it covers all three kinds of similarity: one can apply a mathematical theory a) elsewhere in mathematics, b) to scientific contexts, and c) as part of a mathematico-scientific theory, to reality. (For most purposes, the word 'applied' is perhaps *too* wide in its use, and one could argue for a return to the older (near-)synonym 'mixed' to designate the second and third kinds.) In all kinds the possibility of *non*-similarity is to be borne forcibly in mind, and indeed the word 'similarity' is to be taken throughout the chapter as carrying along its opposite also. The first, or intra-mathematical, kind of similarity has many familiar manifestations, and I shall not dwell too long on them in this chapter, for my chief concern is with the other two kinds, which bear upon a major point of interaction between the philosophy of science and the philosophy of mathematics: the use of mathematics in scientific theories.

Various issues in the philosophy of science are involved. I leave most of them to one side here, referring to [Grattan-Guinness 1986 = §2], to which this one is a sequel and a development. However, three features will be useful. Firstly, the remark that scientists 'reify' objects of concern when forming theories: they *suppose* that certain kinds of entity or process exist for the purpose of theory-building (and, for the concerns of this chapter, may use mathematics in the process). Secondly, when testing theories they check if some of these reified entities and processes actually exist in reality: if so, then reification becomes reference. It may be that a full theory refers, in which case it becomes 'ontologically correct', a category I use to replace some of the misuses of truth. Finally, there is the notion of 'desimplification', in which a scientific theory is formulated in which various effects pertaining to the phenomena are *knowingly* set aside as negligible or at least too complicated to deal with in the current state of knowledge, but then are reinstated later in desimplified forms of the theory, when the measure of scientific (and maybe also ontological) structure-similarity is thereby increased. I propose this notion for the philosophy of science as an improvement upon the category of ad hoc hypotheses, for there is no component of ad hocness in desimplification: one might even *hope* to show that the neglected effect is indeed small enough to be set aside in the measure of accuracy and fine detail within which the current scientific and experimental activity is set.

Our concern in this chapter is with *how* a mathematical theory can mean, not *what* it may mean: the paper [Grattan-Guinness 1987b] is based upon the theme 'How it means' within a particular historical time-period. Here is one respect in which the differences of intent from much modern philosophy (of mathematics) are evident.

2. Plan and purpose

Part Two of the chapter contains a selection of case studies. After the intra-mathematical considerations of the next two sections, in which the main novelty is the distinction between icons and representations, scientific and empirical similarities dominate. ¶4 and ¶5 contain some examples from the uses of the calculus in mechanics and mathematical physics, dealing in turn with the formation and the solution of differential equations. Some of them involve Fourier series, which also raise the question of linearisation of scientific theories; this matter is discussed in more general terms in ¶7. Then In ¶8 the focus turns to mathematical psychology, in the form of Boole's algebra of logic and its early critics.

The examples used in Part Two happen to be historical, and involve cases that I can explain without difficulty to the reader and with which I am sufficiently familiar to draw on with confidence for current purposes. But the points made are equally applicable to modern mathematical concerns, and I hope, therefore, that they will catch the interest of mathematicians as well as historians and philosophers.

In Part Three more general and philosophical considerations are presented. ¶9 contains remarks on the limitations of axiomatised mathematical theories. In the same spirit, doubts are expressed in ¶10 about philosophers' 'philosophy of mathematics', which is centered on logics and set theories but rarely get further; and mathematicians' 'philosophy of mathematics', where the breadth of the subject to noted (to some extent) but logical theory and related topics are disregarded. An alternative philosophy is outlined in ¶11-¶13, in which both the range of the subject and logic (and related topics) are taken seriously.

PART TWO   SOME CASE STUDIES

3. Intra-mathematical similarity between algebra and geometry

It is a commonplace that different branches of mathematics relate by similarity to each other. The example of $a + b$ as numbers and as lengths, just mentioned, is a canonical example of the numerous cases that apply to the complicated relationships between geometries and algebras: my use of plurals here is deliberate). It can be extended to the possibility of non-similarity, for at various times objections have been made to the legitimacy of negative numbers [Pycior 1987], so that their interpretation as suitably directed line-segments was not adopted.

Other such cases include the interpretation of powers of variables relative to spatial dimensions: it is striking to note that when François Viète [1591] advocated the new 'analytic art' of algebra he used expressions such as 'squared-cube' to refer to the fifth power, with similar locutions for all powers above the third, and thereby to draw on structure-similarity between algebra and geometry, and (by an implication that involves a huge burden of questions concerning our theme), onto space. By contrast, a few decades later René Descartes [1637] showed no qualms in his *Géométrie* when advocating higher powers in his algebrisation of geometry, and in writing z^4 just like z^3, and thereby discarding this similarity; in the same way he regarded negative roots of equations as 'false' and so dispensed with that possible link also.

In some respects, therefore, Descartes was a non-similarist. Yet at the beginning of Book 3 he announced that 'all curved lines, which can be described by any regular motion, should be received in Geometry', a typical example of an isomorphism between one branch of mathematics and another. Notice, however, that the similarity often does not go too far. For example, there are no obvious structural similarities between basic types of algebraic expressions and fundamental types of geometrical curve; hence classification has been a difficult task for algebraic geometry (or should it sometimes be thought of as geometrical algebra?) from Isaac Newton onwards. Again, dissimilarity is evident in problems involving the roots of equations, where the algebra is unproblematic but a root does not lead to a geometrically intelligible situation (an area becomes negative, say). The occurrence of complex zeroes in a polynomial can be still more problematic as geometry, since their presence is not reflected in the geometrical representation of the corresponding function in the real-variable plane, and a pair of complex-variable planes will represent the argument and the value of the function but necessarily do not reflect the function itself. Other cases include a situation where the algebraic solution of a geometrical problem supplies a circle (say) as the required locus whereas only an arc of it actually pertains to the problem.

In an undeservedly forgotten examination of 'the origin and the limits of the correspondence' between these two branches of mathematics Cournot [1847] explores in a systematic way these and other cases and sources of such structural non-similarity. The mathematics is quite elementary; the philosophy is far from trivial.

4. Representations and icons

Sometimes these intra-mathematical structure-similarities are put forward more formally as representations, such as in the geometrical characterisation of complex numbers in the plane, or A.-L. Cauchy's definition by such means of infinitesimals in terms of sequences of real values passing to zero (§9.2). There is no basic distinction of type in these cases, but considerable variation in the manner and detail in which the structure-similarity or representation is worked out: Descartes's case was to cause him and many of his successors great difficulty concerning not only over the detail but also the generality of the translation that was effected. In ¶9 we shall note an approach to mechanics in which it was explicitly avoided.

A particular kind of intra-mathematical structure-similarity worth emphasising is one in which the mathematical notation itself plays a role, and is even one of the objects of study. Within algebra matrices and determinants are an important example, in that the array can be subjected to analysis (by graph-theoretic means for large sparse matrices, for example). Following C. S. Peirce, I call them 'icons', and draw attention to one of his examples, in algebraic logic, where systems of connectives were set up in squares and other patterns in ways that reflect their significations [Zellweger 1982].

Another type of example is shown by algebraic logic: the principle of duality, which was exploited by Peirce's contemporary Ernst Schröder. He stated theorems in pairs, deploying a formal set of (structure-conserving) rules of transformation of connectives and quantifiers in order to get from one theorem to the other. He consciously followed the practise of J. V. Poncelet and J. D. Gergonne, who had laid out theorem-pairs in projective geometry following a set of rules about going from points, lines and planes to planes, lines and points. A current interest in logic is a type of generalisation of duality into analyses of proof-structures in order to find structure-isomorphisms between proofs and maybe to classify mathematical proofs into some basic structural categories. In cases like these, the mathematical text itself is the (or at least an) object of study, not (only) the referents of the mathematical theory. The situation is not unlike the use of a laboratory notebook in empirical science, where the matters of concern can switch back and forth between the laboratory experiments and the contents of the notebook itself.

5. Modelling continua: the differential and integral calculus

The calculus has long been a staple method in applied mathematics, especially in the differential and integral form introduced by G. W. Leibniz. Here the principal device was represent the (supposed) continuity of space and time by the theory of differentials, infinitesimally small forward increments dx on the variable, with its own second-order infinitesimal ddx, and so on [Grattan-Guinness 1990a]. (Newton presented his second law of motion under the same regime, in that he stated it as the successive action of infinitesimally small impulses.) The key to the technique lay not only in the smallness of the increment (a controversial issue of reification, of course) but especially in the preservation of dimension under the process of differentiation: if x is a line, then so is dx (a very short line, but infinitely longer than ddx). The various orders of infinitesimal were used to reflect individually the increments on the variables of the chosen problem, and literally 'differential equations' were formed: that is, equations in which differentials were related according to some physical law. The measure of structure-similarity is quite substantial: for example, a rate of change dx/dt was the ratio $dx \div dt$ of two infinitesimal increments, a property lost in an approach such as Cauchy's or Newton's based on limits, where the derivative $x'(t)$ does not reflect its referent in the same way.

The integral branch of the calculus also can exhibit issues pertaining to our theme when interpreted as an area or as an infinitesimal sum. In energy mechanics, for example, the work function $\int P\, ds$ designates the sum of products of force P by infinitesimal distance ds of traction, and founders of this approach in the 1820s, such as Poncelet and G. G. Coriolis, explicitly made the point. Before them, P. S. Laplace had inaugurated in the 1800s a programme to extend the principles of mechanics (including the use of the calculus) to the then rather backward discipline of physics by modelling "all" physical phenomena on cumulative inter-molecular forces of attraction and repulsion [Grattan-Guinness 1987a]. The key to his mathematical method was to reflect the cumulative actions by integrals; but his follower S. D. Poisson modified this approach in the 1820s by representing the actions instead by sums. Cauchy did the same in his elasticity theory of the same time, and for a clearer reason than Poisson offered, more clearly involving structure-similarity: the magnitude of the action on a molecule was very sensitive to locations of the immediately neighbouring molecules, and the integral would not recognise this fact with sufficient refinement [Grattan-Guinness 1990b, 1003-1025]. [Cournot 1847, chs. 13-15] is again worth consulting, for a range of examples concerning not only the basic aspects of the calculus but also rectification and the definability of functions as definite integrals.

6. Linearity or non-linearity: the case of Fourier series

> The elegant body of mathematical theory pertaining to linear systems (Fourier analysis, orthogonal functions, and so on), and its successful application to many fundamentally linear problems in the physical sciences, tends to dominate even moderately advanced University courses in mathematics and theoretical physics. [... But] nonlinear systems are surely the rule, not the exception, outside the physical sciences.
>
> R. M. May [1976, 467]

Let us take now a major example of ontological similarity. One of the main periods of research that led to the great importance of linear modelling of this non-linear world was linked to the rise of classical mathematical physics on the early years of the 19th century. The note just taken of the elasticity theory of that time was part of this adventure, and Cauchy is a prominent example of a linearist. Another major figure was J. B. J. Fourier, who introduced the mathematical theory of heat diffusion and radiation from the 1800s onwards. His deployment of linearity embraced not only the assumptions used in forming the diffusion equation but also in solving it by Fourier series.

These series form an excellent example of ontological non-similarity and incorrectness: not only the (non-)similarity of these linear solutions with the phenomena to which they refer but also the question of their manner of representing (in Fourier's case) heat diffusion. Mathematically speaking, they comprise a series of terms exhibiting integral multiples of a certain periodicity specified by the first term, sometimes prefaced by a constant term; in addition, sine/cosine series exhibit evenness/oddness. They describe diffusion at the initiation of the time t: for later values of t exponential decay terms are multiplied into the time terms.

Now the ontological structure-similarity of these time terms is clear (and indeed rather important for the legitimacy of the analysis, especially for its non-linear critics!); but the trigonometric terms raise questions. Does heat have to be interpreted as waves, in families of corresponding periodicities? The nature of heat was an important question at that time, with debates as to whether it was a substance (caloric), a product of molecular action, an effect of vibratory motion (that is, the waval theory just mentioned), or something else; but Fourier himself did not like the question, preferring to treat heat and cold as opposites without reifying their intimate structure. His most explicit approach to a structure-similar reading occurred in a treatment of the solution of Laplace's equation (as the case of the diffusion equation for steady state in the lamina in the Oxy plane) of the form

$$\sum_{r=0}^{\infty} a_r \exp(-rx) \cos ry \; ; \qquad (1)$$

here he spoke of each component term 'constitut[ing] a proper and elementary mode', such that 'there are as many different solid laminae that enter in to the terms of the general surface, that each of these tables [laminae?] is heated separately in the same manner as if there were only a single term in the equation, that all these laminae are superposed' [Fourier 1807, 144]. Even here he went so far only to affirm superposition; the periodic character of the functions involved was not affirmed ontologically as heat behaviour.

Fourier series are well known also in acoustics, and indeed Daniel Bernoulli had already proposed them in this context. Here structure-similarity between periodicities and pitch level were proposed, together with further similarities with pendular motion [Cannon and Dostrovsky 1981]: 'for the sounds of horn, trumpets and traverse flutes follow this same progression 1, 2, 3, 4,..., but the progression is different for other bodies' [Bernoulli 1755, art. 3]. He did not have the formulae for the coefficients of the series; by contrast, Fourier did know them, and also the manner of representing the function by the series outside its period of definition, but when discussing his predecessors on the matter he did not follow Bernoulli's advocacy of structure-similarity when vindicating his case against the criticisms offered at the time by Leonhard Euler and Jean d'Alembert.

Bernoulli's stance was adopted much later by G. S. Ohm, a famous German scientist not oriented towards Fourieran positivism although much influenced by Fourier's methods in his own earlier work on electromagnetism. In his paper [Ohm 1843] he rejected the usual view of sound as composed of a small number of simple components and proposed instead that (infinite) Fourier series could be so structurally interpreted: 'If we now represent by F^1 any sound impulse striking the ear at time t, then Fourier's theorem says that this impulse is analysable into the [trigonometric] components', where the constant term 'corresponds to no oscillation but represents merely a displacement of the oscillating parts of the ear. The other components, however, all correspond to oscillations that take place around the displaced position'.

In this way Ohm imposed the structure-similarity of Fourier series onto acoustic theory in a way that Bernoulli had envisaged a century earlier but without the details of the mathematical theory that Fourier was to provide between them. From then on this interpretation of series became well known, and even of mechanical representation in Lord Kelvin's harmonic. Kelvin is a wonderful subject for our theme, for he was an advocate of the method of analogy from one theory to another; for example, his early work in electromagnetism was based on a 'flow analogy' from Fourier's theory of heat [Wise 1981]. His machine evaluated integrals of the form

$$\int f(x) \, {\sin \atop \cos} \, nx \, dx \quad \text{for} \quad 0 \le n \le N, \text{ with } N \text{ small}, \qquad (2)$$

via the motions of rotating cranks linked in series to imitate the sum of the trigonometric terms. In the particular case of tidal theory, the terms corresponded to the various 'tidal constituents' that he asserted were caused by various solar and lunar effects; the formula was also applicable to the important problem of compass deviation on iron ships [Kelvin 1881]. Later workers developed desimplified versions of this machine that could calculate the Fourier coefficients for arbitrarily large values of n, for a variety of purposes [Henrici 1892].

Here we see scientific structure-similarity in a stark form. Kelvin calculated the tidal level as literally the sum of the component trigonometric terms, but he also recognised that these terms only approximated to the periodicity of those planetary effects; thus scientific *non*-similarity is also evident, like Fourier advocating superposition but not committing himself to a waval theory of heat.

This practice goes back at least to Euler's celestial mechanics, when he took Newton's second law to express both planetary-solar and interplanetary actions; the latter group of effects involved powers of the appropriate distance functions, which were stated via the triangle formula. Now this expression took the form $(a + b \cos \alpha)^{-3/2}$, where α was the angle between the radius vectors of the two planets involved; and to render this expression in more amenable form he used De Moivre's theorem to convert it into a series of the form $\sum_r (a_r \cos r\alpha)$ (see, for example, [Euler 1749]). But this is trigonometric series once again, in a different context. However, this time they arose out of a pure-mathematical artifice, *bereft* of the structural interpretation that was then being advocated by Bernoulli for acoustical theory (and being *rejected* by the same Euler in favour of the functional solution of the wave equation).

7. Linearity or non-linearity: the general issue

I remember being quite frightened with the idea of presenting for the first time to such a distinguished audience the concept of linear programming.

After my talk the chairman called for discussion. [... Hotelling] said devastatingly: '*but we all know the world is non-linear*'. [...]

Suddenly another hand in the audience was raised. It was von Neumann. [...] 'The speaker called his talk "Linear programming". Then he carefully stated his axioms. If you have an application that satisfies the axioms, use it. If it does not, then *don't*', and he sat down.

George Dantzig [1982, 46]

If scientific and ontological structure-(non)-similarities are not incorporated into the philosophical scenario of developments such as the ones described in the previous section, then the full richness of the intellectual issues that were at stake cannot be appreciated. And the concerns are not only historical: indeed, the rise of computers and the new levels of efficacy achievable in non-linear mathematics has raised the issue of linear versus non-linear reification and reference to a new level of significance. Acoustics and related branches of science themselves form a fine example of this trend, for especially since the 1920s attention has focussed on non-linear oscillations of all kinds, especially the so-called 'relaxed' variety, and upon associated phenomena known as 'irregular noise' some decades ago and now carrying the trendy name 'chaos' (see the survey in [West 1985, ch. 3], a profound study of 'the importance of being nonlinear').

Competition between utility of linear or non-linear versions of theories in same range of concern can ensue: an interesting example is linear and non-linear programming, in which the latter was instituted in the early 1950s soon after the establishment of the former, but took rather different origins [Grattan-Guinness 1994]. Hotelling's reservation about linear programming, quoted at the head of this section, was soon to be dealt with, at least in part.

From the point of view of the philosophy of mathematics, the extension of structure-similarity is a central feature of the questions raised so far. For the philosophy of science, desimplification of theories, and the realm of legitimate reification, are correspondingly central concerns. For science itself, the manner of extending theory from mechanics and physics to other branches has been a major issue.

8. Mathematical psychology and the algebra of thought

My last context seems to be quite different; yet we shall meet some striking metasimilarities. It concerns Boole's formulation of an algebra to represent normatively 'The laws of thought' (to quote the title of his second book on the matter, of 1854). (This occurred before Peirce's concerns with semiotics mentioned in ¶4.) The basic law $x^2 = x$ of 'duality' obeyed by 'elective symbols' x (which selected the individuals satisfying a property such as 'European') differentiated the algebra of his logic from the common algebras, which were structurally inspired by arithmetic; but Boole maintained similarity by deploying the other three connectives, and made use of subtraction, division and addition as well as the multiplication involved in the law of duality.

The chief object of concern here is Boole's definition of '+'; for convenience I shall use his interpretation in terms of classes. $(x + y)$ was defined so that 'the symbol +' was 'the equivalent of the conjunctions "and", "or"': however, 'the expression $x + y$ seems indeed uninterpretable, unless it be assumed that the things represented by x and the things represented by y are entirely separate; that they embrace no individual in common' [1854, 55, 66]. The specification of the unions of non-disjoint classes x and y, corresponding to the exclusive and inclusive senses of 'or', required the intermediate definitions of disjoint classes:

$$x(1 - y) + y(1 - x) \quad \text{or} \quad x + y(1 - x), \qquad (3)$$

as required [p. 57].

Thus we see that the symbol '$x + y$' actually reflected a process structurally similar to addition, although it was defined under the hypothesis of the disjointness of the components. Boole tried to argue for the necessity of this restriction; but his arguments were hardly convincing, and they were rejected by his first major commentator, W. S. Jevons, in his little book *Pure logic* [Jevons 1864]. His sub-title, 'the logic of quality apart from quantity', expressed his desire to reduce the structural links with mathematics as espoused by Boole (although this and other work by Boole himself and others showed that the distinction between quality and quantity did not capture the essentials of the mathematics of that time). Jevons worked with 'terms', that is, 'any combination of names and words describing the qualities and circumstances of a thing' [1864, 26, 6]; a proposition *'is a statement of the sameness or difference of meaning between two terms*, to be written A = B with '=' read as 'is' (pp. 8-9). He followed Boole in 'combining' terms A and B in a Boolean manner to produce AB, and accepted the law of duality (as a law satisfied by terms, giving it the name 'the law of simplicity'); but he rejected entirely the restriction laid upon the definition of addition, arguing that the natural use of language permitted the definition of full union of intersecting classes (or overlapping terms). 'B or C is a plural term [...] for its meaning is either that of B or that of C, but it is not

known which' ([p. 25]: later examples show that inclusive disjunction was intended). His 'law of unity' for a term A allowed that
$$A + A = A; \qquad (4)$$
that is, two (and therefore any number) of self-disjunctions of A may be reduced to a single A without change of meaning. Hence logical alternation was different from mathematical addition: intra-mathematical structure-similarity was being rejected. Jevons distanced himself further from Boolism by dispensing entirely with division and subtraction.

Boole and Jevons corresponded on these matters in 1863 and 1864, some months before Boole's unexpected death [Grattan-Guinness 1991]. They took as a "test case" the example of $(x + x)$. For Jevons it satisfied (4); for Boole it was not interpretable at all, although in his system it followed that the equation $x + x = 0$ was reducible to $x = 0$ via one of his general expansion theorems for a general logical function. Thus the differences in ontological structure-similarity ran quite deeply, and Jevons found Boole to be ontologically incorrect.

These changes required Jevons to develop different methods of obtaining consequences (or, as he called them, 'inferences'). 'Direct inference' worked in effect on the transitivity of terms equated in the premises; for example, from A = B and B = C, A = C could be inferred [Jevons 1864, 10-13]. In the more powerful method of 'indirect inference' he formed all possible logical combinations of the simple terms involved in the premises, together with the contrary terms, combined these compound terms with both members of each premise, and retained only those terms that either were consistent with both members of at least one premise or contradicted both members of all of them. The consequences were drawn by taking any simple or compound term (C, say) and equating it to the sum (in his sense of '+') of all the retained terms of which it was part. In other words, he found the plural term to which C was equal ('=') under the premises. Basic laws such as duality and simplicity, and various rules of elimination, simplified the resulting propositions (pp. 42-53).

This procedure of selecting and inspecting was rather tedious: in order to render it more efficient [Jevons 1870] introduced his 'logical machine'. The similarity of function with the harmonic analysers of Kelvin and his successors in ¶7 is worth noting: in both cases a structure was carried through from a theory and its referents to a mechanical imitation.

PART THREE
TOWARDS A GENERAL PHILOSOPHY OF MATHEMATICS

> If I speak very decidedly about the consequences of the neglect of pure logic by mathematicians, as I have done elsewhere about the neglect of mathematical thought by logicians, I shall not be supposed to have any disrespectful intention [...]
> Augustus De Morgan [1869, 180],
> alluding to George Peacock

9. Prologue: the limitations of axiomatisation

Under a deeply unfortunate educational tendency, itself partly inspired by the so-called 'philosophy of mathematics', many theories are formulated in a neo-axiomatic way; and the impression accrues that the subject is deeply unmotivated. In many cases this impression arises because the axioms are used precisely because of their epistemological status, *as* starting-point for deductions: they have little or no intuitive character (which is why they are so often hard to find or identify as axioms in the first place!). In particular, structure-similarity is rarely evident: even the intuitive feel in Boole's algebra is largely lost in the axiomatisation of the propositional calculus (which is not due to Boole or Jevons, incidentally). Further, the axiomatised states have nothing to say about the desimplification of theories — ironically, not even about the enrichment of axiomatised versions! — and thus do not focus upon scientific knowledge as a process of growth.

Take classical mechanics, which was and is an especially rich branch of mathematics from our point of view, with its variety of formulations [Grattan-Guinness 1990c]. The most axiomatised version is the variational tradition, of which two main forms developed: Lagrange's, based on the principles of virtual work and of least action; and the extended version based on Hamiltonians. In both cases the aim was to develop a very general theory from few assumptions: the aim of those times corresponding (but *not* in close detail) to our conception of axiomatisation. But, as in modern cases, the price of (alleged) generality is intuition: one cannot claim the Lagrange equations as an *evident* way of founding dynamics. The reason can be stated in terms of structure-non-similarity: the various terms do not imitate in an way the phenomena associated with their referents, and neither do their sums or differences.

Allied with variational mechanics is potential theory, which has a curiously ambiguous place in our context. On the one hand some potentials, such as the velocity potential, for example, do not have a clear referent (so that structure-similarity is ruled out); and the point is important from an epistemological angle, since the potential of some category X could replace X in the list of ontological commitments; for example, if the work expression always admits a potential, does 'force' become only *façon de parler*? However, equipotential surfaces can, and indeed are intended to, take a direct reification and even reference — as streamlines in hydrodynamics, or as the defining surfaces for lines of optimal electromagnetic flow.

10. Between two traditions

It is clear that by 'mathematics' I intend to refer to "all" of the subject, both ancient and modern, and also the modes of reasoning that attend it. Thus I try to bridge the following lamentable gap.

On the one hand, there is the "philosophy of mathematics" that starts out from logics, set theories and the axiomatisation of theories, but rarely gets much further. It has flourished since the 1920s, mostly in the hands of philosophers. Without doubt important and fruitful insights have come out of this tradition, often with consequences beyond the fragments of mathematics studied; but they belong only to a corner of the wide range of questions that the philosophy of actual mathematics excites. One might as well think that music is the same as piano sonatas.

On the other hand, there is the opposite absurdity practised by mathematicians, which respects (much of the) range of their subject but tends to adopt the metaphilosophy of ignoring the logico-philosophy of the subject. It takes logics and proof theory for granted even if axiomatisation is (over)-emphasised; for the logical issues at hand are often poorly understood [Corcoran 1973].

This separation has long been in place, unfortunately, in one form or another. A paper [Grattan-Guinness 1988] on the contacts between logics and mathematics between the French Revolution and the First World War took the title 'Living together and living apart' to underline the modest degree to which contacts functioned.

Both traditions are quite often formalist in character, either in the technical sense of the word (falsely) associated with Hilbert or in a more general way as unconcerned with reference [compare Goodman 1979]. Further, while structure-(non)-similarity is a major component, this philosophy has little purpose in common with the structuralist philosophies, which are often involved only with set-theoretic formulations and/or abstract axiomatisations of theories, with associated model theory [Vercelloni 1988]. It is time to bring these two traditions together, in philosophy that takes 'how it means' as the prime question rather than the 'what it means?' of normal modern philosophy (of mathematics). In the next three sections I shall show how *both* philosophies can and must be accommodated.

11. The philosophy of forms

There is nothing at all new in my emphasis on the use of one theory (T_1, say) in another one (T_2). But new to me, anyway, is the importance of structure-*non*-similarity between theories, and the consequent idea of *levels of content* that T_2 may (or may not) exhibit in T_1. From this start I propose the following formulation, in which intra-mathematical and scientific structure-similarity (and non-similarity) are borne in mind.

Mathematics contains *forms,* which may be expressions, equations, inequalities, diagrams, theorems with proofs, even whole theories. Some forms are atomic, and can be concatenated together to produce *superforms*, or *compound forms*. The level of atomicity can be varied, depending on the need and context: thus in hydrodynamics the integral may be used as an atom, but in the foundations of analysis it would be dissected into its components. (An example for this case is given in ¶13 below.) The forms themselves characterise mathematics, and distinguish it from, say, chemistry.

The assembly of forms is very large {Grattan-Guinness 1999}. It includes not only those from abstract algebra such as group and field but also integral (as just mentioned), exact differential, least squares, addition, neighbourhood, limit, and so on and on. To each form there are sub- or special forms: double integrals and Abelian groups, for example. An important special case of intra-mathematical structure-similarity was stressed in ¶4 under 'icons': mathematical notations, which themselves can play the role of T_2.

When T_1 is applied to T_2, the repertoire of forms of T_1 works in T_2, but with differing levels of content (some clear examples were given in Part Two). Thus one can understand *how* the 'unreasonable effectiveness of mathematics in the natural sciences' occurs: there is no need to share the perplexity of [Wigner 1960] on this point if one looks carefully to see what is happening (§12). Further, the notion of desimplifying scientific theories can be used here; for two of its sources are the increases in structure-similarity and in levels of content. The *genuine* source of perplexity that the mathematician-philosopher should consider is the *variety* of structures and of levels of content that can obtain within one mathematico-scientific context.

12. The philosophy of reasonings and structures

So far no notice has been taken of logics and allied theories. I bracket them together under the word 'reasonings'. This word is chosen without enthusiasm, but every candidate synonym or neonym is defective: 'logic(s)', 'deduction' and 'proof' are definitely too narrow, and 'argument' and 'connection' not sufficiently specific. In this category I place logics, both formal/axiomatic and natural deduction, bivalent and also non-classical forms, including predicate calculi with quantification and set theories; the associated metatheories are also on hand. In addition, place is granted to rules for valid and invalid inference, and the logic of necessary and sufficient conditions for the truth of theorems; and for proof methods such as mathematical induction, by reduction to the absurd, by modus ponens, and so on, again with special kinds (first- and higher-order induction, for example). There is quite a bit of repetition in this catalogue, for many of these methods could be expressed within some logical frameworks; but the agglomeration is not offensive to this sketch.

In addition, we need definitional theory, a neglected area [Gabriel 1972] where mathematics often comes adrift [Dugas 1940, Rostand 1960]. Topics include rules for well formation of nominal definitions in formal but also non-formal mathematical theories; formation to meet given criteria (such as the correlation coefficient in statistical regression, or G. U. Yule's structurally similar Q-parameter for association); definitional systems and their relationship to axioms; and creative and contextual definitions. Some of these types of definition involve model-theoretic notions such as (non)-categoricity [Corcoran 1980], so that they sit here also. In addition, philosophical ideas about existence and uniqueness of defined terms will need consideration. Finally, some philosophy of semiotics will be needed to appraise the use of iconic forms.

The main distinction between forms and reasonings is that the former pertain to mathematics itself while the latter are suitable in other areas of thought. But they have an important common factor: like forms, reasonings have structures, which are *not* objects in the way that forms and reasonings are (at least, as they are in the liberal ontology that I have admitted into this sketch). Take the example mentioned several times in Part Two of the integral, treated as a superform under the structural concatenation of the forms limit, function, sum, difference and product, together with the reasonings of nominal definition (':=') and existence:

$$\int f(x) \, dx := \lim \sum_r [f(x_r) \Delta x_r] \text{ as } \Delta x_r \Rightarrow 0 \text{ if the limit exists.} \qquad (5)$$

The structure *glues this concatenation together,* but it is not itself an object of the concatenation. Further, this structure is distinguished from other structures by the glueing, which tells the integral as this sum apart from the integral as an infinitesimal sum or as the inverse of a derivative. An analogy may be drawn between a book, any of its chapters, and its price or weight:

the book and the chapters are objects in a sense denied to the price and the weight {see §12.3-§12.4 on parts and moments}.

13. Mathematics as forms, reasonings and structures

Several features of mathematical development and progress can be illuminated by the philosophy proposed here. The important notion of analogy [Knobloch 1989] falls into place: T_1 has been applied with success to T_2, and task of analogy is to assess the similarity of structure between T_1 in T_2 and T_1 in some new context T_3 (as Kelvin did with the success noted in ¶6). Again, the process of 'having a new idea' and thereby advancing mathematical knowledge can be given a more precise characterisation in those cases (which will be the vast majority) in which the idea is *not* completely novel in itself. The famous phrase of the novelist E. M. Forster, 'only connect', is apposite: *new* combinations of forms, structures and reasonings are made, and a theory developed from there. Similarly, missed opportunities are situations where the connections are not made [Dyson 1972].

This philosophy also has the advantage of conveying the great multitude of complications, of all kinds, that attend mathematics. As was mentioned in ¶10, the number of forms is very great, and a rich basket of reasonings is involved also; thus the assembly of structures is commensurately large. In addition, the chains of reference (¶1) are varied in their kinds: from mathematics to scientific theory and maybe on to reality. Finally, there is a range of possibilities in a theory for structure-similarity to be upheld between some components but denied to others (for example from ¶6, Fourier on superposition but not on the waval interpretation of heat).

When this philosophy is used historically (as just now), the usual caveats about anachronism would have to be watched with especial care; in particular, the ignorance of logic among mathematicians will require the historian to deploy reasonings with especial delicacy. But, as the examples of Part Two show, this philosophy has much to say about the past. Here is another important source of difference of this philosophy from traditions that take no notice of the evolution or development of a theory: for example, here is a further manner of expressing the reservations of ¶9 on exaggerating the place of axiomatisation. I am much more in sympathy with the view of [Polya 1954 and 1962-1965] and of followers such as [Lakatos 1976], and also the neglected [Rostand 1962], on the role of modification of proofs to generate mathematics — which, among other things, is an historical process. A few other modern philosophies of mathematics take history seriously in some way [see *passim* in Tymoczko 1985].

When one takes on also the use of probability and statistics within this context, a range of *quite basic* additional issues concerning purpose of theory are raised: probability as a compensation for ignorance or as a genuine category for reference or reification, and the various interpretations of probability and their referentiability. It is a great pity that questions of these types do not occupy an important place in the practice of the philosophy of mathematics, for they occupy comparable positions in mathematics itself.

Yet even now not every question involved in the philosophy of mathematics has been raised. For example, the creative side of the subject is not basically touched, although, for example, the questions concerning mathematical heuristics may be tackled from the point of view of maximising levels of content among the various presentations available [compare Polya 1954]. As a special case this remark restates once again my criticisms of axiomatisation made in ¶9.

To conclude, in this philosophy mathematics is seen as a group of problems, topics and branches in which forms and reasonings are chosen and deployed in a variety of structures exhibiting differing levels of content. This, briefly, is *how* this philosophy of mathematics means, and also how this philosophy of mathematics *means*.

A developed version of would be highly taxonomic in character. What are the atomic forms, the reasonings, and structures? How do they relate to each other; via (meta)structural isomorphism for example? Does it matter that versions of set theory occur both as forms and in reasonings? What, if anywhere, is the place of a priori knowledge? I am not at all sure of the answers to these questions; but I am sure that they are *fruitful* questions, and examining them could close the lamentable gap that exists between the practise of mathematics and the reflections upon it that mathematicians and philosophers make (§21). It is a great pity that the philosophy of mathematics is always bogged down in preliminaries.

Bibliography

Bernoulli, D. 1755. 'Réflexions [...] sur les nouvelles vibrations des cordes [...]', *Mémoires de l'Académie des Sciences de Berlin, 9* (1753), 147-172.

Boole, G. 1854. *The laws of thought* [...], London: Walton and Maberley. [Repr. New York: Dover, 1958.]

Cannon, J. T. and Dostrovsky, S. 1981. *The evolution of dynamics: vibration theory from 1687 to 1742,* Heidelberg: Springer.

Corcoran, J. 1973. 'Gaps between logical theory and mathematical practise', in M. Bunge (ed.), *The methodological unity of science,* Dordrecht: Reidel, 23-50.

Corcoran, J. 1980. 'Categoricity', *History and philosophy of logic, 1,* 187-207.

Cournot, A. A. 1847. *De l'origine et des limites da la correspondance entre l'algèbre et la géométrie,* Paris and Algiers: Hachette.

Dantzig, G. B. 1982. 'Reminiscences about the origins of linear programming', *Operations research letters, 1,* 43-48. [Slightly revised version in A. Bachem, M. Grotschel and B. Corte (eds.), *Mathematical programming. The state of the art,* Berlin: Springer, 1983, 78-86.]

De Morgan, A. 1869. 'On infinity and the sign of equality', *Transactions of the Cambridge Philosophical Society, 11* (1864-69), 145-189.

Descartes, R. 1637. 'La géométrie', In *Discours de la méthode* [...], Leiden: Jan Maire, 297-413. [Various reprints and translations.]

Dugas, R. 1940. *Essai sur l'incompréhension mathématique,* Paris: Vuibert.

Dyson, F. 1972. 'Missed oppportunites', *Bulletin of the American Mathematical Society, 78,* 635-652.

Euler, L. 1749. *Recherches sur la question des inégalités du mouvement de Saturne et de Jupiter,* Paris: Académie des Sciences. [Repr. in *Opera omnia,* ser. 2, vol. 25, Zürich: Orell Füssli, 1960, 45-157.]

Fourier, J. B. J. 1807 'Mémoire sur la propagation de la chaleur', ms. in the *Ecole Nationale des Ponts et Chaussées,* ms. 1851. Published in [Grattan-Guinness and Ravetz 1972] *passim.*

Goodman, N. 1979. 'Mathematics as an objective science', *American mathematical monthly, 86,* 540-551.

Gabriel, G. 1972. *Definitionen und Interessen. Über die praktischen Grundlagen der Definitionenlehre,* Stuttgart: Frommann.

Grattan-Guinness, I. 1986. 'What do theories talk about? A critique of Popperian fallibilism, with especial reference to ontology', *Fundamenta scientiae, 7,* 177-221. [§2]

Grattan-Guinness, I. 1987a. 'From Laplacian physics to mathematical physics, 1805-1826', in C. Burrichter, R. Inhetveen and R. Kötter (eds.), *Zum Wandel des Naturverständnisses,* Paderborn: Schöningh, 11-34.

Grattan-Guinness, I. 1987b. 'How it means: mathematical theories in physical theories. With examples from French mathematical physics of the early 19th century', *Rendiconti dell'Accademia del XL, (5)9,* pt. 2(1985), 89-119.

Grattan-Guinness, I. 1988. 'Living together and living apart: on the interactions between mathematics and logics from the French Revolution to the First World War', *South African journal of philosophy, 7,* no. 2, 73-82.

Grattan-Guinness, I. 1990a. 'Small talk in Parisian circles, 1800-1830: mathematical models of continuous matter', in G. König (ed.), *Konzepte des mathematisch Unendlichen,* Göttingen: Vandenhoeck und Ruprecht, 47-63.

Grattan-Guinness, I. 1990b. *Convolutions in French mathematics, 1800-1840. From the calculus and mechanics to mathematical analysis and mathematical physics,* 3 vols., Basel: Birkhäuser; Berlin: Deutscher Verlag der Wissenschaften.

Grattan-Guinness, I. 1990c. 'The varieties of mechanics by 1800', *Historia mathematica, 17* , 313-338.

Grattan-Guinness, I. 1991. 'The correspondence between George Boole and Stanley Jevons, 1863-1864', *History and philosophy of logic, 12,* 15-35.

Grattan-Guinness, I. 1994. '"A new type of question": On the prehistory of linear and non-linear programming, 1770-1940', in E. Knobloch and D. Rowe (eds.), *History of modern mathematics,* vol. 3, New York: Academic Press, 43-89.

Grattan-Guinness, I. 1999. 'Forms in algebras, and their interpretations: some historical and philosophical features', in L. Albertazzi (ed.), *Shapes of forms. From Gestalt psychology and phenomenology to ontology and mathematics,* Dordrecht: Kluwer, 177-190.

Grattan-Guinness, I. with the collaboration of Ravetz, J.R. 1972. *Joseph Fourier 1768-1830. A survey of his life and work, based on a critical edition of his monograph on the propagation of heat, presented to the Institut de France in 1807,* Cambridge, Mass.: MIT Press.

Henrici, O. 1892. 'Über Instrumente zur harmonischen Analyse', in W. von Dyck (ed.), *Deutsche Mathematiker-Vereinigung. Katalog mathematischer und mathematisch-physikalischer Modelle, Apparate und Instrumente*, Munich: Wolf, pt. 1, 125-136.

Jevons, W. S. 1864. *Pure logic, or the logic of quality apart from quantity*, London: Stanford. [Repr. in his [1890], 1-77.]

Jevons, W. S. 1870. 'On the mechanical performance of logical inference', *Philosophical transactions of the Royal Society of London, 160*, 497-518. [Repr. in his [1890], 137-172.]

Jevons, W. S. 1890. *Pure logic and other writings* (ed. R. Adamson and H. A. Jevons), London: Macmillan.

Kelvin, Lord 1881. 'The tide gauge, tidal harmonic analyser, and tide predictor', *Minutes of the proceedings of the Institute of Civil Engineers, 65*, 3-24. [Repr. in *Mathematical and physical papers*, vol. 6 (ed. J. Larmor), Cambridge: Cambridge University Press, 1911, 272-305.]

Knobloch, E. 1989. 'Analogie und mathematisches Denken', *Berichte zur Wissenschaftsgeschichte, 12*, 35-47.

Lakatos, I. 1976. *Proofs and refutations* [...], Cambridge: Cambridge University Press.

May, R. M. 1976. 'Simple mathematical models with very complicated dynamics', *Nature, 261*, 459-467.

Ohm, G. S. 1843. 'Über die Definition des Tons [...]', *Annalen der Chimie und Physik, 59*, 513-566. [Repr. in *Gesammelte Abhandlungen* (ed. E. Lommel), Leipzig: Teubner, 1911, 587-633.]

Polya, G. 1954. *Mathematics and plausible reasoning*, 2 vols., Oxford: Clarendon Press.

Polya, G. 1962-1965. *Mathematical discovery*, 2 vols., New York: Wiley.

Popper, K. R. 1972. *Objective knowledge*, Oxford: Clarendon Press.

Pycior, H. 1987. 'British abstract algebra: development and early reception', in I. Grattan-Guinness (ed.), *History in mathematics education. Proceedings of a workshop held at the University of Toronto, Canada, July-August 1983*, Paris: Belin, 152-168.

Rostand, F. 1960. *Souci d'exactitude et scrupules des mathématiciens*, Paris: Vrin.

Rostand, F. 1962. *Sur la clarté des démonstrations mathématiques*, Paris: Vrin.

Tymoczko, T. 1985. (Ed.), *New directions in the philosophy of mathematics*, Basel: Birkhäuser.

Vercelloni, L. 1988. *Filosofia delle strutture*, Florence: La Nuova Italia.

Viète, F. 1591. *In artem analyticem* [sic] *isagoge* [...], Tours: Mettayer. [Repr. in *Opera mathematica* (ed. F. Schooten), Leiden: Elzevir, 1646, 1-12. English trans.: J. Klein, *Greek mathematical thought and the origin of algebra*, Cambridge, Mass.: The MIT Press, 1968, 315-353.]

West, B. J. 1985. *An essay on the importance of being nonlinear*, Berlin: Springer (Lecture Notes in Biomathematics, no. 62).

Wigner, E. 1960. 'The unreasonable effectiveness of mathematics in the natural sciences', *Communications in pure and applied mathematics, 13*, 1-14.

Wise, M. N. 1981. 'The flow analogy in electricity and magnetism. Part I. William Thomson's reformulation of action at a distance', *Archive for history of exact sciences, 25,* 19-70.

Zellweger, S. 1982. 'Sign-creation and man-sign engineering', *Semiotica, 38,* 17-54.

12 Solving Wigner's Mystery: the Reasonable (Though Perhaps Limited) Effectiveness of Mathematics in the Natural Sciences

In 1960 the physicist Eugene Wigner published an article on 'The unreasonable effectiveness of mathematics in the natural sciences' that has been very influential. I counter the claim stated in its title with an interpretation of science in which many of the uses of mathematics are shown to be quite reasonable, even rational, though maybe somewhat limited in content and indeed at times ineffective. The alternative view emphasizes two factors which Wigner largely ignores: the effectiveness of the natural sciences in mathematics, in that much mathematics has been motivated by interpretations in the sciences, and still is; and the central place of theories in mathematics and the sciences, especially theory-building, in which analogies drawn from other theories play an important role. A major related feature is the desimplification of theories, which attempts to reduce limitations on their effectiveness. Significant also is the ubiquity and/or generality of many topics and notions in mathematics. It emerges that the connections between mathematics and the natural sciences are, and always have been, rationally though fallibly forged links, not a collection of mysterious parallelisms.

1. Wigner's thesis

Wigner states as his main thesis 'that the enormous usefulness of mathematics in the natural sciences is something bordering on the mysterious and that there is no rational explanation for it'; for example, 'The miracle of the appropriateness of the language of mathematics for the formulation of the laws of physics is a wonderful gift which we neither understand or deserve' [1960, 2, 14]. By way of illustration he recalls a story about two friends studying population statistics by means of the normal (or Gaussian) distribution and being bewildered by the presence in the analysis of π: 'surely the population has nothing to do with the circumference of the circle' [p. 1]. He judges this mystery to be 'plain common sense' and does not discuss it again in the article. {He does not give a source for the mystery, but one exists in [De Morgan 1872, 285-286].}

Wigner's article has been cited especially by scientists and mathematicians on many occasions, with approval or at least without demur;

This article was first published in *The mathematical intelligencer*, *30* (2008), no. 3, 7-17; it is reprinted by kind permission of Springer Science+Business Media. Note the additional considerations of types of generality provided in §21.

some related articles have appeared.[1] Philosophers have also considered the article, and some have largely accepted the force of the argument.[2] One should note that most of the established philosophies of mathematics favoured by philosophers have been developed to grasp mathematical theories *already developed* rather than to address theory-building. There [Polya 1954a, 1954b] is much more promising, with his masterly survey of 'plausible reasoning' and the dynamic relationships between theorems and proofs; however, he focusses largely upon pure mathematics. In my approach, which in general terms follows Polya, the unreasonableness will largely disappear, but doubts are raised over effectiveness. The discussion is set at the level of formed cognition and theory-building; I do not address the interesting subject of the psychology of mathematical creation.

Several of my points have been made in earlier discussions of Wigner's article, but to my knowledge nobody has taken as central the two theses presented in the next section. In general terms I follow the spirit of [French 2000], who nicely defends the reasonableness of one particular kind of application. This paper is noted in ¶6.6, in a section devoted to examples of the methods of theory-building that have been adopted. These examples are largely historical ones, as that is my background, but their potency is not thereby lost; for if mathematics *is* unreasonably effective in the natural sciences, then it always has been so, or at least for a long time. In any case, we can surely learn from our past masters. A solution of the mystery about π follows in ¶9.

2. Two counters

2.1 Firstly, in a part of his article called 'What is mathematics?' Wigner asserts that while elementary concepts in mathematics (especially geometry) were motivated by 'the actual world, the same does not seem to be true of the more advanced concepts, in particular the concepts which play such an important role in physics' [1960, 2]. In reply I build upon a

---

[1] For example, the rather ineffective [Hamming 1980]. In a review of Wigner's article for the *Zentralblatt für Mathematik* [Kiesow 1960] welcomed a 'brilliantly written essay'. *Mathematical reviews* did not cover it. I do not attempt a full bibliography of reactions to Wigner's article, but see the Wikipedia online entry on it.

[2] [Colyvan 2001] sees Wigner's 'puzzle' as a conundrum for some prevailing philosophies of mathematics, within which mathematics is 'developed primarily with aesthetic considerations in mind' [p. 267]. [Sarukkai 2005] emphasizes the language of mathematics as such rather than mathematical theories, which of course need language for expression. His account of intuitionism is not happy, and both authors misrepresent Hilbert as a formalist.

In a study of the epistemology of questions and answers, [Hintikka 2007] sees Wigner's thesis as exemplifying *a priori* knowledge, and associates mathematics especially with his 'function-in-extension', which plays the central role of linking who/what/where/… questions with the proposed answers. While supporting his philosophical enterprise, I am not persuaded that apriority captures Wigner's thesis, nor that the function need be placed in mathematics rather than in the pertaining logic just because functions (and functors) play major roles there.

large truth coming strongly from the history of mathematics, quite counter to his claim; not only elementary theories and branches of the subject were (and are) motivated by some problems found in the actual world, including on occasion sciences outside the physical ones, but so *equally* were (and are) the more advanced theories. Much mathematics, at all levels, was brought into being by worldly demands, so that its frequent effectiveness there is not so surprising.

It is necessary to emphasise this feature of mathematics, because especially since the middle of the 19th century a snobbish attitude gradually developed among substantial parts of the growing mathematical community to prefer pure over applied mathematics ('dirty mathematics' to Berliners, for example). As a consequence the impression has grown that mathematics is always, or at least often, developed independent of the natural sciences, or indeed anything else; given this presumption, its undoubted effectiveness is indeed mysterious.

2.2 Secondly, in a part of his article called 'What is physics?' Wigner emphasises the role of observing regularities in the world for formulating 'laws of nature' (using Galileo Galilei's law of fall as an example) which nevertheless are subject to 'probability laws' due to our incomplete knowledge [1960, 3-6]. This point about regularities is valuable, and should form part of a wide-ranging analysis of theories *as such*, especially their initial formation and later elaboration. These processes are *central* features of the development of mathematics pure or applied, and indeed of any science, and so they form the basis of my own approach.

The status of theories depends upon whether one subscribes to a philosophy of science that treats theories as mere devices for calculation or prediction (instrumentalism, conventionalism, some kinds of positivism) or to a philosophy that pays attention to the (apparent) explanatory power of theories (inductivism, fallibilism, some kinds of Platonism).[3] These differences matter, because the criteria for (in)effectiveness vary between the two kinds of philosophy. The discussion below will apply to both of them, as obtains also in Wigner's article.[4]

3. Developing theories in the presence of other theories

In science as in everyday life, when faced by a new situation, we start out with some guess. Our first guess may fall wide of the mark, but we try it and, according to the degree of success, we modify it more or less. Eventually, after several trials and several

---

[3] On the different philosophies of science see, for example, [Dilworth 1994].

[4] Wigner ends with a rather strange section on 'the uniqueness of the theories of physics' in which he stresses 'the empirical nature' of laws of nature and considers cases where '"false" theories' give 'alarmingly [sic] accurate descriptions of groups of phenomena' [1960, 11, 13]. Since much of that discussion focuses upon some specific physical phenomena and the possibility of reconciling quantum mechanics with relativity theory, it does not centre on mathematics; so I leave it alone.

modifications, pushed by observations and led by analogy, we may arrive at a more satisfactory guess.

Georg Polya [1954b, 158]

3.1 *Parts and moments.* When forming a problem and attempting to solve it, a scientist is not isolated: he operates in various contexts, philosophical, cultural and technical, in some cases consciously recognised while in others intuitively or implicitly adopted. *Thinkers develop theories in the presence of other theories already available as well as by observations of the actual world, and can be influenced positively or maybe negatively by these theories.* My approach complements the theory of 'abduction' due to C. S. Peirce, where theory-building is considered largely in terms of reactions to (new) observations.[5] In both studies *it is the world of human theories that is anthropocentric, not the actual world.*

In the discussion below 'notion' is an umbrella term covering not only objects such as function and matrix but also concepts such as convexity, systems of symbols, and proof methods, that occur in mathematical theories; these latter are often called 'topics' when they include individual theorems or algorithms as well as larger-scale bodies of results. The distinction between topic and notion resembles that made by phenomenologists between a part and a moment of a whole; for example, between the third chapter of a certain book and the price of that book [Smith 1982]. {Both parts and moments may have parts and moments, and a part of a moment may also be a moment of a part (the first few pages of that chapter; the price of those pages; my surprise at that price). The difference between parts and moments may be expressed in terms of *separation*; I can detach the pages of the chapter from the book, but I cannot detach its price. Some parts or moments may be empty; for example, the non-existent index of that book, or the blueness of its green dust-jacket.}

3.2 *Relationships between theories.* Assume that the creator of a new theory S_2 was aware of another theory S_1 already available and drew upon it in some way; this does not forbid the possibility that he independently re-created S_1 on his way to S_2. Four categories of relationship may obtain between S_2 and S_1. Analogies are mentioned here, and analysed in some detail later.

Category 1: Reduction. S_1 not only actively plays a role in the formation and development of S_2, but the theorist also hopes to *reduce* S_2 to the sphere of activity of S_1. Analogies now become special cases of S_1 in S_2; and S_1 may be seen as an extension of S_2, maybe even a generalisation of it. There are also reductions within a mathematical topic, when it is shown that a particular kind of object may be reduced to a special kind of itself without loss of generality.

---

[5] Wigner notes Peirce in [1960, 1, 4]. {On Peirce's theory of abduction see, for example, [Tursman 1987, ch. 1].}

Category 2: Emulation. S_1 actively plays a role in the formation and development of S_2, with resulting structural similarities, but reduction is not asserted or maybe even sought. Analogies are just similarities; for example, S_2 uses (close versions of) some of the mathematical notions already deployed in S_1.

Category 3: Corroboration. S_1 plays little or no role in the formation and development of S_2; but the theorist draws upon similarities to S_1, maybe including structural ones, to develop S_2 further and thereby enhance the measure of analogy between S_2 and S_1.

Category 4: Importation. S_1 is imported into S_2 basically intact, to serve as a mathematical tool. Thereby S_1 and S_2 have certain notions in common, creating analogies; and if some of them are of sufficient generality to surpass the spheres of activity of both S_1 and S_2, then they are *instantiated* in S_1 and S_2.

Theory S_2 may well have several S_1s of various kinds in its ancestry. What relationship does it hold to its principal parents? The word 'revolution' is often used to refer to substantial changes of theory, but in my view excessively and without adequate allowance for the different kinds of relationship that may obtain. I propose the following tri-distinction [Grattan-Guinness 1992a].

Category 5: Revolution. Adoption of S_2 means that S_1 is replaced, perhaps discarded or at least much reduced in status to a prediction device, with much of its explanatory power abandoned.

Category 6: Innovation. S_2 says some quite new things about which S_1 was silent, or at most treated only some special cases. Replacement will occur, for example when S_2 is preferred over S_1 in certain circumstances, but it is not the main feature.

Category 7: Convolution. In its development from S_2 exhibits both old and new (sub-)theories; S_1 and S_2 wind around each other, showing both old and new connections and thereby mixing elements of replacement and innovation.

It seems that convolutions are the most common relationship to be found between old and new theories, with innovations and revolutions as opposite extremes; thus the tri-distinction is more of degree than of kind. A very widespread use of convolution occurs when a mathematician takes some existing piece of mathematics (of any kind) and modernises parts of it in some ways before embarking on his new work or while doing so; I call this use of old mathematics 'heritage', to distinguish it from its historical analysis [Grattan-Guinness 2004; and §13]. A nice example is the 'genetic approach' to the calculus given in [Toeplitz 1963], heritage that also exhibits historical sympathy.[6]

4. Some basic topics and notions in theories

> In the demonstrative sciences logic is used in the main for proofs — for the transmission of truth — while in the empirical sciences it is almost exclusively used critically — for the retransmission of falsity.
> Karl Popper [1972, 305]

We consider now some of the main topics and notions that are invoked in the application of mathematics to the natural sciences. They can obtain also within mathematics, between different branches of the discipline and/or parts of the same one; I shall not pursue this feature here, but I note that it increases the content of the mathematical theories involved, and thereby the potential measure of their effectiveness in applications. A significant part of so-called 'pure' mathematics is applic*able,* done without any stated applications but with a clear potential there: the various kinds of solution of differential equations are a prominent example.

Table 1 provides some significant topics, notions and strategies that help in theory building to produce some sort of convoluted theory out of previous theories. None of the lists in the three columns is meant to be complete (especially not the first one), though every item is noteworthy. Apart from a few groupings in the columns, the order is not significant; and only one connection by row obtains. In several cases the opposite notion is also to be noted (for example, disequilibrium from equilibrium). Those in the third column can be manifest within mathematics also.

5. Ubiquity and the role of analogies

Analogies (and disanalogies) between theories play a very significant role in the reasonable effectiveness of mathematics in theory-building; in particular, in the second way (emulation) of deriving S_2 from S_1 listed above.[7] Two such theories have some mathematical notion M in common, which therefore is an invariant relative to S_1 and S_2 ; for example (which is

---

[6] However, for Wigner 'It is absurd to believe that the existence of mathematically simple expressions for the second derivative of the position is self-evident, when no similar expressions for the position itself or for the velocity exist' [1960, 11]. Is this strange remark some allusion to Newton's second law?

[7] The philosophy of analogies is not yet well developed. The most extensive account is given in [Kaushal 2003, chs. 3-6]; see especially his synoptic table illustrating 'the contents of a structural analogy' on p. 93. In the rest of his book he considers their use in the humanities and in the Hindu religion. [Polya 1954a] stresses analogies, mainly in pure mathematics. [Steiner 1998] draws quite a lot on them, partly as a reply to [Wigner 1960]; he also advocates an anthropocentric standpoint. My own approach, based upon 'structure-similarity', is sketched in [Grattan-Guinness 1992b = §11].

given its historical context later), both heat diffusion and acoustics use Fourier series.

A major source of the importance of analogies is that all of these topics and notions are *ubiquitous*, in mathematics and/or in the actual world; hence *lots* of analogies may be tried, and the successful ones help to explain the 'uncanny usefulness of mathematical concepts' [Wigner 1960, 2]. We can also assuage the puzzlement of Steven Weinberg that mathemat-

Table 1
Some topics, notions and strategies used in mathematics and the natural sciences.

Topics from mathematics	Notions from mathematics	Notions from the sciences and/or the actual world
Matrices	Linearity	Space, time
Determinants	Generalisation	Force
Arithmetic of real numbers	Convexity	Energy
Common algebra	Equality, inequality	Mass, weight
Complex numbers, analysis	Ordering	State of a system
The calculus	Partitioning	Causality
Functions, functors	Approximation	Continuity
Series	Invariance	Optimisation
Differential equations	Duality	Regularity
Theory of limits	Boundary	Notion of lim
Set theory, the infinite	Recursion	Conservation
Potentials	Operators	Equilibrium, stability
Mathematical statistics	Combinations	Discreteness
Stochastic processes	Bilinear and quadratic forms	Symmetry
Probability	Dispersion, location	Analogy
Topology	Regression, correlation	Periodicity
Mechanics	Nesting	Simplicity, complexity
Theory of equations	Mathematical induction	Generality
Group theory	Proof by contradiction	Randomness
Fields (and other abstract algebras)	Superposition	Identification
(Non-)Euclidean geometries	Structure	Abstraction
Vector algebra, analysis	Axiomatisation	Taxonomy

icians have often produced theories before the physicists [Mathematics 1986, 725-728, mentioning Wigner]: the mathematicians thought up these theories in specific contexts using various ubiquitous topics and notions, which physicists *then* found also to be effective elsewhere.

In addition to analogies between S_1 and S_2, each theory (I take S_1) will have analogies with the pertinent mathematical notion M. A dual role obtains for M: both to be correctly developed as mathematics, and to make sense at some level of detail in S_1. The level to which the similarity holds between M and S_1 measures their common *analogy content*; for example, it increases if S_1 not only uses integrals M but also interprets them as areas or as sums. Analogy content can be modest; for example, when an abstract algebra (lattices, say) M is imported into S_1, the analogy content between M and S_1 may be limited to the lattice structure.

Kaushal nicely exhibits ubiquities with lists of scientific contexts in which certain mathematical equations and functions arise: for example, the exponential decay function, and the form $(a - b)/c$ [2003, 60, 75; see also pp. 52-57, 67, 85]. Polya gives simple examples from applications that draw upon analogy [1954a, chs. 9-10]. [Knobloch 2000] notes cases of analogies in the early modern period.

6. Examples of theory-building

Let us now take some further examples of these seven categories and the table of notions working in harness, not necessarily oriented around analogies.

6.1 *Among importations* of elementary mathematical theories, arithmetic has been deployed since ancient epochs, trigonometry and Euclidian geometry for a long while, and common algebra since its innovation by the medieval Arabs. The examples below come from more modern times and mostly from more advanced mathematics: I have chosen ones with which I am fairly familiar, and which collectively illustrate the variety as much as possible in a limited space. It is impossible to cite the original sources for these examples or give their full contexts in detail: short surveys of all the branches and topics of mathematics involved are to be found in the encyclopaedia [Grattan-Guinness 1994]. The reader will be able to construct lots of further examples from his own knowledge.

In these examples enough of the pertaining science was already available when the mathematicisation described took place, and the mathematics and science were competently hawndled. Neither property holds in general; in particular, the *simultaneous* development of mathematics and science in a theory-building context is a central feature of mathematical modelling. To reduce complications in the presentation, I have reluctantly avoided cases where major roles are played by notations and notational systems, or by diagrams; they deserve studies of their own. Out of respect for my ignorance, I have not offered examples from the life sciences or medicine.

6.2 *Among the notions*. Inequality has been much under-rated as an importation [Tanner 1961]. It is at the centre of theories such as thermodynamics, mathematical economics, (non-)linear programming, certain foundations for mechanics, and it also underlies many of the principal definitions and proofs in real- and complex-variable analysis and their uses, in connection with the theory of limits. In contrast, the high status of symmetry is well recognised by, for example, [Wigner 1967], and also [Weyl 1952] in general and [Mackey 1978] in the context of harmonic analysis.

Simplicity has obvious attractions to reductionists, and it grounds conventionalist philosophies; but when two notions are not close together in kind, the relation 'simpler than' between them requires complicated (sic) analysis.[8] (The use of 'simple' in 'desimplification' *is* of this close kind.) Sometimes it is also used to back up the empiricists, who cut their philosophical throats with Ockham's razor.

Linearity has been of especial importance, even though most of the phenomena observed in the actual world are non-linear. It covers all manifestations of the linear form aA +bB + ..., finitely or infinitely. An example of a general kind is forming a problem as a linear differential, or difference, or difference-differential, equation, for many forms of solution are available or may become so; by around 1900 linearisation had become something of a fixation [Grattan-Guinness 2008a]. Linear algebra also brought with it, and to some extent motivated, a further wide range of applications, partly overlapping with that of the calculus.

6.3 *Perturbation theory*. An important example of both strands was initiated by Isaac Newton's innovative insight in celestial mechanics that the planets were 'perturbed' from their basic orbits around the Sun by their mutual attractions. The mathematics to express this situation was not difficult to state but horrible to manipulate, until in the 1740s Leonhard Euler had the superb insight that the distance (and other) astronomical variables could be converted into infinite trigonometric series of appropriate angles, which increased a uniformity of approach [Wilson 1980a]. A major use of this method occurred in *proving* that our planetary system was stable; that is, no planet would ever fly out of the system like a comet, or way off out of the ecliptic plane. Euler (and Newton before him) had been content to rely on God as the guarantee of stability; but in the 1770s J. L. Lagrange secularised the problem by truncating the expansions to their first terms, thereby expressing the motions in a system of linear ordinary differential equations with constant coefficients, which took finite trigonometric series solutions. By a marvellous analysis he made great progress towards establishing stability [Wilson 1980b]. Later work by others (including, surprisingly, A.-L. Cauchy and Karl Weierstrass) played major roles in establishing the spectral theory of matrices (the theory of their latent roots and vectors) [Haw-

---

[8] For example, ponder whether analytical mechanics is simpler or more complicated than Newtonian mechanics, and note that many pertinent points of view are involved. What sort of useful answer would result?

kins 1975, §13.4] — a fine example of the reasonable effectiveness of the natural sciences in mathematics.

This example also exhibits both kinds of generality mentioned above. Firstly, Lagrange's analysis formed part of his development of analytical mechanics, in which he claimed, controversially, that dynamics could be reduced to statics. Secondly, it hinged on a brilliant transformation of the independent variables that (to use matrix theory, heritage style) reduced the matrix of the terms in the differential equations to an antisymmetric one; the task was then to show that all the latent roots and vectors were real.

6.4 *Contributions from Fourier.* Euler's trigonometric series are not be confused with Fourier series, which came back into mathematics in the 1800s (compare §11.6). The context was heat diffusion, where Joseph Fourier innovated the first large-scale mathematicisation of a branch of physics outside mechanics, in a fine display of convolution [Grattan-Guinness and Ravetz 1972].

Importing the differential and integral calculus in its Leibniz-Euler form, Fourier went for linearisation in forming his differential equation to represent the phenomenon. But in adopting the series as the preferred form of solution for finite bodies he revolutionised the understanding of a mathematical theory that had been known before him but were disparaged for reasons (especially concerning their manner of representing a function) that he showed to be mistaken. However, he did not apply analogy to carry the periodicity of the trigonometric terms over to a waval theory of heat and promote a superposition of basic states, although such a theory was being advocated at that time; for him heat was exchanged with cold, each notion being taken as primitive, and he rejected explaining their nature in other terms such as waves or a substance (caloric). The term 'positivism' can fairly be applied here, as in the late 1820s his work was to be a great influence upon the philosopher Auguste Comte. For diffusion in infinite bodies Fourier innovated around 1810 the integral named after him, where the waval reading does not obtain anyway. The physical interpretation of each term of the Fourier series was due especially to G. S. Ohm in the 1840s, in the context of acoustics; it marks an increase in analogy content relative to Fourier.

6.5 *Contributions from Thomson.* Another of Fourier's early foreign supporters was the young William Thomson, later Lord Kelvin (compare §11.6). In his teens in the early 1840s, he not only studied heat diffusion and the series method of solution but also quickly moved on to electricity and magnetism, and then to hydrodynamics [Grattan-Guinness 2008b]. He is a particularly interesting case to note, since he explicitly invoked analogies when passing from one topic to another. The similarities carried over not only at the mathematical level (similar differential equations and methods of solution) but also as physics (for example, from isothermal surfaces to equipotential ones). He was a prominent pioneer in potential theory, not only because of his own contributions but also for popularising George Green's innovative theorem of 1828 relating the internal organization of a solid body to its surface potential. Thanks to these and others' endeavours, potential theory came a massive source of emulation, analogies, instantiations and importations across many branches of mechanics and classical mathematical physics [Bacharach 1883].

Thomson was also a major figure in the mid-century advocacy of the principle of the conservation of energy in mechanics and physics, or 'energetics', which became another major source of emulation, importation, instantiation and corroboration across many sciences. But its parent, energy/work principles in mechanics, had already provided a striking example of corroboration, in the wave theory of light. Its main pioneer from the mid 1810s was A. J. Fresnel, whose theories used a variety of emulations from mechanics, such as assuming the simple harmonic motion of the molecules in his punctiform aether and a cosine law for the decomposition of intensities. The corroboration occurred over his 1821 analysis of Huygens's law of double refraction, that a ray of light of unit intensity at incident angle I in crystals such as Iceland spar split into two rays of intensities $\sin^2 I$ and $\cos^2 I$. After carrying out this analysis he realized from the trigonometric version of Pythagoaras's theorem that his theory conformed to the principle of the conservation of energy if he presumed the aether to be transparent for its transmission; so he annotated some older manuscripts to this effect [Fresnel 1866, 472, 483, 496].

6.6 *Quantum mechanics*. Thomson died in 1907, just when his empire of classical (mathematical) physics was being replaced by new scientific regimes. One of them was quantum mechanics, especially the emulation by Niels Bohr and others of celestial mechanics with his 1913 planetary-like model of the hydrogen atom as a nucleus surrounded by a charged electron orbiting in a circle (or, for the desimplifying Arnold Sommerfeld, in an ellipse) [Hermann 1971].[9] Given this approach, the governing differential equation, Ernst Schrödinger's, was linear as usual, and for it a wide repertoire of solutions was available. But the physics, especially the notions of atomic states and quanta of light and other phenomena to which Planck's constant had become associated, dictated that analogy should *not* guide the choice of solutions; to be reasonable the mathematics had to follow routes different from Fourier series (although Werner Heisenberg's first theory of the atom drew upon them), special functions and the like. Instead Hilbert spaces, infinite matrices and integral equations played prominent roles; and as all three mathematical topics were still rather new at the time (the 1910s onwards) to some extent we see again the effectiveness of a natural science upon their development (and conversely, their applicability). Two main forms of quantum mechanics developed in the 1920s, matrix mechanics and

---

[9] Much of [Steiner 1998] is taken up with quantum mechanics, but unaccountably he omits the contributions of John von Neumann. For a desimplified version of my summary history, see [Beller 1983].

wave mechanics; in the latter development Schrödinger closely emulated the analytical mechanics and optics of W. R. Hamilton. Schrödinger and others showed in 1926 that the two versions were mathematically equivalent; however, their physical differences remained rather mysterious. Paul Dirac came up with a third candidate in his quantum algebra; then he embraced all three in his 'transformation theory'.

Another importation into quantum mechanics was group theory, which had developed over the previous 70 years or so, initially in other specific mathematical contexts and then as a general and abstract theory [Wussing 1984]; several basic kinds of group proved to be effective, especially rotation, unitary, continuous and permutation [Mackey 1978, 1985]. This example is especially striking to note because a significant pioneer was one Eugene P. Wigner; his book on the matter [Wigner 1931] is surely a fine counter-example to his thesis of 1960 [French 2000].

6.7 Statistical mechanics. The relevance of anthropocentrism, mentioned above, is nicely exemplified in the survey [Ruelle 1988] of equilibrium statistical mechanics. Early on he is willing to 'define mathematics as a logical construct based on the axioms of set theory' (oh Gödel, where art thou at this hour?), and praises Wigner's 'beautiful' article without 'concern[ing] ourselves with this mystery'. Then, to outline his theory of indeed 'human mathematics' he not only invokes equilibrium but also imports parts of point set topology, the integral calculus, operator algebra and mathematical statistics; he even stresses that 'the intrusion of physics therefore changes the historical development of mathematics' [p. 265], and indicates uses of his subject elsewhere. That is, he does much to dissipate the mystery that he claims to be ignoring!

6.8 Complex numbers and variables. Like many areas of pure and applied mathematics, quantum mechanics also imported complex-variable analysis. Wigner points to the 'formal beauty' in the mathematics of complex numbers [Wigner 1960, 3: compare p. 7]; they may possess it, but it does not begin to explain their importance. For that we need to distinguish the data, in this case the positive and negative real numbers, from the theories about them, of which the first were the formulae for the resolution of quadratic, cubic and quartic polynomials.[10] The complex number field has to be invoked because the operations of taking square and higher roots are closed in it but not in the real number field: for example, $\sqrt{(a + ib)}$ is always complex for real a and b while $\sqrt{a}$ is real only if a is not negative.[11] Com-

---

[10] The derivation of the formulae depended upon the insight of Scipione del Ferro around 1500 that a cubic polynomial could be reduced without loss of generality to one lacking the quadratic term, and similarly with Lodovico Ferrari on the quartic about 40 years later.

[11] This reading of complex numbers belongs to the advent of structural algebras during the early 20th century [Corry 1996]. An earlier reading deployed the complex plane; but as it depends upon geometry, it might be seen as more of a heuristic aid than an epistemological ground.

plex-variable analysis is a remarkable but reasonable extension, innovated by Cauchy from the 1810s onwards in very close analogy with his concurrent exegesis of real-variable analysis based upon the theory of limits [Smithies 1997].

6.9 *Tweedledum and Tweedledee.* Finally, there is the extreme case of analogy, namely identification: this = that, maybe modulo a Gestalt switch. A remarkable instance occurred in October 1947 when John von Neumann and George Dantzig shared their respective interests in economic behaviour and linear programming; they found that they (and, it turned out, a few others) had been using planar convex regions, but for von Neumann against a background of fixed-point theorems while for Dantzig concerning the performance of objective functions [Dantzig 1982, 45; §11.7]. The effect of the resulting union of theories was a rapid expansion in work in both subjects.

7. Increasing effectiveness: desimplification and the science of small effects

The discussion above should suggest grounds for finding reasonable the impressive utility of mathematics in the natural sciences. The question of its effectiveness, however, requires further consideration. It depends in part upon the demands made of the scientific theory involved, or the expectations held for it; *how* general, for example, or how numerically accurate?

It is a commonplace, but a significant one, to notice that the actual world is a very complicated place; Wigner himself does so in [1960, 4]. Thus the scientist, whether mathematical or not, is forced to simplify the phenomena under study in order to render them tractable: 'the art of the soluble', to quote the artful title of [Medawar 1967]. The long-running preference for linearity noted earlier is a prominent example of such simplifications; in reaction, a notable feature of recent mathematical physics has been a great increase in non-linear methods and models [West 1985].

Among branches of mathematics, mechanics is notorious for the adoption of light strings and inextensible pulleys, the assumption that extended bodies have constant density, the routine ignoring of air resistance, friction, and/or the rotation of the Earth about its axis, and so on. The assumption is fallibly made that in the contexts under study the corresponding effects are small enough to be ignored; but part of the reasonableness of theory-building is to check whether or not such assumptions are justified. I called such checks 'desimplification': putting back into the theory effects and factors that had been deliberately left out.

For example, Lagrange consciously simplified the stability problem in that in truncating the expansions to their first terms he formed a linear approximation to the motions of the planets by taking only the first-order terms in their masses. Thereby he assumed that the terms in higher orders were small enough to be ignored; but should this assumption be checked for reasonableness? In the late 1800s, under the stimulus of a recent analysis by

P. S. Laplace, the young S. D. Poisson and the old Lagrange studied the second-order terms and found a mathematical expression that was of interest in its own right. Thus their study of a particular problem led unexpectedly to a much more general one. For once in the history of mathematics the names attached to the resulting theory, in analytical mechanics, are correct: 'the 'Lagrange-Poisson brackets'.[12] A version of it was to appear in Dirac's algebra noted in ¶6.6 in connection with quantum mechanics.

The longest-running catalogue of desimplifications of which I am aware concerns the so-called 'simple' pendulum. The adjective seems reasonable, for the instrument consists only of a bob swinging on a wire from a fixed point. However, especially from the late 18th century onwards, pendula were observed very exactly for making precise calculations in connection with the needs of geodesy, cartography and topography. This was small-effects science par excellence, literally preoccupied with decimal places. Many scientists studied a wide range of properties [Wolf 1889-1891]. Is the downswing *exactly* equal to the upswing? Does the bob make a little angular kick at the top of its upswing or not? What about the effects of Lunar attraction, the spheroidicity of the Earth, air resistance, the possible extension of the wire under the weight of the bob, and the effect of the bob rotating about its own axis? Do possible movements of the supporting frame affect the swinging of large pendula? What special factors attend the use of a hand-held pendulum [Kaushal 2003, 160-172]? These and various other questions made the simple pendulum a rather complicated instrument! However, all the desimplifications were carried out fallibly but reasonably, for they attempted to establish guides on the orders of smallness of the effects upon the motion of the pendulum.

Some of these strategies involved quantitative approximations to the relevant theory. This invoked numerical analysis, which, when taken with numerical linear algebra, forms a branch of mathematics of special pertinence to our theme [Chabert 1999].

8. Some comments on ineffectiveness, including its own possible ineffectiveness

The account above paints a picture of the development of applied mathematics in sequences of fallible but steadily successful actions. However, it is itself simplified, and needs supplementing with some consideration of types of failure over and above incompetence.

8.1 *Numerical utility.* Some examples are rather slight, even amusing, such as astronomers sometimes calculating values of their variables to ridiculous numbers of decimal places, far beyond any scientific need of their time. However, this action raises the reasonable question, somewhat akin to the considerations of numerical methods just aired: given the in-

---

[12] On Lagrange's and Poisson's work see [Grattan-Guinness 1990, 371-386]. On the place of the theory in analytical mechanics see, for example, [Whittaker 1927, ch. 11].

struments available in some scientific context, what is a/the reasonable number of decimal places to aim at in the theory? More generally, which mathematics goes reasonably and effectively with measurement, both in the natural sciences and elsewhere? For Thomson and many others, it should happen as often and as accurately as possible in science; others have been more cautious.[13]

There are also situations where a genuine problem is addressed, but the theories proposed as solutions are of no practical use whatever; I call this type 'notional applications'. A striking example is [Poisson 1823] on the cooling of an annulus in the desimplified situation when the temperature of the environment was *not* constant and so was itself represented in the diffusion equation by a Fourier series. The consequences for the resulting analysis can be imagined; but what was the motivation? He mentioned the predicament of a sailor using a sextant at sea in a (variably) sunny environment, when the rays from the Sun strike the instrument itself and so cause it to distort out of shape. This is a genuine problem; but how do the parades of sines and cosines resolve it, especially in any calculable manner?

8.2 *Mathematics in economics.* One subject where the use of mathematics has been questioned in a fundamental way is economics. In particular, accepting Wigner's thesis, [Velupillai 2005] entitled his attack 'The unreasonable *in*effectiveness of mathematics in economics'. The criticisms are wide-ranging: 'the mathematical assumptions are economically unwarranted' and often dependent upon weak analogies with other subjects. For example, several main figures in the early stages of neo-classical economics in the second half of the 19th century emulated mechanics with enthusiasm, especially the notion of equilibrium, and deployed major assumptions such as d'Alembert's principle; but the resulting theories were not very effective [Grattan-Guinness 2007, {2010}]. What, for example, corresponds in economics to the continuous and uniform force of gravity? There is still a very wide spectrum of views on, for example, the effectiveness in economics of the notion of equilibrium [Mosini 2007].

Velupillai specifically finds mathematics in economics 'ineffective because the mathematical formalisations imply non-constructive and uncomputable structures'; as medicine he recommends constructive mathematics, especially in the import of number theory and recursion theory into economics when its data have been expressed as integral multiples of some basic unit. One would certainly have a lot of sophisticated theories to deploy (he explicitly recommends Diophantine analysis); but it is a moot point as to whether the great complications that attend constructive mathematics in general would render economics more effective (or alternatively, whether they can be avoided). There is a widespread practice of mathematicising the proposed theory whatever its content — 'bad theory with a mathematical

---

[13] For a history of the mathematics of measurement largely linked to physics, see [Roche 1998]. For a wide-ranging review of measurement, in various scales and many contexts, see [Henshaw 2006].

passport', according to [Schwartz 1962, 358] — but much less concern for bringing it to test. Some branches do exhibit effective testing; for example, financial data subjected to time series analysis, and not just to find correlations for their own ineffective sake.

8.3 *Beyond the physical sciences.* The failure of the mechanisation of economics shows that the gap between the physical and the social sciences is very wide. How about the gap between the physical and the life sciences? [Lesk 2000] takes up the matter in connection with molecular biology, copying Wigner's title; so we expect to learn of some more unreasonable effectiveness. However, he is very cautious, stressing disanalogies between the physical and the life sciences, especially over matters concerning complexity; and indeed in a follow-up letter [Lesk 2001] reports that his hosts asked him to speak of 'effectiveness' rather than 'ineffectiveness' in a lecture of which his paper is the written version! There are topics that can be handled effectively within and without the physical sciences; for example, adaptation in control engineering and biology when oriented around optimisation [Holland 1975]. However, one is tempted to think that desimplification will not be radical enough, and that the non-physical sciences — life, mental, social — may need *fresh kinds of mathematics.*[14] Relative to Table 1, perhaps we should retain the notions (or most of them) and build different topics around them.

Lesk's remark also draws attention to the limitations of human mental capacities; maybe some phenomena are just too complicated or elusive for effective theorizing, whatever the science. Wigner himself raised this striking point when he noted 'the two miracles of the existence of laws of nature and of the human mind's ability to divine them' [1960, 7; see also p. 5]; I strengthen it by regarding the laws as existing only because of human effort in the first place. There may indeed be limitations on the human capacity to formulate a problem clearly, and/or think up theories to solve it (a possibility that worried philosophers such as Kant, Whewell and Peirce); but the means for theory-building laid out in this essay suggest that there is still plenty of room for human manoeuvre!

8.4 *Want of spirit.* Let us finally note three kinds of ineffectiveness due to human frailty. The first is vanities such as the generalisation racket, where a mathematician takes a theorem involving (say) the number -2 and generalises it to all negative even integers -2n, where however the only case of any interest is given by n=1.

The second kind can involve narcissism, where a mathematical theory is applied in a scientific context in an inappropriate form *because* its

---

[14] An example of ineffectiveness is the attempt in [Matte Blanco 1975] to construe the unconscious in terms of set theory, which however is not well handled; for example, paradoxes are admitted, seemingly unintentionally. The *principle* of applying set theory to the mental sciences may be in question, as well as this particular practice.

pure practitioners prefer it. Then indeed the influence of mathematics upon that science is 'pernicious' [Schwartz 1962] — that is, worse than ineffective. For example, since Cauchy's time in the 1820s the mathematically superior version of the calculus has been based upon a theory of limits; but the older Leibniz–Euler theory using the dreadful differentials often has a better analogy content to the scientific context (especially if the latter involves continua) and so should be given its due [Thompson 1910]. Thomson's life-long career reveals many examples of heresy, including those mentioned above.

The third kind obtains in any science: oversight! Mathematics has eventually exhibited some nice 'missed opportunities' [Dyson 1972]; what will turn out to be the good ones of today?

9. Concluding remarks

It may be that Wigner was drawn to his thesis by his experience with quantum mechanics; he gives some examples from there [1960, 9-12]. Perhaps its first practitioners struck lucky in analogising from the experiential celestial heavens to the highly non-experiential atom, and enjoying some remarkable later successes; but for those who follow [Popper 1959] in seeing science as guesswork, then sometimes it is bull's-eye time, and quantum mechanics was one of them — for a time, anyway. For a *general explanation* of mathematics Wigner appeals to its beauty and to the manipulability of expressions [1960, 3, 7]: as with complex numbers above, such properties may be exhibited on occasion, but surely they cannot *ground* mathematics or explain its genesis, growth or importance.

Wigner's thesis about unreasonableness is philosophically ineffective, partly because he neglected numerous clear indications from history of sources of both reasonableness and effectiveness of the natural sciences in mathematics. Yet not only were various histories of applied mathematics available by 1960; some eminent mathematicians had published relevant texts. [Polya 1954a, 1954b] has already been cited; it was followed by [Polya 1963] on 'mathematical methods in science', mostly elementary mechanics, and one could add, for example, [Enriques 1906, chs. 5-6] sketching in some detail the history of how physics convoluted out of mechanics, and [Weyl 1949, 145-164] providing an historico-philosophical review of 'the formation of concepts' and 'theories' in connection with mechanics.

Wigner also under-rated the central place of theories being formed in the presence of other theories, and being desimplified when necessary and where possible. In addition, the ubiquity of the topics and notions elucidated in Table 1, and others not listed there, should be emphasized.

The alternative picture that emerges is that, with a wide and ever-widening repertoire of mathematical theories and an impressive tableau of ubiquitous topics and notions, theory-building can be seen as reasonable to a large extent; however, the effectiveness of the output may need some enhancement through (further) desimplifications, if they can be realized. Instead of 'effective but unreasonable', read 'largely reasonable, but how effective?'. This slogan can also guide appraisals of (un?)reasonable (in?)effectiveness in contexts overlapping with the one studied here: for example, notations and notational systems (places where mathematics meets semiotics[15]), graphical and visual techniques, pure mathematics, numerical methods, logics, and probability theory and mathematical statistics. There are consequences to explore concerning the use of the histories of mathematics and of the natural sciences in theory-building, and the content of mathematics and science education.

By the way, π turns up in ¶1 in the statistical *theory* that is applied to the population *data*. {De Morgan explained it in terms of probabilistic expectation in actuarial mathematics; or one may point to normalising the Gaussian distribution.} Wigner does not give any explanation of the mystery in his article.

Acknowledgements

This chapter is based upon a talk given in December 2007 at a conference held at the London School of Economics as part of a research seminar on 'Dissent in science' that was supported by the Leverhulme Foundation. The influence of Popper's philosophy is evident, though it is rather weak on the formation of problems. For comments on the draft I thank Vela Velupillai, Niccolo Guicciardini, Michel Serfati, Chiara Ambrosio, and R. S. Kaushal.

Bibliography

Bacharach, M. 1883. *Abriss zur Geschichte der Potentialtheorie,* Würzburg: Thein.

Beller, M. 1983. 'Matrix theory before Schrödinger', *Isis, 74,* 469-491.

Colyvan, M. 2001. 'The miracle of applied mathematics', *Synthese, 127,* 265–277.

Chabert, J. L. and others 1999. *A history of algorithms. From the pebble to the microchip,* Berlin: Springer.

Corry, L. 1996. *Modern algebra and the rise of mathematical structures,* Basel: Birkhäuser.

Dantzig, G. B. 1982. 'Reminiscences about the origins of linear programming', *Operational research letters, 1,* 43-48. [Slightly rev. ed. in A. Bachem, M. Grotschel and B. Corte (eds.), *Mathematical programming. The state of the art,* Berlin: Springer, 1983, 78-86.]

De Morgan, A. 1872. *A budget of paradoxes* (ed. S. De Morgan), London: Longmans. [Ed. cited: vol. 1 of (ed. D. E. Smith), Chicago: Open Court, 1915; some reprints. Ms.: Smith Collection, Columbia University, New York.]

Dilworth, C. 1994. *Scientific progress,* Dordrecht: Kluwer.

---

[15] Of especial interest is Peirce's theory of icons, the relationships between (families of) signs, their referents, and the cognitive means of correlating the two{; see, for example, [Tursman 1987, ch. 3]}.

Dyson, F. 1972. 'Missed opportunities', *Bulletin of the American Mathematical Society, 78*, 635-652.

Enriques, F. 1906. *The problems of science*, Chicago and London: Open Court.

French, S. 2000. 'The reasonable effectiveness of mathematics: partial structures and the application of group theory to physics', *Synthese, 125*, 103-120.

Fresnel, A. J. 1866. *Oeuvres complètes*, vol. 1, Paris: Imprimerie Impériale.

Grattan-Guinness, I. 1990. *Convolutions in French mathematics, 1800-1840. From the calculus and mechanics to mathematical analysis and mathematical physics*, 3 vols., Basel: Birkhäuser; Berlin: Deutscher Verlag der Wissenschaften.

Grattan-Guinness, I. 1992a. 'Scientific revolutions as convolutions? A sceptical enquiry', in S. S. Demidov, M. Folkerts, D. E. Rowe and C. J. Scriba (eds.), *Amphora. Festschrift für Hans Wussing zu seinem 65. Geburtstag*, Basel: Birkhäuser, 279-287. [*ROL*, §8.]

Grattan-Guinness, I. 1992b. 'Structure-similarity as a cornerstone of the philosophy of mathematics', in J. Echeverria, A. Ibarra and T. Mormann (eds.), *The space of mathematics. Philosophical, epistemological, and historical explorations*, Berlin and New York: de Gruyter, 91-111. [§11]

Grattan-Guinness, I. 1994. (Ed.), *Companion encyclopaedia of the history and philosophy of the mathematical sciences*, London: Routledge. [Repr. Baltimore: Johns Hopkins University Press, 2003.]

Grattan-Guinness, I. 2004. 'The mathematics of the past. Distinguishing its history from our heritage', *Historia mathematica, 31*, 161-185. [*ROL*, §2.]

Grattan-Guinness, I. 2007. 'Equilibrium in mechanics and then in economics, 1860-1920: a good source for analogies?', in [Mosini 2007], 17-44.

Grattan-Guinness, I. 2008a. 'Differential equations and linearity in the 19th and early 20th centuries', *Archives internationales d'histoire des sciences, 58*, 343-351.

Grattan-Guinness, I. 2008b. 'On the early work of William Thomson: mathematical physics and methodology in the 1840s', in R. G. Flood, M. McCartney and A. Whitaker (eds.), *Lord Kelvin: life, labours and legacy*, Oxford: Oxford University Press, 44-55, 314-316.

Grattan-Guinness, I. 2010. 'How influential was mechanics in the development of neo-classical economics? A small example of a large question', *Journal of the history of economic thought*, to appear.

Grattan-Guinness, I. with the collaboration of Ravetz, J. R. 1972. *Joseph Fourier 1768-1830. A survey of his life and work, based on a critical edition of his monograph on the propagation of heat, presented to the Institut de France in 1807*, Cambridge, Mass.: The MIT Press.

Hamming, R. 1980. 'The unreasonable effectiveness of mathematics', *The American mathematical monthly, 87,* 81-90.

Hawkins, T. W. 1975. 'Cauchy and the spectral theory of matrices', *Historia mathematica, 2,* 1-29.

Henshaw, J. M. 2006. *Does measurement measure up? How numbers reveal and conceal the truth,* Baltimore: Johns Hopkins University Press.

Hermann, A. 1971. *The genesis of quantum theory,* Cambridge, Mass.: The MIT Press.

Hintikka, J. 2007. *Socratic epistemology,* Cambridge; Cambridge University Press.

Holland, J. R. 1975. *Adaptation in natural and artificial systems. An introductory analysis with applications to biology, control and artificial intelligence,* Ann Arbor: The University of Michigan Press.

Kaushal, R. S. 2003. *Structural analogies in understanding nature,* New Delhi: Anamaya.

Kiesow, H. 1960. Review of [Wigner 1960], *Zentralblatt für Mathematik, 102,* 7.

Knobloch, E. 2000. 'Analogy and the growth of mathematical knowledge', in E. Grosholz and H. Breger (eds.), *The growth of mathematical knowledge,* Dordrecht: Kluwer, 295-314.

Lesk, A. 2000. 'The unreasonable effectiveness of mathematics in molecular biology', *The mathematical intelligencer, 22,* no. 2, 28-36.

Lesk, A. 2001. 'Compared with what?', *The mathematical intelligencer, 23,* no. 1, 4.

Mackey, G. W. 1978. 'Harmonic analysis as the exploitation of symmetry: A historical survey', *Rice university studies, 64,* 73-228. [Repr. in [1992], 1-158.]

Mackey, G. W. 1985. 'Herman Weyl and the application of group theory to quantum mechanics', in W. Deppert and others (eds.), *Exact sciences and their philosophical foundations,* Kiel: Peter Lang, 131-159. [Repr. in [1992], 159-188.]

Mackey, G. W. 1992. *The scope and history of commutative and noncommutative harmonic analysis,* [No place]: American Mathematical Society and London Mathematical Society.

Mathematics 1986. 'Mathematics: the unifying thread in science', *Notices of the American Mathematical Society, 33,* 716-733.

Matte Blanco, L. 1975. *The unconscious as infinite sets: an essay in bi-logic,* London: Duckworth.

Medawar, P. B. 1967. *The art of the soluble,* Harmondsworth: Pelican.

Mosini, V. 2007. (Ed.), *Equilibrium in economics: scope and limits,* London: Routledge.

Poisson, S. D. 1823. 'Sur la distribution de la chaleur dans un anneau homogène et d'une épaisseur constante [...]', *Connaissance des temps,* (1826), 248-257.

Polya, G. 1954a, 1954b. *Mathematics and plausible reasoning,* 2 vols., 1st ed., Princeton: Princeton University Press. [2nd ed. 1968.]

Polya, G. 1963. *Mathematical methods in science,* Washington: Mathematical Association of America.
Popper, K. R. 1959. *The logic of scientific discovery,* London: Hutchinson.
Popper, K. R. 1972. *Objective knowledge,* Oxford: at the Clarendon Press.
Roche, J. J. 1998. *The mathematics of measurement. A critical history,* London: Athlone Press.
Ruelle, D. 1988. 'Is our mathematics natural? The case of equilibrium statistical mechanics', *Bulletin of the American Mathematical Society, 19,* 259-267.
Sarukkai, S. 2005. 'Revisiting the "unreasonable effectiveness" of mathematics', *Current science, 88,* 415-422.
Schwartz, J. 1962. 'The pernicious influence of mathematics on science', in E. Nagel and others (eds.), *Logic, methodology and philosophy of science,* Stanford: Stanford University Press, 356-360. [Repr. in R. Hersh (ed.), *18 unconventional essays on the nature of mathematics,* New York: Springer, 2005, 231-235.]
Smith, B. 1982. (Ed.) *Parts and moments. Studies in logic and formal ontology,* Munich: Philosophia.
Smithies, F. 1997. *Cauchy and the creation of complex function theory,* Cambridge: Cambridge University Press.
Steiner, M. 1998. *The applicability of mathematics as a philosophical problem,* Cambridge, Mass.: The MIT Press.
Tanner, R. C. H. 1961. 'Mathematics begins with inequality', *Mathematical gazette, 44,* 292-294.
Thompson, S. P. 1910. *Calculus made easy,* 1st ed., London: Macmillans. [Deservedly numerous later eds.]
Toeplitz, O. 1963. *The calculus. A genetic approach,* Chicago: The University of Chicago Press. [German original 1949.]
Tursman, R. 1987. *Peirce's theory of scentific discovery. A system of logic conceived as semiotic,* Bloomington and Indianapolis: Indiana University Press.
Velupillai, K. V. 2005. 'The unreasonable *in*effectiveness of mathematics in economics', *Cambridge journal of economics, 29,* 849-872.
West, B. J. 1985. *An essay on the importance of being nonlinear,* Berlin: Springer.
Weyl, H. 1949. *Philosophy of mathematics and natural science,* Princeton: Princeton University Press.
Weyl, H. 1952. *Symmetry,* Princeton: Princeton University Press.
Whittaker, E. T. 1927. *Analytical dynamics,* 3rd ed., Cambridge: Cambridge University Press.
Wigner, E. P. 1931. *Gruppentheorie und ihre Anwendung auf die Quantenmechanik der Atomspektren,* Braunschweig: Vieweg. [Rev. English trans.: *Group theory and its application to the quantum mechanics of atomic spectra,* New York: Academic Press, 1959.]
Wigner, E. P. 1960. 'The unreasonable effectiveness of mathematics in the natural sciences', *Communications on pure and applied mathematics, 13,* 1-14. [Repr. in [1967], 222-237; and in *Philosophical re-*

flections and syntheses (ed. G. Emch), Berlin: Springer, 1995, 534-548; and elsewhere.]

Wigner, E. P. 1967. *Symmetries and reflections: scientific essays,* Bloomington: Indiana University Press.

Wilson, C. A. 1980a, 1980b. 'Perturbation and solar tables from Lacaille to Delambre: the rapprochement of observation with theory', *Archive for history of exact sciences, 22,* 53-188, 189-304.

Wolf, C. J. E. 1889-1891. *Mémoires sur le pendule* [...], 2 pts., Paris: Gauthier-Villars.

Wussing, H. 1984. *The genesis of the abstract group concept,* Cambridge, Mass.: The MIT Press.

13 History or Heritage?
An Important Distinction in Mathematics and for Mathematics Education

During recent decades there has been a remarkable increase in work in the history of mathematics, including its relevance to mathematics education. But at times considerable differences of opinion arise, not only about its significance but even concerning *legitimacy* — that is, whether or not an historical interpretation counts as history at all. This chapter considers the latter issue, and notes some consequences for education.

1. Interest and disagreements

The disagreements about using history in mathematics education are general, in that they may arise for any branch of mathematics in any period or culture; so they need a general resolution. I offer one in the form of a distinction in the ways of interpreting a piece of mathematics of the past. Take such a mathematical notion N; it could be anything from one notation through a definition, proof, proof-method or algorithm to a theorem, a wide-ranging theory, a whole branch of mathematics, and ways of teaching it. By its 'history', which becomes a technical term, one considers the development of N during a particular period: its launch and early forms, its impact, and applications in and/or outside mathematics, and so on. It addresses the question 'what happened in the past?' by offering descriptions. Maybe some kinds of explanation will also be attempted, to answer the companion question 'why did it happen?'.

History should also regard as important two companion questions, namely 'what did not happen in the past?' and 'why not?'. The reasons may involve the other side of this distinction, which I call 'heritage'. There one is largely concerned with the effect of N upon later work, during any relevant period including that of its launch. Some modernised versions of N are likely to be taken, for heritage is largely concerned with the question 'How did we get here?', that is, to some current version of the context in question.

The distinction between history and heritage is often sensed by people who study some mathematics of the past, and feel that there are fundamentally different ways of doing so. Hence the disagreements can arise; one man's reading is another man's anachronism, and his reading is the first

---

This article was first published in the *American mathematical monthly*, *111*(2004), 1-12, and again in G. van Brummelen and M. Kinyon (eds.), *Mathematics and the historian's craft. The Kenneth O. May lectures*, New York: Springer, 2005, 7-21. It is reprinted by kind permission of the Mathematical Association of America, which retains copyright. Note that the word 'notion' is used more broadly than in §12.

one's irrelevance. The discords often exhibit the differences between the approaches to history usually adopted by historians and those often taken by mathematicians.

The claim put forward here is that *both history and heritage are legitimate ways of handling the mathematics of the past; but muddling the two together, or asserting that one is subordinate to the other, is not*. Many consequences flow from this stance, which will be treated in ¶3 and ¶¶f4; first let us take a simple and well-known example, from the distant past.

2. Pythagoras's theorem, Euclid style

One of the best-known theorems in Euclid's *Elements* (-4th century) concerns the sides of a right-angled triangle ABC in Figure 1. We recognise it as saying of the sides AB, AC and BC that

$$AB^2 + AC^2 = BC^2 ; \qquad (1)$$

but Euclid actually says something quite different (Euclid *Elements*, Book 1, Proposition 47): 'In right-angled triangles the square on the side subtending the right angle is equal to the squares on the sides containing the right angle'. There is an attached diagram, of which Figure 1 is part, and the differences between it and (1) are basic.

Figure 1

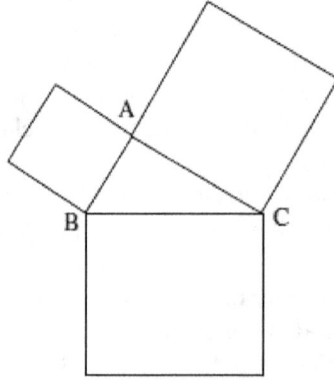

Not only is (1) algebraic while the Figure is geometric; the diagram shows the squares outside the triangle, which (1) does not convey. Were any of the squares to lie over the triangle, then both (1) and the theorem would still be true; but the complicated proof, not shown in the Figure, could not be effected. The algebraic character of (1) emerges further when, as was and is commonly done, and the letters 'a', 'b' and 'c' are used for the sides; for algebra is the branch of mathematics in which special words and especially symbols are used to a significant extent to represent constants, unknowns, variables and operations.

Another important difference concerns the word 'on'. Euclid never used the phrase 'side squared', for in his geometrical Books he never multiplied geometric magnitudes together, either in the statement of theorems or (more importantly) in any proof — for example in this theorem, neither the complicated proof just mentioned, which uses congruence, nor a more elegant one for the more general theorem about rectangles with the same ratio of sides set upon the sides of the triangle, where the proof deploys similar triangles and ratio theory [Book 6, Proposition 31]. Thus 'BC^2' is already a transgression from his geometry (and the frequent use in diagrams of small letters such as in 'a^2' even more so). Instead Euclid constructed a square *on* a given line — indeed, in the Proposition immediately preceding Pythagoras's theorem [Book 1, Proposition 46].

The issue is more profound than it may seem. Both here and everywhere else in the *Elements* Euclid works with *lines* rather than *lengths*, the latter being lines upon which some arithmetical measure has been imposed. Euclid presented geometry without *arithmetic* in the sense just explained; numbers are also present, but for other purposes, such as saying that this line is twice that line, or that the ratio of two lines is the same ratio as 5:7. In the same way he worked with planar regions but not (measured) areas, with solids but not volumes, with angles but not in degrees. By contrast, which is sometimes overlooked, in the arithmetical Books 7-9 multiplication of integers themselves occurs as usual [Grattan-Guinness 1996 = *ROL*, §11].

These remarks concern the history of Euclid. When one moves to its heritage, then a quite different situation arises, in which (1) and many other such equations are prominent. For the *Elements* played a major role in the development of common algebra among some of its Arabic initiators, and a still greater one when Europe at last woke up during the 12th century and began to elaborate that algebra with symbols introduced both for unknown quantities and for operations. *Both* (1) and Pythagoras's theorem with the Figure are legitimate readings of Euclid, but are quite different from each other.

The *Elements* is a particularly interesting historical example, because common algebra as in (1) became the dominating reading of Euclid (including in mathematics education) to such an extent that during the 19th century it also became the normal historical interpretation; apparently Euclid had been a 'geometric algebraist', talking geometry but really practising common algebra. A supporter of this reading was T. L. Heath, whose English edition and translation, first published in 1908, is still the most widely used, usually now in the second edition [Euclid 1926]. Greek specialists tell me that his translation is very reliable both to the language and to the mathematics; in particular, for Pythagoras's theorem and all other contexts he says there 'square on the side', not 'square of the side' as many earlier translations had rendered (the word 'apo' can admit both translations) but which can easily lead to the algebraic 'side squared'. Nevertheless, Heath added to his translation many algebraic versions of the propositions without seeming to notice the differences entailed.

While some historians of that time did not follow the algebraic interpretation of Euclid — for example, the Dutchman E. J. Dijksterhuis [van Berkel 1996, ch. 5] — the standard view came under severe challenge only from the 1960s. In particular, in the mid 1970s the historian Sabatei Unguru attacked it strongly, to the opposition of some mathematicians interested in history. Unguru's charges of anachronism and ahistory are largely vindicated; his mathematician opponents were inheritors [Rowe 1996].

We shall take another Euclid example in ¶8 below. First, though, let us explore some general consequences of the distinction.

3. Some principal differences between history and heritage

The distinction between the history of a notion N and its heritage obviously involves its respective pre- and post-histories; but much more is at hand, for history has to use post-history also. To see this, let us consider the advice, which is quite often put forward for history of all kinds, about a way of being 'history-minded' about N (say, Pythagoras's theorem in Euclid); namely, forget everything that has happened since N was formed, and read Euclid with the eyes with which he wrote it. But this advice *begs the question at hand*. For in order to forget everything E that has happened since N, then one has to know E already; however, to do that one needs to be able to distinguish E from the history and pre–history of N; but this is the task to be attempted.

Thus the distinction between history and heritage rests in part upon the *ways* in which notions later than N are to be used. When they are determined to be later notions, the view urged here is this: by all means bring them to bear, and deploy them to understand the heritage from N, but *avoid* feeding them back to appraise its history (such-and-such did not happen). Further, when considering periods intermediate between that of N and some later ones such as now, apply the distinction carefully. Thus, in our example the equation (1) is not only part of the history of René Descartes and the heritage of Euclid but also belongs to the heritage from, among others, the algebraist François Viète in the 16th century, whose work also belongs to the history of Descartes. Note also that history is usually *a history of heritages*; it is a tale of mathematicians taking and modifying notions from the past (often pretty recent) without enquiring about the history of those notions.

Various other matters can be explored; a more detailed discussion, largely focussed upon history, is given in a companion paper [Grattan-Guinness 2004 = *ROL*, §2]. Table 1 summarises the main features of handling past notions N in the two different ways suggested.

An apparent contradiction between the third and fourth rows needs to be addressed. When the historian reconstructs past muddles, he will conflate notions that we now know to be different, a feature which the inheritor will stress. But the difference that the reconstruction exposes is that between past ignorance of the distinction, which is different from our (and the inheritor's) present knowledge of it.

4. Breaking habits

A further type of issue, which is not susceptible to tabular expression, concerns the use of notions that have become standard and therefore are now used habitually. Such habits may well help in determining the heritage from N: but historical anachronism can easily arise, which needs to be controlled. I note three important examples.

Table 1
History versus heritage.

Feature	History	Heritage
Motivation(s) to N	Important issue; maybe hard to find (for example, Euclid's *Elements*)	Probably only of minor interest
Types of influence	Can be negative as well as positive; both should be noted	Likely to draw only upon the positive cases
Relationships of N to earlier and to later notions	Major issue; differences stressed as much as similarities, maybe more	Important issue; similarities stressed more than differences
Handling unclarities evident in N	Reconstruct them, and as clearly as possible	Recognise them, but clean them up
Successful developments	Very important; but also study failures, delays, missed opportunities, and late arrivals	Likely to be the main concern
Role of chronology	Usually important; can be hard to establish	Beyond broad details, may well not matter so much
Historical consequences	May try to reconstruct the *foresight* (hopes, and so on) for N held by the historical figures	May try to construct *hindsight* and historical *perspective* of developments after N
Determinism?	Preferably not claimed: the actual developments were so-and-so, but not necessarily so-and-so	May carry a determinist flavour; we *had* to get here (but see the history column!)
Foundations of a theory	Dig down to them, and build upon a swamp	Lay them down and build upon them, as if on solid ground
Level of importance or popularity of N	Can vary over time independently of content; should be noted (and maybe explained)	Not normally considered; current importance assigned

Firstly, after an interesting history of its own from the 1870s [Dauben 1979], Georg Cantor's set theory has been part of our mathematical furniture for just over a century; so for the mathematics of this period its use may well be faithful. Now collections of things have been handled in mathematics since at least Greek antiquity; but the earlier theory of so doing was part-whole theory, where (say) British women form part of the class of women, membership is not distinguished from inclusion, and an object is not distinguished from its unit class. The differences between part-whole and set theories are considerable, both technically and philosophically, and the historian needs to mark them carefully; by contrast, the inheritor can deploy set theory with little chance of deception.

Secondly, while the influence of Euclid was great in Western mathematics, his stress on axioms and common notions was rarely imitated (though to some extent Newton's *Principia* is an example). The axiomatisation of mathematical theories became more prominent only during the late 19th century, especially in connection with the axioms of Euclidian and non-Euclidian geometries, and the emergence of abstract algebras [Cavaillès 1938]. Both developments attracted the attention of David Hilbert, and led him to launch the wide-ranging use of axiomatisation during the first half of this century, an attitude which has now become pretty standard: a clear path of heritage can be traced up to present-day practises. But the historian should be careful when looking at the structure of earlier mathematical theories, for axiomatisation may well not be prominent beyond specifying basic principles or laws. Cantor's set theory is a good example; while it too was developed during the late 19th century, he showed little interest in the axioms that it may require.

Thirdly, vector and matrix theory have become standard fare in mathematics, though (especially in the second case) only from the 1930s onwards and after rather scrappy historical developments in various contexts during the 19th century. Once again, care should be exercised in applying them to earlier work. For example, much of the mechanics as developed by figures such as Leonhard Euler, J. L. Lagrange and P. S. Laplace can be rewritten in vectorial and matricial forms. But historical understanding will not profit; none of these figures knew that their theories could be developed in terms of strings or arrays of scalar elements; they worked instead in terms of collections of simultaneous linear or differential equations, or quadratic and bilinear forms. The introduction of vectors or matrices is *not* merely a matter of changing notation; new theories are involved. It is of course nice to save such space, for one thing; but if the historian does deploy these theories, then a chronological health warning should be appended.

By contrast to all these cautions to the historians, the inheritors can execute all these reformulations of theory quite legitimately; indeed, much nice heritage mathematics may emerge. Further, some history of mathematics produced after the initial period under study might be created; for, as was mentioned in ¶1, mathematicians normally read the past in a heritage spirit.

As an example, take Lagrange and others in mechanics. A major problem, which he created in the 1770s, was to prove mathematically that the planetary system was stable. (Previous figures such as Newton and Euler had relied on God to watch out for danger; that is, a religion influenced mathematics [*ROL*, §15].) In terms of matrix theory, Lagrange's brilliant theory sought proof of the reality of all the latent roots and latent vectors of a certain matrix. But he had no such theory, and worked with the corresponding quadratic forms; so did Laplace, who adapted his results to some extent; neither man found a watertight proof. The next major contribution came in 1829 from (surprisingly) A.-L. Cauchy, and in 1829 he *did* formulate 'tableaux' of scalar entries in his own work on this problem [Hawkins 1975]). Thus matrix theory may — indeed, should — be used to describe Cauchy's contribution, and indeed grasp to an important part of his heritage from his predecessors. And we also have a nice example of the 'what did not happen?' question; for Cauchy never realised the significance of his achievement and rarely used it later, so that unfortunately he was not a influential founder of the spectral theory of matrices.

5. Some philosophical background

It is obvious that this talk of earlier and later notions, the development of theories, and so on, is not confined to mathematics: such features occur also in the histories of other sciences (including technology, engineering and medicine), and indeed elsewhere (for example and nice one, practices to be adopted and avoided in the 'authentic performance' of older music). The main general principles that underlie the discussion above are as follows.

Firstly, history is *unavoidable*, whether one likes it or not. A mathematician who presents his theory without concern with history is not thereby immune from it. For example, an enthusiast for axiomatics mentioned in ¶3 will lay out his theory in a very formal way without reference to predecessors or precedents; but they will be there, including previous formal theories laid out by preceding axiomatists without reference to their own predecessors or precedents. Thus the question of whether or not one can use history in mathematics is *miscast*: it is rather the question of whether it is done consciously or not. Indeed, independent of the content of this chapter, it is useful to have some general historical idea of a topic of interest, whatever it may be.

Secondly, knowledge and ignorance go together. This symbiosis has not received the general philosophical attention that it deserves. In particular and of special significance for mathematics, there is *knowledge of ignorance,* especially when one formulates a problem: when, for example, J. P. G. Dirichlet studied the convergence problem of Fourier series in the late 1820s, he knew that he did not know sufficient conditions on a function to establish convergence to it: finding some was precisely his problem. Having done so, he knew that he did not know whether or not they could be weakened, thereby setting the next problem in this chain (to which the first answer was the Lipschitz condition, by the way). One can also have ignorance of ignorance, or unawareness, where people do not know that they do not know something because the required connections between notions have not yet been laid down. Thus Dirichlet did not know that he did not know how his proof bore upon the specification of function spaces, because that notion did not emerge the late 19th century [Siegmund-Schultze 1982].

Thirdly and following from the above, knowledge of all kinds is stratified into theory, metatheory, ... For mathematics this means not only metamathematics of the technical kind that Hilbert launched but also informal kinds. In particular, the history of notion N is one kind, its heritage is another, manners of its possible teaching a third, heuristic strategies to explain its significance a fourth, and there may well be others. The relationship between knowledge and ignorance just outlined lie in the metatheory of the notions involved. Similarly, metatheory requires metametatheory for its own forum for discussion, and so on upwards as far as is needed (§2.10.4). An example of metametatheory is the history of the history of mathematics, an interesting story recently recorded in detail in [Dauben and Scriba 2002]; the comments on Heath in ¶2 form an example of it; and this chapter itself is a self-referring one, with its heritage (if any) awaited!

The recognition of history and heritage as metatheoretic also releases both historians and inheritors from the need to like what they find in the past that they study. Why should they? After all, they were not there (as a rule). The point seems obvious enough; after all, one can be a good historian (or heritage) of, say, military history without being a militarist. Yet not infrequently historians and inheritors become overly attached to their objects and figures of study, in any kind of history, and feel that they have to defend what they find. While of course such attachment can be felt if it arises naturally, no compunction to it should even be encouraged.

The generality of stratification is an insight forged in connection with symbolic logic in the early 1930s, thanks principally to Kurt Gödel and Alfred Tarski. In logic the distinction of (object-level) logic itself from metalogic is especially tricky but thereby all the more important; as was known already in Greek times, failure to make a distinction of some kind admits nasty paradoxes. Gradually stratification spread into other disciplines, especially mathematics and some types of philosophy. One follower, inspired by Tarski in the mid 1930s, was Karl Popper (§6). Several parts of his philosophy of fallibilism are metaphilosophical; for example, his preference for indeterminism over determinism [Popper 1982]. Of particular relevance to this chapter is his essay 'On the sources of knowledge and ignorance' [Popper 1963, introduction], for it contains an insight largely missing from other kinds of philosophy; that *ignorance is nice,* for it is the site (in metatheory) of our problems as knowledge of ignorance. In most other philosophies ignorance is a disease to be cured by the acquisition of knowledge however that acquisition is claimed to occur (see [Unger 1975, chs. 1-6] for the various forms of this view maintained within the sceptical tradition of philosophy). So far explicit use of stratification has not been widely canvassed among prevalent philosophies of history (which are well surveyed in [Stanford 1997]); but it seems worthy of further elaboration.

6. General remarks about history in mathematics education

In recent decades a considerable and international increase has developed in the use of history in mathematics education, in order to temper and challenge the normal picture of mathematics as a human-free zone, all answers but no questions, all solutions but no problems. Several edited or authored books and special issues of journals have appeared containing material of various kinds: textbooks significantly informed by the relevant history; summary histories of particular developments; surveys of the lives and works of important historical figures; international and/or multicultural comparisons of the development of (more or less) the same theories; translations of original texts with commentary; and suggested strategies for using history in teaching practice, both in specific contexts and in general. The emphasis often falls upon motivation and context, on showing that mathematics is after all human activity despite appearances, and moreover that much of it is not Western in origin. The range of concerns is well captured in [Fauvel and van Maanen 2000]. Most attention seems to have fallen on teaching at school and college level, but the university level has also been addressed. Much more work has been done on pure mathematics than on applied or applicable mathematics, or on probability and statistics; a redress of balance would be most welcome. I shall not attempt to review this literature here, but I consider the place and utility of the distinction between history and heritage in mathematics education in general.

As with researchers in history mentioned in ¶1, there is evident sense of the distinction in this kind of educational literature, or at least that the mathematics of the past can be used in different ways. Where is mathematics education to be found between history and heritage? My answer is that *that is exactly where* it *should be found*, for that it can profit from *both* sides. In particular, if notion N is to be taught, then both its history and its heritage can be used. Euclid's *Elements* is a good example: the inherited use of algebra has been used quite frequently; but the historical Euclid also deserves attention, with its geometry presented without *arithmetic* with lines rather than lengths, and the beautiful theory of ratios used in both his geometry and his arithmetic.

7. History-satire and the calculus

In [Grattan-Guinness 1973 = §14.3] I introduced the term 'history-satire', to characterise a way in which history and also heritage can be used in mathematics education. Under it the broad features of the historical record are respected and used; but usually many detours and complications occurred which, while they of attraction for the historian, will impede teaching and so should be set aside or at most treated only in passing. The 'genetic method' of [Toeplitz 1963], introduced initially in the late 1920s in connection with teaching the differential and integral calculus, is similar in sentiment. More recently the Mathematical Association of America published a novel and important textbook in real-variable mathematical analysis, in which the main developments especially of the 19th century, such as Fourier series, are given prominent places [Bressoud 1994].

As Bressoud duly notes, a major innovation of the century was the founding of analysis in the 1820s by Cauchy. His approach was based upon a newly sophisticated *theory* of limits, not merely as an intuitive notion. Undoubtedly it was much superior to the preceding versions in the organisation of the subject and statements and proofs of the theorems; however the loss in heuristics was heavy, and both his colleagues and students objected forcefully to it [Grattan-Guinness 1990, chs. 10-11 *passim*].

For an explicit example, here is a use of history-satire that I found helpful in my own teaching. In a remarkable analogy, Cauchy adapted his real-variable analysis to complex variables and their functions and thereby introduced a major new subject into mathematics. But it seems a strange subject when first learnt: it uses the corresponding expressions as in real-variable analysis, but there are no curves, tangents to them, or areas underneath them to think about or look at. Among the many theorems that he proved, a main one is now named after him: namely, that the integral of a single-valued and differentiable function with a continuous derivative around and inside a closed contour C is zero. To students, including me long ago, it seemed to be a peculiar result; and a quick and doubtless valid proof using the Cauchy-Riemann equations and Green's theorem did not assuage the perplexity.

Cauchy developed his theory fitfully from the 1810s to the 1840s [Smithies 1997], and this version of his theorem is the last one, with the complex plane available as the site for C. I found that an earlier stage of his theory helped understanding of the theorem. In his treatment of the real-variable integral of f(x) over the range $x_0 \leq x \leq X$ (I use his symbols) he formed the area sum S for a partition of values of x over the range, took successive sub-partitions and formed the corresponding sums, and defined the integral as the limiting value of the sequence *if* it existed at all. This manner of defining the integral has long been standard, and his version is still worth reading and teaching [Cauchy 1823, lecture 21].

Soon afterwards Cauchy deployed his analogy. He defined the integral of a finite-valued and continuous complex-variable function '$f(x + y\sqrt{-1})$' by forming the expression corresponding to S for f(x) but with $x + y\sqrt{-1}$ taking a sequence of values between the limits A = '$x_0 + y_0\sqrt{-1}$' and B = '$X + Y\sqrt{-1}$' for which both x and y were continuous functions of a parametric variable t. Then, drawing upon integration by parts and the calculus of variations, he proved that the value of this integral between A and B '*is independent of the nature of the functions*' involved [Cauchy 1825, sect. 3]. The closed contour theorem then follows by taking the integral along one sequence of values between A and B and back along another sequence under which the required conditions obtained; the two integrals for the two sequences cancel out, so that the value of the integral over C is zero.

Working through the theorem this way certainly takes more than a few lines; but the understanding increased substantially, especially as the definition of the real-variable integral had already been taught elsewhere. My account above follows Cauchy historically to the extent of deliberately avoiding diagrams, for both types of integral; at that stage in his career he regarded geometric notions as unrigorous and so wished to avoid them. The status of geometry makes a nice point to debate in the classroom, and in fact I increased the measure of satire by using diagrams myself. I also ignored several special cases and other details of the theory as Cauchy was then developing it. But I raised questions such as whether or not Cauchy assumed the derivative of F to be continuous (yes, but implicitly); and I also taught his 1825 version of the residue theorem, noting that, contrary to most later practice, he allowed $x + y\sqrt{-1}$ to go *through,* and not just *round*, a pole of $F(x + y\sqrt{-1})$ [1825, sect. 8].

The considerations of this section have used the calculus and mathematical analysis because these case studies happen to come from it.

But genetic approaches and history-satire can be applied to any mathematical notion or level of teaching.

8. The proposals of Bashmakova

The relationships between knowledge and ignorance outlined in ¶5 deserve serious consideration, including the niceness of ignorance as the source of problems (big or small) for tackle. One important area of education where these relationships are prominent is the design of a of a course syllabus and the manner and order of teaching the topics proposed, when in effect the designer is considering the stage at which the pupils or students should cease to be ignorant of some specific notions.

Let us take an example, examining the historiography proposed in recent years by the Russian historian I. G. Bashmakova; for two of her books have recently been translated into American and published by the Mathematical Association of America for their utility in mathematics education. While dealing with the history of common algebra, her position is put forward in a general way, most explicitly in the joint paper [Bashmakova and Vandoulakis 1994]. For them, there are two main stages in handling an historical text [p. 251]:

> First the text should be "translated" into the [sic] contemporary mathematical language, i.e. an adequate model for it should be constructed. This is absolutely necessary in order to *understand* the text, to reveal its mathematical content.

In the next stage 'it is necessary to *embed* the considered work in the context of science of its day' [p. 252].

The authors state that the second stage is 'more difficult' than the first: for me it might well be impossible, since the first stage will have put so much heritage in place that the historical context could be masked. They state the aim of heritage very clearly: 'the mathematicians of every new age reconsider the previous material and restate it in new terms, thus making readily available and applicable for the contemporary scientist' [p. 250].

The examples given in Bashmakova's writings seem to exhibit the conflation of history and heritage, without the stress on the distinction between them that was argued in ¶7. For example (Figure 2), she takes Book 2, Proposition 4 of Euclid's *Elements* to express the quadratic identity

$$(a + b)^2 = a^2 + 2ab + b^2 \qquad (2)$$

as a legitimate prime reading [Bashmakova 1990, 88; Bashmakova and Smirnova 2000, 165]:

Figure 2

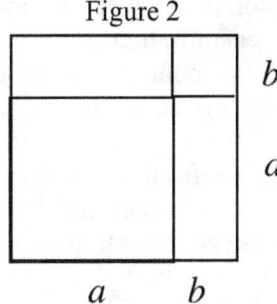

In a more recent co-authored short history of algebra, (2) is held to be 'equivalent' to the diagram [Bashmakova and Smirnova 2000, ch. 2]; and throughout that book the modern notations dominate, although the older terms and symbols are also presented in some detail [chs. 4-5]. The dominance of heritage is clear in a general historiographical appendix, where in specifying the term 'geometric algebra' the authors characterise algebra as an historical category drawn from 'the class of problems associated with algebra today' [p. 164]. For them, therefore, Euclid's Book 2 is concerned with algebraic identities such as (2) (see especially [Bashmakova 1997]): indeed, her most recent stance is to impose algebraic readings onto ancient arithmetic and geometry for *all* cultures, and to vote for the mathematicians and against Unguru in the disagreement noted above in ¶2 [Bashmakova and Smirnova 2000, 163-172].

The preference for modern notations in the book fits its primarily educational purpose well, exposing an important chain of heritage influences. But the general statements of *historical interpretation* quoted above seem to be heritage mistaken for history. For me, in Book 2 Euclid presents theorems relating sub-regions of planar rectilinear constructions involving rectangles, squares and triangles (as in the above example, where the relative locations of the sub-squares and rectangles are lost in the ubiquitous sign '+'); the algebraic content is empty, as also in all his other geometry Books. By contrast, algebra looms very large in the post-Grecian heritage from Euclid's geometry. Both readings are valuable to mathematics education, but presented as *distinct* sources; indeed, like Euclid the history of the theory of polynomial equations is especially suitable for historical satire.

9. Concluding remark

In this chapter, and in somewhat more detail in its companion [Grattan-Guinness 2004 = *ROL*, §2], I assert that the history of mathematics differs fundamentally from heritage studies in the use of the mathematics of the past, and that both are beneficial in mathematics education when informed by the mathematics of the past. The majority of the examples presented come from fairly modern periods; not by accident, they are my own specialist areas, and so the examples have no special significance. Indeed, since the distinction between history and heritage is held to be general, then indefinitely many more examples could be presented; the reader is invited to construct some of his own. A rich resource comes from considering the many ways in which notions are changed, especially when they are (major) theorems or theories. These include the alteration of known results by extension, generalisation and/or abstraction; reaction to counter-example; the exposure as axioms or as procedures of assumptions previously taken for granted; the adaptation of algorithms; the introduction, or maybe removal, connections between branches (such as geometry with or without arithmetic); classifications into kinds of objects in a theory; switches between axiom, theorem and definition; and new applications, both within mathematics and to other disciplines.

More attention has been paid in this chapter to history and historiography than to heritage and heritage studies; but no value judgement is involved, for, as stated in ¶1, neither activity is subordinate to the other. A companion paper concentrating on good and bad practices in heritage work could be written. The two activities are distinct; but they interact in fruitful ways, each posing questions for the other to address.

Acknowledgement

This chapter is based upon a plenary lecture delivered to the joint annual meeting of the Mathematical Association of America and the American Mathematical Society, which was held in Baltimore in January 2003. Thanks are offered to the former organisation for the invitation.

Bibliography

Bashmakova, I. G. 1990. 'Diophantine equations and the evolution of algebra', *Translations of the American Mathematical Society, (2)147,* 85-100.

Bashmakova, I. G. 1997. *Diophantus and Diophantine equations* (trans. A. Schenitzer), Washington: Mathematical Association of America.

Bashmakova, I. G. and Smirnova, G. 2000. *The beginning and evolution of algebra* (trans. A. Schenitzer), Washington: Mathematical Association of America.

Bashmakova, I. G. and Vandaloukis, I. M. 1994. 'On the justification of the method of historiographical interpretation', in K. Gavroglu and others (eds.), *Trends in the historiography of science,* Dordrecht, Boston and London: Kluwer, 249-264.

Bressoud, R. 1994. *A radical approach to real analysis.* Washington: Mathematical Association of America.

Cauchy, A.-L. 1823. *Résumé des leçons données à l'Ecole Polytechnique sur le calcul infinitésimal,* vol. 1 [and only]. Paris: de Bure. [Repr. in *Oeuvres complètes,* ser. 2, vol. 4, Paris: Gauthier-Villars, 1898, 5-261.]

Cauchy, A.-L. 1825. *Mémoire sur les intégrales définies, prises entre des limites imaginaires,* Paris: de Bure. [Repr. in *Oeuvres complètes,* ser. 2, vol. 15, Paris: Gauthier-Villars, 1974, 41–89.]

Cavaillès, J. 1938. *Méthode axiomatique et formalisme,* 3 pts., Paris: Hermann.

Dauben, J. W. 1979. *Georg Cantor,* Cambridge, Mass.: Harvard University Press. [Repr. Princeton: Princeton University Press, 1990.]

Dauben, J. W. and Scriba, C. J. 2002. (Eds.), *Writing the history of mathematics: its historical development.* Basel: Birkhäuser.

Euclid *Elements*. [Edition used:] Ed. and trans. T.L. Heath, *The thirteen Books of Euclid's Elements,* 2nd ed., 3 vols., Cambridge: Cambridge University Press, 1926. [Repr. New York: Dover, 1956. First ed. 1908.]

Fauvel, J. and van Mannen, J. 2000. (Eds.), *History in mathematics education. The ICME study.* Dordrecht: Kluwer.

Grattan-Guinness, I. 1973. 'Not from nowhere. History and philosophy behind mathematical education', *International journal of mathe-*

matics education in science and technology, 4, 421-453. [Parts in §14.]

Grattan-Guinness, I. 1990. *Convolutions in French mathematics, 1800-1840.* 3 vols., Basel: Birkhäuser; Berlin: Deutscher Verlag der Wissenschaften.

Grattan-Guinness, I. 1996. 'Numbers, magnitudes, ratios and proportions in Euclid's *Elements*: how did he handle them?', *Historia mathematica, 23,* 355-375. [Printing correction: *24* (1997), 213. *ROL,* §11.]

Grattan-Guinness, I. 2004. 'The mathematics of the past. Distinguishing its history from our heritage', *Historia mathematica, 31,* 161-185. [*ROL,* §2.]

Hawkins, T. W. 1975. 'Cauchy and the spectral theory of matrices', *Historia mathematica, 2,* 1-29.

Popper, K. R. 1963. *Conjectures and refutations,* London: Routledge and Kegan Paul.

Popper, K. R. 1982. *The open universe. An argument for indeterminism* (ed. W. W. Bartley, III), London: Hutchinson.

Rowe, D. 1996. 'New trends and old images in the history of mathematics', In R. Calinger (ed.), *Vita mathematica. Historical research and integration with teaching,* Washington: Mathematical Association of America, 3-16.

Siegmund-Schultze, R. 1982. 'Die Anfänge der Functionalanalysis', *Archive for history of exact sciences, 26,* 13-71.

Smithies, F. 1997. *Cauchy and the creation of complex function theory,* Cambridge: Cambridge University Press.

Stanford, M. 1997. *An introduction to the philosophy of history,* Oxford: Blackwell.

Toeplitz, O. 1963. *The calculus. A genetic approach,* Chicago: University of Chicago Press.

Unger, P. 1975. *Ignorance. A case for scepticism,* Oxford: Clarendon Press.

van Berkel, K. 1996. *Dijksterhuis. Een biografie,* Amsterdam: Bert Bakker.

14. Not from Nowhere.
History and Philosophy Behind Mathematical Education

I discuss some of the historical and philosophical aspects of mathematical education, especially in foundational subjects where many of the principal issues seem to lie. ¶1 is motivated partly by problems arising in school mathematics, although the subject matter is treated from a more advanced viewpoint. ¶2 develops some general methodological considerations. ¶3 and ¶4 contains a number of broad conclusions, including the question of the role of the history of mathematics in mathematical education, at university and school levels. The notion of 'history-satire' is introduced.

After an unbroken tradition of many centuries, mathematics has ceased to be generally considered as an integral part of culture in our era of mass education. The isolation of research scientists, the pitiful scarcity of inspiring teachers, the host of dull and empty commercial textbooks and the general educational trend away from intellectual discipline have contributed to the anti-mathematical fashion in education. It is very much to the credit of the public that a strong interest in mathematics is none the less alive.
                          Richard Courant [Kline 1953, foreword]

The same inhuman — in fact anti-humanistic — trend pervades the climate in which science is taught, the classrooms and the textbooks. To derive pleasure from the art of discovery, as from the other arts, the consumer — in this case the student — must be made to re-live, to some extent, the creative process. In other words, he must be induced, with proper aid and guidance, to make some of the fundamental discoveries of science by himself, to experience in his own mind some of these flashes of insight which have lightened its path. This means that the history of science ought to be made an essential part of the curriculum, that science should be represented in its evolutionary context — and not as a Minerva born fully armed [...] The traditional method of confronting the student not

---

This chapter comprises parts of a paper that was first published in *International journal of mathematical education in science and technology,* 4(1973), 421-453; they are reprinted by permission of Taylor & Francis (http://www.informaworld.com). Omitted are a general outline of theories and their metatheories that has already been rehearsed in earlier chapters, now using the distinction between history and heritage; and most of the summary histories of mathematical analysis and set theory, which have already been presented in §9.2 and §11-§13 (and also in *ROL,* §12-§13). The two quotations at the head are new.

with the problem but with the finished solution, means depriving him of all excitement, to shut off the creative impulse, to reduce the adventure of mankind to a dusty heap of theorems.

Arthur Koestler [1966, 260-261]

1. In and around the 'New Mathematics'

1.1 *Eddington's moral.* Sidney Chapman died in the summer of 1970, to the great sadness of those who knew or had met him. Like some other great scientists, he left behind him not only a mass of important writings but also a fund of stories and anecdotes about his life and personality. One of them serves as an admirable prologue to this chapter, as well as a tribute to him; its truth seems assured, for he told it to me himself.

When he first went to Cambridge as a young man, he was shown around the laboratory buildings by Arthur Eddington. As they walked together down a corridor, they came across an open door. Chapman looked through it and saw a large lecture theatre filled with students listening intently to the lecturer at the front. 'What's going on in there?' asked Chapman. 'That's a meeting of mathematicians', replied Eddington. 'But I didn't know that there were so many mathematicians in the whole world!' cried Chapman. Eddington pulled him away from the door. 'Between you and me', he answered, 'there aren't'.

'Eddington's moral', as I shall call it, is simple but important; there are not a lot of mathematicians in the world, but many people are going to mathematics lectures and taking degrees in mathematics or some related subject. Indeed, the moral may be extended; there are even more people studying mathematics in school and institutes, and therefore an even smaller proportion of them possess substantial mathematical ability.

Yet mathematics is not only for mathematicians; everybody requires a certain amount of mathematical knowledge for their daily lives, and a perceptible minority, such as those who attend mathematics lectures during their degree courses, need considerably more. So the educational questions are, basically, what should be taught? To whom? How? When? And above all, why?

The programmes of 'New Mathematics' in school education have sharply divided the mathematical profession in recent years; indeed, there seems to be a regrettable polarization of viewpoints towards either strong commitment or sweeping criticism. In this chapter I have tried to explore some of the underlying philosophical questions concerning mathematical education at both school and university levels, bearing in mind the force of Eddington's moral.

1.2 *Are purple ducks red herrings?* To begin at the beginning: once upon a time there were three ducks swimming in line down a river. Leading was a purple duck, followed by a green one and finally a red one. All at once and all together each duck turned round in its own water and swam off in the opposite direction. The red duck was now in the front and the purple one at the back.

The purpose of this story is to convey the idea that sets can be ordered in different ways: in particular, that when the ducks turn round and swim in the other direction, they exhibit the inverse of the original ordering of a set. It is an example of the type of 'New Mathematics' being offered nowadays at the primary level; indeed, it comes (in an illustrated form) from a book I saw a few years ago in a primary school.

Mathematics of this type has been characterized recently by [Hammersley 1968] as 'soft intellectual trash' (although his detailed remarks are directed mainly at the university level). While accepting in general the force of his criticism, I would question the use of the little word 'soft'. It refers to the trivialization of mathematical ideas by excessively simple examples, but this estimation concerns mathematics as a body of knowledge. From the point of view of teaching and learning the subject, especially for young children with little general experience, the use of such examples seems to me to make the mathematics *excessively hard* for those to whom it is directed.

The difficulty is conceptual, and arises from the very ordinariness of the example that may seem to be its virtue. There are many ways of thinking about ducks swimming on the water, and the one that interprets them as a set of objects in the world susceptible of a variety of different orderings is about the *last* of which anyone, of any age, would conceive. The fact that such examples *do* belong to most children's experiences only serves to stress the additional fact that they never interpret them in this abstract and peculiar fashion. Thus mathematics becomes associated for them with idiosyncrasy, with looking at ordinary things in a funny way.

To grasp the purpose and generality of this interpretation requires precisely those qualities of *maturity* and of *self-analysis* in the learning process that children usually lack. With their natural tendency to fancy, they are all too likely to chase completely different but more accessible aspects that are in fact quite irrelevant, such as the fact that they have never seen ducks in these colours or (if they have noticed it) that when ducks turn round they swim in an arc following their leader so that the ordering in the reverse direction is the same as it was before.

In other words, the game and story approach to mathematics has dangers and limitations. Between ordered set and purple duck there are many red herrings; each of them can lead to confusion and irrelevance, and each is likely to be followed, especially by children of originality and enthusiasm who might eventually qualify for Eddington's approval. For example, Jean Piaget's very original researches [1952] into the child's conception of number exhibit many of these red herrings and show how children pursue them. He and others have demonstrated, time and again, that our daily-life theories of number, substance, weight, mass, force, time and so on are in fact extremely subtle and complex theories of invariants; the categories with respect to which these artefacts are dependent or independent are numerous, and children constantly make mistakes about them. The prophets of the New Mathematics could take a note of caution from Piaget's work, instead of rushing past the sorts of difficulty that he has studied so carefully.

Set theory does not normally fill the pages of the New Mathematics books; indeed, often the page count is small. But, along with algebraic structures, it is very important in embodying a kind of goal towards which mathematics learning should approach. It has achieved this status in education in broad imitation of its progress in research.

Let us turn now to the more general questions that have arisen.

2. Methodological issues for mathematical education

This section is largely philosophical and methodological, and its general points are not necessarily restricted to mathematical education. It is a difficult part, and I would have tried to simplify it were it not clear to me that in a number of respects (especially in passing over problems of language) it was already over-simplified. It is also very abstract, and may seem remote from practical teaching; yet the classroom situation is methodologically very subtle and complicated, involving many matters other than the subject material being taught. I feel that teachers should try at least to recognize the character of these seemingly abstract ideas if they are to obtain genuine insights into their teaching situation.

2.1 *Justificationism and fallibilism.* Whether explicitly or implicitly, teachers usually believe in the solidity of foundations and the correctness of the knowledge built upon them. For centuries this spirit was represented by a dogmatic belief in Euclidean geometry. There one found the apotheosis of mathematics, the paradigm of the Immutable Fact, the tree of secure knowledge, the cathedral of rigour to which all other parishes must look for inspiration and enlightenment.

Now one of the principal claims of the New Mathematics is that it has swept this type of thinking away. Nobody believes in the primacy of geometry any more, or in the teaching of artificial presentations of knowledge: heuristic motivation, straight to the heart of mathematical thinking, is the chief aim. My own view is that the old Euclideanism has been replaced by a new one of a different type. Instead of the Immutable Fact, we have the Immutable Structure; the tree of secure knowledge has become the tower of basic artefacts; the church of rigour has been taken over by a Corporation of Fundamental Concepts, whose products we all need in order to think. The foundations lie no more in the truths of Euclidean definitions, but in the 'generality and simplicity' (to use a phrase of Georg Cantor, the chief founder of set theory) of these fundamental components.

The sense in which this is a new Euclideanism needs to be described carefully, for it lies not in mathematics itself but in methodology. Perhaps the most immediate way to describe it is to consider the status and role of foundational knowledge, for it is here that the most pressing issues concerning the New Mathematics lie.

Among the most pervasive assumptions in all human thought are doctrines that assert the need for certainty, for things to believe in. The nuggets of truth have been sought everywhere: in personal commitments, religions, sense data, language and many other sources. Basically the doctrines fall into two categories; the rationalists, who hope for the attainment of certainties through reason (though not necessarily formalized deductive reason); and the irrationalists, usually acting in reaction against the failure of rationalist programmes, who search their souls and their psyches for the ultimates.

All these doctrines have been characterized by Karl Popper as 'justificationist' [1963, introduction]. The common feature is their interpretation of some category of knowledge as authoritative over all (or at least several) other categories. But their crucial failing lies in the need for a guarantee that the authoritative category really is authoritative; for this guarantee must also be authoritative and therefore requires its own guarantee; and thus an infinite regress is instituted.

Popper's solution to this difficulty is to urge a fallibilist methodology [1963, introduction and ch. 10]. Here any sort of knowledge, foundational or otherwise, is 'acceptable', but only on a tentative basis. The prime criterion for the assessment of a theory is its success and failure to solve problems, and criticisms of it are directed at its capacity to do so. The problem above of the infinite regress is avoided, since the regress is not allowed to start in the first place.

The fallibilist view of foundations, therefore, is that one digs down to them rather than builds up from them; they are set in mud rather than concrete. More precisely, there are no foundations to knowledge unless there is technical knowledge requiring foundations in the first place. Thus at any stage the technical knowledge will be informal wherever it reaches beyond currently known foundations. Conversely, any foundational format will lag behind the progress of the technicalities built upon it (although new foundations may suggest some types of technicality not previously explored), and sometimes to such an extent that completely fresh foundations will have to be designed to cope as well as possible with the latest technical developments.

Thus there is *a double growth of knowledge*, a simultaneous development of both technical and foundational knowledge, with each acting upon the other. The growth leads to theories at different depths, both of foundations and of technicalities built upon them, and the passage from one depth to another is often more important than the depth either left or attained.

The distinction between fallibilism and justificationism rests not only on the knowledge itself, but also on certain interpretations of it; on whether or not some part of it can be authoritative, for example. The point is important, for it raises the question of the status of methodological knowledge. At first sight it appears to be metatheoretic, in that it is a theory that talks about theories. But then we notice that there are metatheories that not only are not methodological in this sense but in fact have methodological features of their own for example (to be discussed in more detail below), educational metatheories of learning mathematical theories.

2.2 *Theories and their sub-structure.* By a *category* I refer to some general area of knowledge whose form and limits are not clearly specified. Within each category there are various *theories*, each of which is a version of the category. A theory may take various different *presentations*, of which the following types are most pertinent to the present context:

An *epistemological presentation*, where the theory is laid out simply as a set of statements from basic principles. In mathematics, axiomatic formulations are examples, although informal theories have them also.

An *historical presentation*, in which is described the historical development of a theory through various phases. It may include elements from the recent past or even the present day. There are two main aspects, the creative and the doctrinal. The creative aspect deals with the historical discovery or invention of the various stages of the theory by the persons involved, while the doctrinal aspect describes its influence or lack of influence as a doctrine on the scientific community.

A *heuristic presentation*, in which the theory is laid out with the aid of motivations or examples from familiar contexts. It is especially relevant to educational problems, for the teaching and learning of theories are usually done via their heuristic presentations.

These three types are not exclusive; in particular, heuristic presentations are quite likely to contain elements of the epistemological and historical presentations. But the difference between the latter two is often very marked, especially in foundational theories. For since in the historical process one digs downwards to successively deeper foundations, then at any stage the epistemological presentation will be approximately the reverse of the historical, since it starts with the ideas most recently produced and then proceeds through a sequence of results that were discovered in chronologically reversed order.

Of course this picture is oversimplified; stages of foundational development do not often embrace their predecessors so neatly (and technical developments still less so). But the force of the distinction is still significant, and it provokes the question of the position of heuristic presentations relative to history and epistemology. Briefly, a heuristic presentation will be in between the two; but to be successful *it must surely imitate history more than epistemology*. For, like history, heurism is fallible; while epistemology, though fallible also, is justificationist in appearance.

Presentations concern only the layout of a theory; its possible commitments are the concern of its features. The words 'philosophical' and 'methodological' are sometimes used to describe features, but they have been avoided here as they are already in use in much more general senses. Instead we have the following distinction: *ontological features* of a theory are those of its components and techniques with which it is asserted to be ultimately concerned, while *procedural features* of a theory are those components and techniques that are used simply as convenient devices; no ontological commitment to them is involved.

Thus features are quite different from presentations; in fact, they are connected with epistemological standpoints concerning the theory as well as with the theory itself. Any presentation can contain both ontological and procedural features, while any feature can occur in more than one presentation. Thus a theory is viewed as a collection of presentations together with a specification of ontological and procedural features. Let us take a few examples of these distinctions.

2.3 *The foundational crisis in mathematics around 1900* showed that the whole edifice of mathematical knowledge was in danger — or did it? Much of the relevant writing of the time showed that attitude; for example, when Gottlob Frege learnt of Bertrand Russell's paradox in 1902 he wrote back to say that 'not only the foundations of my arithmetic, but also the sole possible foundations of arithmetic, seem to vanish' [van Heijenoort 1967, 127-128]. This is a typical justificationist reaction; the concrete has given way. Yet in fact it is only this particular version, this theory, of arithmetic that is so imperilled; if we move back our foundational commitment so as to avoid using set theory, then we will have a different version of arithmetic that is safe — or at least, that is not subject to those paradoxes. A conspicuous feature of the mathematics of that time is that most mathematicians took no part in the foundational discussions at all, and proceeded quite successfully with their own researches at a less deep level. Further, the very fact that the foundational studies produced several alternative programmes suggests that none of them was the Solution, in the justificationist spirit; rather, that each led to a mathematical system in which (hopefully) the difficulties were avoided.

Thus the history of set theory, far from recounting the discovery of certain truth, is in fact an excellent example of the tentative growth of knowledge from one collection of shaky foundations to another. It also shows that, after a long period of successful progress in a particular line, the foundations obtained are 'simple' only with respect to an epistemological presentation, 'simple' in the sense of 'primitive'; from the point of view of heuristic understanding, they are exceedingly difficult in that their superiority over other versions is almost impossible to appreciate without a considerable acquaintance with the intermediate versions.

This is my original criticism of the New Mathematics; that the purple duck can be a red herring since the ideas that it is used to exemplify are so remote and abstract. In terms of our distinctions, one could say that the mathematics involved is unsuitable for educational purposes, since it uses abstract and conceptually sophisticated versions of general categories of knowledge, whose significance is not conveyable to untutored minds. Nevertheless, it is urged upon the young because educators do not distinguish sufficiently between epistemology and heurism and therefore convert the prima facie justificationist layout of an epistemological presentation into an explicitly justificationist heuristic presentation, complete with exact definitions, basic concepts and so on. In other words, heurism imitates epistemology much too closely and usually does so in a justificationist spirit.

Earlier, I referred to the New Mathematics as a 'New Euclideanism'. Nothing could highlight the analogy better than the fact that these criticisms apply equally well to the traditional doctrine of Euclidean geometry. Here again we have a sophisticated and complicated theory, in this case assembled by Euclid from previously known informal theories into an epistemological presentation. The presentation is justificationist; the definitions are essentialist, the axioms (which include the underlying logic as well as the geometrical assumptions) are immutable, and the theorems flow from them impeccably.

Of course the traditional teaching method followed the epistemological layout; thus the limitations of the Euclidean axiomatic system were not stressed. Yet they are among the most enlightening aspects to teach. For the axioms are largely irrelevant to the ordinary study of geometrical figures, not only because they are inessential to the production of geometry but also because in any case they are often incapable of validating the topological aspects of diagrams, such as points being inside or outside figures. Such properties become provable only in a version such as David Hilbert [1899]; his version of geometry is deeper than Euclid's.

These various distinctions apply not only to mathematical theories but also to educational theories, to which we now turn.

2.4 *Possibilities for theories of learning.* There is a fundamental new difficulty in theories of learning mathematics, which may be expressed as follows. Let us suppose that a teacher has a theory of how a student learns some mathematical theory. Then the mathematics being learnt by the student is at the object-theoretic level; the student's own understanding of the mathematics is metatheoretic, for it is his theory about the object theory; and the teacher's theory of learning is metametatheoretic, since it is a theory about the student's metatheory about the mathematics being learnt.

Thus there are three interacting levels of theory in the structure. The metatheories will proliferate, for there will be a separate one for each presentation of a theory; for example, learning an epistemological axiomatic presentation of set theory is different from learning its history. Indeed, such metatheories may consider concurrently several different presentations of the same theory, or even different versions of the same category.

Theories at each level have their own presentations and features; but they will basically be different in kind from each other. If the object theory is set theory, say, then it refers to sets and their properties; the child learning it at primary school will construct theories about purple ducks or whatever as well as about set theory; and the teacher will be considering, say, the child's general level of intelligence as well as ducks and set theory.

Similarly, the ontological and procedural features of the theories at each level will be independent of the features of the theories at other levels, even when there are features in common between them. It is possible — but not necessary — to have the same ontology for the mathematics learnt as well as for the method of learning it and the theory of learning involved. For example, one may have simultaneously psychologistic commitments to set theory, to learning set theory, and to theories of learning. But it is also possible to have some objective ontology for set theory, an empiricist theory of learning set theory and a psychologistic theory of learning.

All social problems seem to have this type of hierarchical structure. Take the simplest possible situation of two friends discussing a problem together. Each one has to deal not only with the problem, which is at the object-theoretic level, but also with his own metatheoretic understanding of it and a metametatheoretic appreciation of his friend's understanding of it. Or take any kind of sociological theory. One factor that it should take into account is itself and its practitioners as a social activity; this is the social generalization of the previous example and is also at the metametatheoretic level. So it is not surprising that in the social sciences, progress has been much slighter than in the natural sciences; the methodological situation is intrinsically much more complicated.

One consequence of this comparatively primitive state of affairs in the social sciences is that the various distinctions discussed earlier tend to be less marked. In particular, the epistemological presentation of a theory is often much the same as the historical presentation, because there have been many fewer stages of development. However, the interpretation of features as ontological or procedural is a more prominent question. For example, statistical techniques are widely used in the social sciences, and the question arises (as it does also in parts of the natural sciences, such as quantum mechanics) whether such theories are really — that is, ontologically — statistical, or whether they are just procedural devices for representing, or approximating to, situations that are really founded in other items.

2.5 *On Piaget.* In the context of mathematical education, distinctions such as these seem to be missed in, for example, Piaget's judgements on the relationship between his discoveries on learning and the mathematical theories being learnt. People such as Russell are criticized because they have 'forcibly separated logistic investigation from psychological analysis, whereas each should be a support for the other in the same way as mathematics and experimental physics' [1952, ix]. Hence Russell's formulation of arithmetic fails for a number of reasons. His definition of the equality of the cardinality of two classes in terms of a one-one correspondence between their members is 'too simple' because it omits the 'qualitative correspondence between the two classes' [Piaget 1952, 182-183]. There are

> various kinds of qualitative correspondence, which depends on the spatial position of the elements and have no numerical significance. [... Hence] number is not a 'class of classes', but the result of a new operation brought in from outside, which is not contained in the logic of classes as such. In fact, this 'quantifying' correspondence is only achieved by disregarding all the attributes in question, i.e. by disregarding the classes.
>
> In order to transform classes F and F_1 into numbers, the first essential condition is that their terms shall be regarded as equivalent from all points of view simultaneously.

Now all these remarks concern perception and experience, and the analysis of sense data in terms of sets and their membership. In other words, the mathematics involved is treated in terms of metatheoretic understanding of heuristic presentations. But the Russell-Frege version of arithmetic referred to is an epistemological presentation; it may have difficulties, but not from heuristic considerations of this type. Even Piaget's valid criticism [1952, 183-184] on the necessity for objects to be distinct is not correctly used. The difficulty here lies in claiming this necessity to be 'logicist'; but it arises within the epistemological presentation. The heuristic or empirical assertion of difference in terms of colours or shapes concerns only the heuristic presentation. Piaget's analogy with mathematics and experimental physics is inexact, since the latter does not concern the different presentations of a theory but the interpretation of a mathematical theory into a physical theory or vice versa.

The writings of educational psychologists often show conflations of different versions of the category, or of different presentations of the same version, or of the same interpretation forced upon different theories. A significant general instance is the confusion between a concept and a conception. It is marked, for example, in the important work of Z. P. Dienes, who has extended the work of pioneers such as Piaget to develop programmes of mathematical education. He often discusses 'concept formation' at length (for example, [Dienes 1964a, ch. 7; 1964b, ch. 1]), but then describes the formation of conceptions of heuristic presentations of mathematical theories.

Concepts are alleged to be the basic notions of theories. The claim for their importance is an offshoot of essentialism: they are, in fact, the entities whose essences are described by exact definitions. In fact children form conceptions, that is, metatheories of understanding (or misunderstanding) of some theory. This theory may be at any level, and the conception of it is one level higher. Hence one may have a conception of some theory of arithmetic, say; one may also have a conception of a theory of learning arithmetic: and so on. Conceptions are quite different in kind from concepts; concepts are the (alleged) atoms of the theory, while conceptions are metatheories of understanding that theory.

A form in which this type of confusion arises is the claim for the New Mathematics that 'children think in sets'. The argument of which this is the conclusion presumably proceeds as follows:
1. Set theory plays a fundamental role in mathematics:
2. Sets are among the basic concepts and ontological features of mathematics;
3. Sets are also among the basic concepts of learning mathematics, and are an ontological feature of learning theories;
4. Hence children think in sets.

Each of the alleged deductions can be challenged: in particular, the moves from the second to the third, and from the third to the fourth statements seem dubious. Like the rest of us, children think neither in sets nor in anything else; they think about problems and theories, and form conceptions of them, that will vary from child to child and may or may not be set-theoretic. The world of theories and ideas is as accessible to them as to anyone else; thus the more deliberately it is represented to them in this spirit, the more successful they will be in trying to enter it.

3. On university mathematical education

3.1 *On possible roles for the history of mathematics.* Throughout the historical element could be significant, though not at all dominant. The teaching would be centred on selected problem areas, of which at least some would be extensive enough to transcend fossilized 'subjects' and could incorporate related branches of science in the course of their solution. These problem areas would be taken through several phases or stages of solution, noting geneses of each from its predecessors as well as from external influences. In some cases it should be possible to reach contemporary problems, or at least comparatively recent work whose structure is not yet settled.

The foundations would deliberately be held at a 'working level' relative to new technical results, whose consequences then entail foundational modifications. Rather than attempting to imitate an epistemological presentation from abstruse foundations to prodigious technicalities, one would start from somewhere 'in the middle' and then move in both directions concurrently. Exhaustive detail would never be attempted; the student would be given sufficient content to grasp the structure and leading points of the subject matter and enable him to look up further details as demands and his later interests dictate.

This methodological procedure, far from being left implicit, would be brought out from time to time quite explicitly in order to convey ideas such as tentativity, double growth of knowledge, and theories at different depths or levels of rigour.

3.2 *Students* tend to be keener to learn about the history of mathematics than teachers are to give it to them; and I can easily understand the teachers' reservation. There is simply not enough usable literature available, least of all concerning the developments of the past two hundred years that are most closely related to currently taught Subjects. Instead, there are books that claim to cover the whole or large parts of the history of mathematics but omit almost every development after 1800, which is where much of modern mathematics — including that on the under- and post-graduate curricula — has its origins (However, a recent shining exception is [Kline 1972].[1]) The result is that mathematicians tend to regard historical work as dilettantish activity for failed mathematicians or real mathematicians when they are old, but not man's work; good for a few anecdotes and 'first-time' identifications, perhaps, but not for anything else. One may not subscribe to this estimation, but it is easy to see how it has arisen.

In addition to such arguments, there are purely methodological reasons for the limited role for history. For if one were to follow only historical progress, then one will finish only with historical questions. Therefore the historical role must be restricted.

Further, one of the reasons that students seek an historical element in their education is that they believe that it will make the subject 'easier' to learn. I do not subscribe to this view; on the contrary, in terms of the

---

[1] {In the nearly 40 years to have passed since this passage was written the interest in the history of mathematics since 1800 has increased considerably, as has its use in mathematics education. The main motivation seems to have come from mathematicians' dissatisfaction with their learning or teaching experiences rather than any general philosophical standpoint, Popperian or otherwise (§13; *ROL,* §3, §9).}

demands made on students, I am sure that an historical approach would be much harder. Under current teaching systems, a student can often survive even if he has learnt the subject matter in only a rote manner; but if confronted with the original problems and the techniques, he will have to spend hours trying to reconstruct the situation from unfamiliar ideas. If a textbook on some branch of mathematics were written following an historical line, then it would be the hardest book on the market.

But the rewards for such efforts are very much greater. Here is the real world of ideas, seen in genesis, ripening and decay, rather than some artificial mock-up with the problem background removed. This is the sense in which learning is 'easier'; a personal sense in which the student relives creative work and imitates the individual creative discovery of the results.

Here we find the obvious but vital factor in favour of historical elements. The fact is that, whatever mathematics we want to learn, it has historical roots and motivations and therefore important and quite possibly essential features caused by its historical background. In this sense, it is more than 'interesting' to know of the differential calculus that 'dy/dx' started life with G. W. Leibniz as the ratio of differentials dy and dx, or that sorting out A.-L. Cauchy's "false" theorem of the 1820s in mathematical analysis required raising the level of rigour of the subject in mid century by distinguishing between various different modes of convergence of infinite series (§9.2.1), or that Cantor was motivated to create set theory in the 1870s by innovating point set topology in the course of studying trigonometric series and thereby raising the level of rigour still higher [Kline 1972, chs. 17, 40]; it is part of the genuine learning of mathematics to know such things [ROL, §13].

3.3 *Attitudes among mathematicians.* Put another way, the distinction between 'mathematics' and 'history of mathematics' is false in principle: *there are only mathematical problems, and they have a history*. Yet in practice the division is regarded as very sharp. The history of mathematics is taught (if at all) in its own lecture course as if it were a separate branch of mathematics instead of an essential part of every branch. Historians tend to work clandestinely at their subject, officially employed to do something else, since few mathematics departments ever appoint staff on the prime strength of their historical work. Indeed, it is well known that in some departments history should not be mentioned apart from honorific pieties to the past, and that the meaning of life is to grind out a few more lemmas, improved inequalities or minor generalizations to add to the trees of mathematical knowledge, rather than to use history to obtain some conception of the forest that the trees comprise. The underlying principle is that, since we definitely know more in mathematics than we did even a few years ago, then the past is dead and can tell us nothing.

This is why the historical element in research and teaching in mathematics (and in science in general) is often slight; it is a penalty of the *success of science*, that because scientific knowledge grows quickly, the past can soon be abandoned. The only sign of a compensating effect is that today the scientific empire is so vast that no one can understand more than fragments of it, and so some scientists find extra nourishment in the historical developments of their own field and hence are a little more interested in history than used to be the case.

But to most mathematicians their subject is a two-dimensional canvas, lacking in three-dimensional perspective. They know well enough that knowledge progressed in the past, but they do not grasp the significance, never 'mind the content, of the progress achieved. To them their predecessors knew so little relative to the present state of knowledge that the differences between the various historical stages are not important. I have heard mathematicians jump backwards and forwards by decades while discussing the development of a subject as if they were passing over adjacent instants in the history of thought.

But some of the Great Men of mathematics knew better. One of the striking features of the mathematical literature is the extent to which some major mathematicians of the past were genuinely interested in the history of mathematics relative to their own time. Their writings sometimes contain quite extensive references to, and even discussions of, the achievements of their own predecessors; and it is commonly the case that the collected or selected works of a Great Man were edited by one or several of his Great successors. If one wants to learn from the past, then this is one of the things to learn; that the Great Men were interested in history, not because they felt obliged to be interested or because it seemed to be a 'good thing to do at the time', but because they knew that they would learn things from history that otherwise were inaccessible. They were not interested in the history of All Mathematics, but in bits of it; usually starting from the history of those branches in which they themselves worked, they proceeded either more deeply into these branches or more broadly into neighbouring branches, or both. They seem to have known their history better than most of their lesser contemporaries; this may well be one reason why they were greater.

So history is there to play an active role in current teaching and research. However, it is not a tool to be used indiscriminately; the Great Men, for example, were often not very competent at handling the specifically historical problems involved. One of them must serve as exemplification: the problem of the use of posterior knowledge. {In the terms of ¶13, mathematicians normally study the heritage of the mathematics of the past rather than its history, which seems to be alien to their training.} Structure, abstraction, generality, rigour — these are the spirit of much mathematical research and degree teaching today, and they infiltrate the attitude of the graduate, whether or not he intends to pursue research thereafter. By contrast, historiography of the kind described above seeks out ambiguity, unclarity, obscurity and sloppiness, and tries to preserve them.

In a phrase, the history of mathematics contains history as well as mathematics, and both elements are vital. It is doing both together that makes it man's work.

4. On school mathematical education

I am not a school-teacher, and I would not dream of telling teachers how to teach their subject. But the historical and philosophical considerations here may be of some interest or help to them, and might even provide fresh arguments to corroborate or criticize their views.

4.1 *Subjects.* Group theory is another prominent part of the New Mathematics. Epistemologically it is an uninterpreted axiomatic structure, with models (sometimes for special types) throughout mathematics. But historically the development was once again the reverse [Wussing 1969], with the subject arising in its models and only a very gradual progress during the nineteenth century to the pure axiomatic form. Once more, if the subject must be studied at early stages this historical process might be imitated more carefully with final abstractions perhaps omitted entirely. It is certainly a more welcome feature of the New Mathematics than the interest in set theory, for it sprung from several different branches of mathematics rather than from a microscopic analysis of one particular part.

Again, one area in which reform has not been as great as one might hope is in elementary arithmetic, in which many young children face their first crises. Rather than continue to use the standard methods, it would be far more desirable to teach some of the powerful and elegant algorithms in which the Eastern mathematicians were so proficient [*ROL*, §12.5]

The decline in the teaching of geometry is especially to be regretted [Thom 1971]. The traditional Euclideanism gave geometry a bad name; but when used heuristically it is a perfect example of mathematics that contains easy empirical reference, subtle structural features and numerous opportunities for cross-reference to other branches of mathematics (§13).

4.2 *History-satire.* Turning to more general matters, one conclusion that I would draw from the above discussions is that in school education the content is not so important; rather, the methodological spirit in which the content is presented can be crucial. For example, some historical element could be fairly significant not only at the university level but also at the later stages of school education, especially if the syllabuses included 'sorties' into some of the larger-scale problem areas that would be done in more detail later. But at earlier stages, and especially in primary education, history is pretty useless. Apart from the inevitable technicalities involved, children have little or no sense of historical progress, least of all in scientific subjects that they associate with immediate things. Nevertheless, historical developments could still be imitated even if the historical contexts are not mentioned. Then, as at the university level, the creative element of science could also be imitated in that the results achieved would be creative for the pupil himself. I use the term 'history-satire' to represent this suggestion.

4.3 *Broad education?* The need for history-satire is all the more urgent when one remembers Eddington's moral (¶1.1): that especially at these early stages *the mathematics is being learnt by budding non-mathematicians,* poised *not* to launch themselves on their mathematical careers. For these children the fundamental concepts on which mathematics is alleged to be based would seem to be especially inappropriate. Yet they are stressed to the exclusion, or at least expense, of more familiar ideas.

The justificationist doctrine on which this attitude is based is an example of a more general principle that I shall call 'the myth of broad education'. The myth claims that education is intended to lay down the broad base of knowledge ready for all applications. The assumption is that the conceptual and structural basis of the knowledge imparted is broad; the possibility is implicitly rejected that children need an elastic and broad methodological basis, drawn from contexts that perhaps are conceptually and structurally limited.

For all learning has important methodological as well as factual and structural aspects; indeed, they may often be the most valuable. Of course it is necessary to know particular facts and rules, theorems and formulae, and to comprehend the structure of theories, but it is equally essential and sometimes more illuminating to detect the methodological aspects involved, even if only implicitly; that a theory solves problems only after a fashion, that a certain type of problem is not reducible to some other type, and so on. This is surely what the old cliché about 'learning from experience' is about; one uses the methodological aspects of knowledge already learnt to solve problems of a quite different character. Such a process starts at the beginning of life, and the more deliberately it is used in school education the more chance it has to develop and mature.

It is this sense in which the methodological spirit of learning at early stages is as important as the content. Indeed, it seems to me to be important enough for the content to be chosen with its latent methodological richness prominently in mind. By contrast, any imitation of epistemology is of limited value; the goal is already achieved, as it were, without indication of why the goal should be chosen or how one might discover it.

5. On education in general

5.1 *On consequences for syllabus construction.* One of the most disturbing practical consequences of justficationism and the myth of broad education is that it leads to impossibly ambitious intentions throughout education, at all levels. All syllabuses are far too full, and their authors cannot see why the content need not be so great, since they assume in some form the aims that the myth dictates. One result is that at present students are tending to be less and less competent at basic techniques, due to the lack of opportunity to practise them. Further, the modern prevailing prejudice in favour of mathematical structures despises both practical techniques and methodological discussions anyway; such things are unnecessary once the structures are grasped.

Another consequence is that to many students mathematics is only 'clever nonsense'. Obviously clever, it is also nonsensical in that it comes from nowhere and leads to little that matters or that cannot be obtained by simpler means, and is full of boring verifications of results that the epistemology-imitating presentations predestine. The problem of classroom boredom, already important and likely to increase still more with the raising of the school leaving age, can surely do without such encouragement.

An absolute reduction in the quantity of mathematics taught seems highly desirable; in particular, any theory containing comprehensive foundational implications should be avoided until well into the university level. Then pupils with whom Eddington's moral is concerned will have a chance of achieving a self-satisfying degree of personal learning. Meanwhile those of original ability who would earn Eddington's approval will be advancing beyond these confines of their own accord. They are the type of problem child, severely abled, who receives all too little official attention nowadays and for whom any educational process is rather a nuisance. Yet for them too such an alternative teaching programme would be more appropriate; it would help them to flower into brilliant human beings rather than into brilliant robots.

5.2 *Coda.* There is a cynical view that asserts that education is damage. It has a corollary for our times that more education is therefore more damage. Cynical though this view may be, it contains a grain of truth that is significant enough to act as a constant warning to teachers.

Of course, education is enrichment also; and perhaps the most important treasure to instil into the young is a deep and methodological lesson for life: *not to be afraid of being uncertain.*

Bibliography

Dienes, Z. P. 1964a. *The power of mathematics*, London: Hutchinson.

Dienes, Z. P. 1964b. *Mathematics in primary school*, Melbourne: Macmillan.

Hammersley, J. M. 1968. 'On the enfeeblement of mathematical skills [...]', *Bulletin of the Institute of Mathematics and it Applications, 4,* 66-85 [with supplements in later issues].

Hilbert, D. 1899. *Die Grundlagen der Geometrie*, 1st ed., Leipzig: Teubner. [Many later eds. English trans: *Foundations of geometry*, 1st ed., Chicago: Open Court, 1902; later eds.]

Kline, M. 1952. *Mathematics in Western culture*, New York: Oxford University Press.

Kline, M. 1972. *Mathematical thought from ancient to modern times,* New York: Oxford University Press.

Koestler, A. 1966. *The act of creation*, London: Hutchinson.

Piaget, J. 1952. *The child's conception of number* (trans. C. Gattegno and F. M. Hodgson), London: Routledge and Kegan Paul.

Popper, K. R. 1963. *Conjectures and refutations*, London: Routledge and Kegan Paul.

Thom, R. 1971. 'Modern mathematics: an educational and philosophic error?', *American scientist, 59,* 695-699.

van Heijenoort, J. 1967. (Ed.), *From Frege to Gödel* [...], Cambridge, Mass.: Harvard University Press.

Wussing, H. 1969. *Die Genesis des abstrakten Gruppenbegriffes*, Berlin: VEB Deutscher Verlag der Wissenschaften. [English trans.: *The genesis of the abstract group concept,* Cambridge, Mass.: The MIT Press, 1984.]

PART 4

SPECULATIONS

I shall not commit the fashionable stupidity
of regarding everything I cannot explain as a fraud.

C. G. Jung, addition to a lecture
given to the Society for Psychical Research in 1919
(*Collected works*, volume 8, London: Routledge and Kegan Paul, 1960, 317)

15 Decline of the Philosophical Spirit

This little swipe at positivism, long the preferred philosophy of most scientists, reviews the position adopted by the physicist, mathematician and philosopher André-Marie Ampère (1775-1836), in the days before positivism began its rise — which was due in n notable part to some of his former students.

Ampère's book *Théorie mathématique des phénomènes électro-dynamiques* (1826) is still worth reading. It is known as the principal founding source of electrodynamics, but other features are just as instructive. It begins with an extensive homage to 'Newtonian philosophy', and continues with a long mixture of physical theory, mathematical analyses and reports of experimental procedures [Grattan-Guinness 1990, ch. 14]. Ampère sought not merely methods of explaining and calculating effects; he wanted to find out how the phenomena actually occurred [Blondel 1982]. Indeed, in the naive tradition sometimes followed at the time, he thought of his theory as a truth, 'uniquement déduite de l'expérience', to complete the title of his book.

Ampère is remembered now only for his work on electrodynamics, but in fact it was a small part of his output [Hofmann 1995]. He was a polymath, whose activities were unified by his philosophical spirit. This spirit informed all his writings and came to its zenith in his *Essai sur la philosophie des sciences* (1834). But he was an outsider in philosophical thought, for the 1830s also saw the start of the rise of positivism in the hands of some of his former students at the *Ecole Polytechnique*: especially Auguste Comte, with a general philosophy of 'positivism' (his word) from the late 1820s onwards [Pickering 1993]; and engineer-scientists such as Charles Dupin and J. V. Poncelet [Grattan-Guinness 1990, ch. 16]. Associated closely at that time with educational and social causes, positivism became one of the dominant philosophies of the 19th century and has maintained its influence, directly and indirectly, until today. Knowledge without metaphysics; rejection of abstract intellectual objects; even some lack of attention to the ways in which mathematics is used in physics. It is a strange contrast to read Ampère's *Essai*, with its Kantian concern with phenomena and their causes, with man's knowledge and his cognitive power to know.

It was through movements such as positivism that philosophy and science became separated. Positivism and its cousins (mechanism, materialism, instrumentalism, behaviourism, and so on) do not solve

---

This passing thought was first published anonymously in *Wireless world*, (July 1981), 29; the references have been added.

philosophical problems so much as ignore them [Brush 1983]. Yet scientists accept positivist tenets without much thought: facts are facts are facts; theories are useful only for predicting new facts; mathematics is just a fiction that in principle has nothing to do with physical reality; the aim of science is consensus (as a noted Fellow of the Royal Society contentedly put it on television recently); the history of science is bunk — and, above all, philosophising about science is time-wasting nonsense. At the same time much of philosophy itself has become an enclosed profession, concerned with footling 'puzzles' in ordinary language; its practitioners rarely know anything beyond the writings of their professional colleagues. There are exceptional figures in the communities of both science and philosophy; but they stand out as such, often nervously.

Meanwhile the real world seems to have remained the same as it was in Ampère's day, especially with regard to the phenomena studied in physics. Thus the objects of scientific study remain basically unchanged, and so docs the need for philosophical as well as technical skill. We now know far more about the technicalities of electricity and magnetism than did Ampère and his contemporaries; but we no longer bring to our theoretical studies the sensitivity to philosophical questions that Ampère, and others of his time, could show. He and his contemporaries were not really scientists in the way that we understand the term; they often called themselves 'natural philosophers', enquirers into the nature around them and into the powers of mankind to think up theories about it. They may have fallen into optimistic naiveties such as the allegedly immediate deduction of theories from facts; but they did not succumb to our reflex dismissals of the non-experiental and our inattention to the place of mathematics in scientific knowledge. The imperatives that informed not only Ampère but also his contemporaries such as Michael Faraday and G. S. Ohm, and successors like Lord Kelvin and James Clerk Maxwell, have faded; the traditions of natural philosophy have long been broken; reflection has given way to 'research'.

Bibliography

Blondel, C. 1982. *A.-M. Ampère et la création de l'électrodynamique*, Paris: Bibliothèque Nationale.

Brush, S. G. 1983. 'Negativism sesquicentennial', in N. Rescher (ed.), *The limits of lawfulness. Studies on the scope and nature of scientific knowledge*, Lanham, MD: University Press of America, 3-22.

Grattan-Guinness, I. 1990. *Convolutions in French mathematics, 1800-1840. From the calculus and mechanics to mathematical analysis and mathematical physics*, 3 vols., Basel: Birkhäuser; Berlin: Deutscher Verlag der Wissenschaften

Hofmann, J. R. 1995. *André-Marie Ampère. Enlightenment and electrodynamics*, Cambridge: Cambridge University Press.

Pickering, M. 1993. *Auguste Comte: an intellectual biography*, vol. 1,. Cambridge: Cambridge University Press.

16 Psychical Research versus the Established Sciences

Psychical research has made important connections with several established sciences, and its respectability has improved somewhat in recent decades, but it still suffers from the extremes of credulous support and equally credulous hostility. Why is this so? What kinds of phenomena does it treat anyway?

> On psy[chical phenomena] I feel confident that such things exist, but the attempt to make them 'scientific' seems to me not only unsuccessful so far, but to lead to a travesty. [...]. I find 'ghost' boring, and their scientific supervision time-consuming; I am not going to waste my time on them: physics is far more interesting. Everybody must make a choice, and it is no good telling us that it is our 'duty' to get involved in 'ghosts'. But good luck to those who make another choice.
> Karl Popper, letter to me of 16 February 1984;
> compare [Popper and Eccles 1977, 117, 146]

> Frank speculation is often the basis for constructive theorising and experimentation. However, the informal dissemination of speculative ideas has been impeded by enormous growth and differentiation in all scientific fields. Recognising this, SPECULATIONS IN SCIENCE AND TECHNOLOGY was launched by the editor in 1978 to provide all scientists with a forum for their speculative papers [...]
> Note: No papers on extrasensory perception or unidentified flying objects will be considered.
> Publicity literature (Elsevier Sequoia Publishers, 1979)

1. Brief

The principal kinds of phenomena that are studied in psychical research include extra-sensory perception, both of other minds (telepathy) and of events or objects (clairvoyance); mental effects upon physical objects, such as fork-bending (psychokinesis); anticipation of the future events (precognition); analysis of the activities of physical and mental

---

This chapter is based upon a chapter in my (ed.), *Psychical research. A guide to its history, principles and practices*, Wellingborough: Aquarian Press, 1982, 344-358; a few references, and the quotation from Popper, have been added. Most of ¶3 is new. In addition, ¶5, and the quotations from A. N. Whitehead, come from my 'Extrasensory perception and its methodological pitfalls', *Methodology and science, 12* (1979), 17-32.

mediums; out-of-the-body experiences; apparitions and ghosts; survival after death; and psychic healing. The critical reader will sprinkle the word 'alleged' around this list.

The question of the non-scientific character of the field could lead to literally dozens of problems about the acceptance of theories, belief systems, and so on. I confine myself to the following five questions, which between them cover quite a lot of useful ground:

What is it about a science that makes it established? (¶2);

What is the rational view to take of anomalous phenomena? (¶3-¶4);

Is psychical research a subject? (¶5);

Should we form theories only in terms of things that we experience? (¶6-¶7);

Why does psychical research sometimes arouse such vehement opposition? (¶8).

Other comparisons of (some) psychical research with various sciences include [Smythies 1967] and [Various 1980]; much more briefly, [Grattan-Guinness 1979], from which ¶7 below comes.

2. Some properties of established sciences

Today we are used to recognizing science (including medicine and technology) as a large-scale professional activity; but this situation has developed only since the early 19th century. Prior to that there were communities of scientists (for example, in astronomy and medicine); but even then they were relatively few in number, and depended for support for their work on the whims, of kings, protectors and (occasionally) governments. The situation in higher education was similarly patchy: in particular, universities were often of poor quality (for example, in the German states) or little interested in science (for example, in England)

The establishment of science, especially as a social factor, really developed in the 19th century. Initially in France after the Revolution, and then in other countries (especially England, Scotland and the German-speaking states), it became possible for a scientist to make his living as a scientist from university or college appointments, industrial or governmental offices, state-supported projects, or possibly a mixture of them all. The increase in scientific research and education between 1800 and 1850 alone is very remarkable; and the further increase over the rest of the century saw in particular the introduction in several countries of many societies and journals for specific sciences.

Hence began the 'enormous growth and differentiation in all scientific fields' that was bemoaned in the second quotation at the head of this chapter, and which of course has continued ever since. The establishment of the sciences themselves was intimately linked with their professionalization. The collection of professions was increased by that of the scientist — a word introduced, incidentally, in the 1830s by William Whewell. Science was to be left to the scientists, just as the law was left to the lawyers. The scientist was trained in his subject at a university (or some equivalent institution), he pursued his studies to obtain his qualifications, and then continued his research as part of his career. Maybe he became an authority in some field, both by his publications (including, quite possibly, textbooks) and by the research students who worked under him and carried his approach over to the institutions where their careers in turn were passed.

An important credential for the status of authority, or at least of professional competence, was knowledge of the 'craft skills' (theoretical, experimental, or whatever) that applied to his particular science at the time in question. Learnt at university, they would be refined and developed during his career, these modifications being introduced into his own later university teaching [Ravetz 1971].

In such a way is 'normal science' practised, where the research follows broad outlines already determined [Kuhn 1970; §9.3]. When critical times arise, the normal science is challenged by observations or experimental results that just do not seem to fit with its predictions. During these times occur major changes, sometimes called 'scientific revolutions' (though the term is over-used). After a rather hectic period a new normal science replaces the old one, and the process of learning craft skills, and so on, continues as before but with different content.

Some comments on this process are worth making. Firstly, a science sometimes gets so over-established as to be rigid in conception; it does not respond to innovations even from its professional members. Secondly, the close link between the establishment of a science and the professionalization of its scientists has entailed both that amateur involvement in science has much decreased and also that phenomena that are not amenable to professional treatment — for example, to successful laboratory work — have been ignored. Thirdly, it has led philosophical concern in science somewhat away from its traditional interest with ontology (that is, what there actually is in the world) and towards consensus (in particular, how scientists to accept or reject theories, what their belief structures and ideologies are, and so on) (§15).

While psychical research draws on techniques, and has organizations, it does not exemplify well the processes described above. For example, as with all immature sciences [Ravetz 1971, ch. 14], it is not clear what its basic principles are, never mind how they should be learnt and developed. Hence the field cannot be associated with a profession and so cannot become an established science. Therefore it is not scientific. Therefore its phenomena are, at best, anomalies.

{3. The predicament of the experients

One major difficulty about psychical phenomena is that the experients themselves have little or no understanding of their own roles. A fine example is the American psychic and magician Joseph Dunninger, with whom I spent several hours in 1979. As a young man he helped make some of the artefacts used by Harry Houdini (for example, fake footcuffs, which I examined), and incorporated some of them in his own act. However, as he assured me with force, he seemed also to have psychic powers that he could use but did not understand. For example, he performed the "trick" of writing down a number on a piece of paper and then asking three members of his audience each to shout out a number between 1 and 100, to find that their sum was the number already written on his paper.

Similarly, even people who hold physical sittings regularly do not understand their own experiences. I have sat with the Scole group in Britain [Solomon and Solomon 1999] and the SORRAT group in the USA [Richards 1982], and conducted long-term distant experiments with the latter involving the apparent extraction of the contents from sealed envelopes and the impression of clear and easily identifiable pictures onto print and slide films without the use of a camera [Grattan-Guinness 1999]. I also seem to be coincidence prone, and have some dowsing ability; but as a healing channel using the Reiki method, I am much more miss than hit. In all these situations, I have no idea what was/is happening on all occasions; in particular, I confess my failure to detect the necessarily attendant frauds.}

4. How should one react to an anomaly?

Obscurantism is the refusal to speculate freely on the limitations of traditional methods. It is more than that: it is the negation of the importance of such speculation, the insistence on incidental dangers. A few generations ago [...] large sections of the clergy were the standing examples of obscurantism. Today their place has been taken by scientists.

A. N. Whitehead [1929, 34-35]

Reactions to anomalies are important factors in science, especially when major changes are in the offing. Well-established theories become so much a part of a scientist's make-up that giving them up would be like losing part of himself (as well, possibly, as involving loss of professional face). In any case, there are all kinds of possibility to account for the mismatch between theory and observation from which the anomaly has arisen: an incompleteness of the theory that can be filled, an error in its development that can be rectified, instrumental faults, experimental gaffes, impure materials, and so on. Critics of psychical research often appeal to such reasons as explanations of the alleged psychical phenomenon. Better to remain cautious, they say, in the interpretation of such evidence. Popper's position, quoted at the head of this chapter, is of this kind.

When one considers the 'logic of caution' (as it may be called), a surprising conclusion turns up, not only for psychical research but also for science in general. *The cautious position accepts that the anomaly exists, rather than rejects it*. For acceptance entails no particular kind of explanation; one of those just listed may turn out to apply, for example. But to deny the existence of the anomaly is bold, even reckless; it rejects a potentially infinite possible range of explanations, even before some have been put forward.

This danger is well exemplified by C. E. M. Hansel. He published a book [1966] with the title *ESP: a scientific evaluation*, with a new edition [1980] entitled *ESP and parapsychology: a critical re-evaluation*. The older title was preferable, as it indicated more clearly the limited range of phenomena covered; for example, he never discusses psychic healing at all, and in his "exposure" of mediums he omits D. D. Home, the greatest to date. Nevertheless, he still feels able to summarise his account of psychical research between 1880 and 1920 as follows: 'The first forty years of psychical research produced nothing that could be regarded as scientific evidence for supernatural processes. It was in the main, a history of fraud, imposture, and crass stupidity' [1980, 73]. If such a statement is meant to be taken seriously, then, by the point made above, every single case has to be *authoritatively* denied. Although Hansel's book provides a useful survey, for newcomers to psychical research, of the types of fraud and experimental gaffe that might be committed, there seems little benefit, for sceptics and adherents alike, to gain from adopting such an extreme view (see the review [Honorton 1981] on this and other points).

Some supporters, and many critics, of psychical research have shared a common hard line against fraud. The Society for Psychical Research has had a tendency to reject all evidence from a psychic if *some* of it turned out to be fraudulent, thus ignoring the argument that since repeatability is so hard to achieve in the field, there is no reason to assume that fraud is always repeated. Among the critics, Hansel follows the same approach in his [1966] and [1980], and otherwise contents himself with construction to show that fraud *could* have been committed in the cases discussed; a good example is thoughtography [1980, 171-184]. If all scientific work were treated this way, then science would disintegrate rather quickly into a collection of scientists rejecting all evidence except their own. This is not a trivial possibility to consider, especially in the life sciences (including psychology, Hansel's own field) where exact repeatability is often impossible or hard to achieve.

A preferable approach is to say: established science is here, anomalous psychical phenomena are over there. Are the phenomena to be brought over here (as the critics require), or is science to modify itself in order to get over there?

5. Is psychical research a subject?

Often people discuss the 'subject' of psychical research. The word 'subject' is used as a convenience, but it may bring difficulties in its trail if it is taken seriously. For it is not obvious that psychical research is a subject in the sense in which (say) astronomy or pharmacy are; it is not even clear that it is a federation of subjects, like physics or medicine.

For psychical research studies psychical phenomena, which are categorised *negatively*, not positively; they are the phenomena that cannot be explained in terms of physics or biology or medicine or ... . But then it need *not* be expected that they will out to be explicable under some *one* new discipline. Rather than concern over questions like 'What is psychical research?', the subject (or non-subject!) should build upon its classification into different branches: of course, some all-embracing theory may emerge later, after a few 'scientific revolutions' have taken place in the interim. As it is, the statuses of some branches as orthodox or paranormal is not clear, or at least have "improved" over time; for example, dowsing, hypnosis and acupuncture.

So is the following ingenious repeated experiment within or without science? During his career as a physics school-teacher, a friend of mine would take a class into his preparation laboratory, and test their powers of co-ordination by making each pupil hold a metal rod in a fixed position in each hand while also carefully walking along a jagged crack line on the concrete floor. At one point some of his pupils would fail to keep the rods in the required position. Afterwards he would take the class into the school yard and show them that that point lay directly above the water inflow pipe of the school, and thus that they had been taking part in a dowsing experiment without knowing it!

6. The problem of interpretation: senses of the word 'existence'

Scientific study of perception is bringing to light an increasingly wide *and varied* collection of sensory means (see, for example, [Wynne-Edwards 1962] on animal perception). Hence it is increasingly likely that different kinds of extra-sensory perception (ESP) will be explicable in terms of different parts of this collection. (Note, incidentally, how these developments make talk of 'the five senses' so outdated; and yet it is still frequently uttered.) But let us suppose that we claim that ESP does not exist. What does this mean? There are various possible answers, and the differences between them affect fundamentally the standpoint adopted.

To say that any phenomena 'exist' or occur could have the following meanings:

OBS: that phenomena occur that are detectable by direct human experience and/or by instrumental registration;

THE: OBS together with the claim that there is a theoretical means of describing at least some of the phenomena and of predicting conditions under which they will occur again;

ONT: THE together with the claim that the theoretical descriptions describe how these phenomena actually occur (that is, things as they actually are, whatever that is).

There is an ascending ladder of philosophical commitments here, and it is essential to make clear which commitment one has chosen. For example, certain kinds of instrumentalism and positivism accept the construction of THE-type theories but regard as illegitimate both the affirmation and the denial of ONT. By contrast, Karl Popper points out that it is possible to achieve ONT in science because one might think up an ontologically correct theory by luck, but that one can never prove that one has been so fortunate as to find such a theory [1963, ch. 3]

Claims both for and against the 'existence' of ESP are often incoherent, for they do not specify the sense of 'existence' that is under discussion. Personally I believe that enough reliable evidence has been collected to assert existence in its OBS sense, but that, despite so much effort, no explanation has yet been offered whose basic concepts can confidently be said to exist in its THE sense. As regards the ONT sense, concerning ontology, the situation is the same as in established science; but that situation involves the following possible confusion of theory construction and ontology.

Scientists often speak of their 'discoveries', in the sense of ONT-type discoveries of ontological fact; but, in fact, most of their activity concerns their decisions about the structure of the world, decisions based on theories that are at the THE sense of existence. Their (claimed) discoveries are involved relatively rarely; for they can only discover the ontologically correct, which, as was pointed out above, cannot be proved to have happened even if it has. (This is what scientists must mean when they say that the truth is unknowable.) Similarly, they also talk of 'facts' as objects on the same ONT level as discoveries; but facts are actually highly theory-laden constructions, on a par with the THE sense of existence (§2.5.1).

This confusion of THE and ONT levels is particularly prominent in the widespread habit in science of appealing to 'Ockham's razor', which requires the elimination of superfluous assumptions from scientific theories. Now this can be quite legitimate for assessing theories at the THE level, but unfortunately it is often carried over into ONT contexts. This is quite illegitimate, since there is no proof that God uses Ockham's razor; explanations for ontology do not have to be preferred on the grounds of intellectual parsimony of assumption.

ESP has been hit very hard by such confusions. Since (some of) the phenomena are explicable without involving extra-sensory factors, then such explanations are preferable, and so 'ESP does not exist'. For example, magicians can effect some kinds of alleged thought transference; therefore anyone who claims to be able to transmit thought is a magician. Q.E.D., by courtesy of Ockham's razor. Unfortunately, the question of ONT-type existence of these phenomena is thereby not even posed.

7. Should theories in psychical research be kickable?
In one sense, Science and Philosophy are merely different aspects of one great enterprise of the human mind.
A. N. Whitehead [1933, 179]

Late Victorian thought in science and philosophy was much occupied with 'psychic science': the higher spirits, manifesting the Creator and his will (§17). It is not surprising that the Society for Psychical Research was founded in Britain then, or that various physicists were interested in its activities. But by and large professionalized science has been opposed to approaches of this kind, especially in France and all English-speaking countries (where they largely died out in the 1900s). The normal view is that theories should be formulated in terms of ideas and concepts that one could directly experience by some means. The term 'kickable' is sometimes used to describe, rather graphically, this desirable property.

One philosophical position of this kind is called 'positivism'. Encouraged by various French engineers in the 1820s and systematized by Auguste Comte (§15), it took 'knowledge without metaphysics' as its slogan and exercised a strong influence through the century in Europe. A broadly similar view was held in the mid 19th century by J. S. Mill, who tried to argue for an empirical basis for all our knowledge, even the laws of logic. He strongly influenced the German psychologist Wilhelm Wundt, who trained up generations of psychologists (including many Americans) to regard theories as inductions (*à la* Mill), built up by associating mental states with situations in the world. A related standpoint, called 'behaviourism', lays the emphasis on people's patterns of behaviour, rather than on their mental states. Although such positions are not completely free from unkickable components, they are removed from the apparent requirements of theories of psychical phenomena; however, compare §18.12.

For psychical research seems to be out of reach of the kickable; certainly it has been criticized on these grounds. It is interesting that some psychical researchers today try to meet this criticism by putting forward the so-called 'observational theories' according to which ESP reduces to psychokinesis. An important type is retroactive psychokinesis, where the psychical phenomenon occurs because the psychic has been told the results of his efforts to date — but possibly long after the events have occurred, in which case he affects their outcome retroactively. (On these theories, see [Millar 1978] and §18.13; and for criticism and discussion, [Braude 1979].) This theory tries to speak only of events and their observation, and so avoid appeal to unkickable forces, fields or radiations. However, it does make strong demands on credulity to think that the psychic will be able to cause a phenomenon after it has happened, by being told that it has happened.

Today scientists seem to be less inclined to desire kickability in their theories than before. Psychical research can claim only a small part of this influence; the reasons lie more within advances in established sciences themselves, and also on the growing realization that the unkickable has a habit of creeping back anyway, particularly for the following reason.

Science is often thought to explain the unknown in terms of the known. This view has some chronological accuracy, for we may transfer a theory from its known areas of relevance to others hitherto regarded as irrelevant (and thus unknown there). But it is meant to describe theories themselves; and there it fails, for the basic entities and concepts of a theory are unknown, by definition. For example, I may explain the fall of bodies by Newton's laws; but it is the bodies that are the known, whereas the basic notions of Newton's laws (force, mass, and so on) are primitive, or else defined in terms of other primitives, which themselves are necessarily unknown. Prejudice in favour of the kickable has made scientists (and many philosophers, too) overlook this point, and assign kickability to ideas where it is not easily found. Force is a good example; some commentators on Newtonian mechanics in the 18th century were puzzled about what sort of "object" force might be, and tried to eliminate it by *defining* it as mass multiplied by acceleration [Fraser 1985].

8. A speculation

Doubtless psychical research has had more than its share of fraud and gaffe; but this surely does not explain the *vehemence*, often emotional in character, which the "subject" arouses. Thus we must look for other factors. Here is a speculation on the question.

During the last 30 years there has been a violent and even emotional controversy over the views of Immanuel Velikovsky that cosmic disturbances have affected the Earth, and indeed the planetary system, within the time of man's residence here. Not being an expert in any of the sciences or the humanities with which his view is involved, I shall not discuss them. But the similarity in character between this controversy and some of those that attend psychical research struck me as so strong that I sought chance for a personal discussion of the matter with Velikovsky. As a result, shortly before his death in 1979 we had two long conversations, of which the relevant parts can be summarised as follows.

Velikovsky did not expect any instant acceptance of his revolutionary views; but he was surprised both at the intensity of the opposition, especially the vendetta planned against his publisher by some scientists, and the rejection of his work by those who were proud to announce that they had not read it [Bauer 1984]. He was also puzzled that the row continued for so long afterwards. The situation interested him professionally, since he had been a practising psychoanalyst in Palestine before concentrating on these studies in 1940. He felt that the reactions of his critics, both in general and even in some details, closely resembled that of hysterical amnesia that he had witnessed in his consulting rooms; for example, the simultaneous horror of discussing the matter and the unavoidable desire to do so. He therefore formulated the following hypothesis. Mankind did indeed suffer the events that he (Velikovsky) has described; they were of truly catastrophic proportions, involving the destruction of sizeable portions of the then existent race; these events entered the "forbidden" part of the race-memory of mankind; and thus the reaction against his work was stimulated by the involvement of this sensitive part of the memory. He wrote a manuscript on mankind's treatment of catastrophe, published as [Velikovsky 1982].

No alleged psychical phenomena are involved in either his theories or the reaction that they excite; but one does not have to be a Velikovskian to know that this planet has been through some rough times, or that mankind has lived in the greatest peril from flood, starvation, wild beasts, and so on, even when the Earth itself has been stable. Thus the speculation arises that, Velikovskian catastrophes or no, at least some psychical abilities developed as *survival mechanisms* for primitive man: psychic healing, telepathy and out-of-the-body experiences would have been particularly useful. As man's control over his environment has increased, so the problems that these abilities were designed to solve have declined in importance. Hence the abilities have become "rusty", as it were, and therefore are now so hard for us to revive (especially when applied to such boring tasks as perceiving shapes in sealed envelopes or bending forks). Further, the very act of reviving them brings recall, by association, of the great dangers that stimulated their development — ancient memories that we want to suppress.

'We want to know in science', Velikovsky said to me, 'but we don't want to know too much'. I shall not be sending this speculation to *Speculations in science and technology*. Unfortunately it seems to be unfalsifiable anyway.

Bibliography

Bauer, H. H. 1984. *Beyond Velikovsky: the history of a public controversy*, Chicago: University of Illinois Press.
Braude, S. 1979. *ESP and psychokinesis: a philosophical examination*, Philadelphia: Temple University Press.
Fraser, C. G. 1985. 'd'Alembert's principle: the original formulation and application in Jean d'Alembert's Traité de dynamique (1743)', *Centaurus, 28,* 31-61, 145-159.
Grattan-Guinness, I. 1979. 'Extra-sensory perception and its methodological pitfalls', *Methodology and science, 12,* 17-32.
Grattan-Guinness, I. 1999. 'Real communication? Report on a SORRAT letter-writing experiment', *Journal of scientific exploration, 13,* 231-256.
Hansel, C. E. M. 1966. *ESP: a scientific evaluation*, Buffalo: Prometheus.
Hansel, C. E. M. 1980. *ESP and parapsychology: a critical re-evaluation*, Buffalo: Prometheus.
Honorton, C. 1981. Review of [Hansel 1980], *Journal of the Society for Psychical Research, 75,* 155-161.
Kuhn, T. S. 1970. *The structure of scientific revolutions*, 2nd ed., Chicago: University of Chicago Press.
Millar, B. 1978. 'The observational theories: a primer', *Journal of parapsychology, 2,* 304-332.
Popper, K. R. 1963. *Conjectures and refutations*, London: Routledge and Kegan Paul.
Popper, K. R. and Eccles, J. C. 1977. *The self and its brain*, Berlin: Springer.
Ravetz, J. R. 1971. *Scientific knowledge and its social problems*, Oxford: Clarendon Press.
Richards, J. T. 1982. *SORRAT. A history of the Neihardt psychokenesis experiments,* Metuchen (NJ) and London: The Scarecrow Press.
Smythies, J. R. 1967. (Ed.) *Science and ESP*, London: Routledge and Kegan Paul.
Solomon, G, and J. 1999. *The Scole experiment*, rev. ed., Waltham Abbey: Campion Press.
Various 1980. Discussion of the status of psychical research, *Zetetic scholar*, no. 6 (July 1980), 1-121.
Velikovsky, I. 1982. *Mankind in amnesia*, New York: Doubleday.
Whitehead, A. N. 1929. *The function of reason*, Princeton: Princeton University Press.
Whitehead, A. N. 1933. *Adventures of ideas,* Cambridge: Cambridge University Press.
Wynne-Edwards, V. C. 1962. *Animal dispersion in relation to social behaviour*, London: Oliver and Boyd.

17 Psychical Research and Parapsychology: Notes on the Development of Two Disciplines

This chapter, based upon a lecture delivered in 1982 to the centenary conference of the Society for Psychical Research, surveys the general historical context in which the Society was founded in 1882. Then it records some of the initiatives taken especially in the USA in the 1920s and 1930s to convert the field into the more scientific activity of 'parapsychology'.

> Is there any room for a speculative philosophy distinct from Psychology and Ethics, and all particular sciences?
> Cambridge University Moral Sciences Tripos, examination paper on psychology and metaphysics, 25 April 1882

1. Introduction

The names of the societies that we celebrate at this conference — the Society for Psychical Research (SPR) and the Parapsychological Association — reflect a change of terms that has occurred over the last 50 years. In this chapter I wish to comment on the early days of psychical research 100 years ago and of parapsychology (using the term in its narrower sense, referring to the experimental and technical parts of psychical research) of about 50 years ago, when the Rhines were beginning to publish their results on card tests. While much excellent historical work has been written on both developments, rather less has been undertaken on the more general philosophical milieux within which they occurred.

This chapter is a tentative exploration of these more general questions. In places I shall draw upon recent work in the history of science, and thus try to draw this discipline to your attention.[1] But my primary purpose is not historical nostalgia, even though we also celebrate in 1982 the centenary of the Inspectorate of Ancient Monuments; I shall suggest at the end that these events of yore seem still speak to us today about the

---

This article was first published in W. G. Roll, J. Beloff and R. A. White (eds.), *Research in parapsychology 1982. Jubilee centenary issue. Abstracts and papers from the Combined Twenty-fifth Annual Convention of the Parapsychological Association and the Centenary Conference of the Society for Psychical Research*, Metuchen, N.J. and London: The Scarecrow Press, 1983, 283-304; a few historical references have been added. The quotation in ¶4 replaces one already used in §2.9.

[1] However, this is not the place for sophisticated analysis of the implications and details of the historical record, nor of all the attendant philosophical niceties. A few of these escutcheons are provided in these footnotes. Specialists in the history and philosophy involved are requested to treat with indulgence the over-simplifications that are inevitably committed.

current state of psychical research and parapsychology, including their relationship to each other.

2. Psychical research and the spirit of inquiry

> The word *spiritual* is one of the most difficult and important terms in nineteenth-century thought.
>
> Frank M. Turner [1974, 3]

2.1 The founding of the SPR. In 1876 the physicist William F. Barrett submitted a paper to the British Association for the Advancement of Science (BAAS). Entitled 'On some phenomena associated with abnormal conditions of mind', it contained a discussion of theoretical and experimental work on hypnotized subjects. There was a row; while he was able to read the paper, only the title was published in the *Reports* of the meeting [Barrett 1877]. The controversy lasted for several weeks in the press [Barrett 1924b], with distinguished objectors (W. B. Carpenter, W. F. Donkin, Ray Lankester) facing equally distinguished supporters (William Crookes, William Huggins, A. Pitt-Rivers, the 3rd Lord Rayleigh, A. R. Wallace).

I suspect that the refusal of publication was fortunate for the development of psychical research. It was largely on Barrett's initiative that a two-day conference on 'Spiritualism and psychological research' was held in the London rooms of the British National Association of Spiritualists on 5 and 6 January 1882. This event led to the formation of the SPR later in the year as a forum that (among other things) would *publish* research in this area; indeed, the bulk of Barrett's 1876 paper appeared in the *Proceedings* [Barrett 1883].[2] Further, the SPR's immediate decision to organize six committees to investigate particular classes of phenomena, a move that doubtless helped it to succeed where earlier organizations had failed, recalls the BAAS's policy of maintaining standing committees on particular subjects.

The description of these committees, which comprises the opening contribution to the SPR *Proceedings*, includes an often-quoted statement of policy:

> The aim of the Society will be to approach these various problems without prejudice or prepossession of any kind, and in the same spirit of exact and unimpassioned inquiry which has enabled Science to solve so many problems once not less obscure nor less hotly debated.

The word 'spirit' is striking; its use was undoubtedly deliberate, for it would have immediately rung its overtones on the educated minds of the time. There was not only the interest in Spiritualism, shared by Barrett and

---

[2] Barrett reprinted his invitation to the conference of January 1882 in [1924a]; 27 letters concerning it are held in the SPR Archives, Barrett mss., file A2. He tried again with the BAAS in 1883 with a paper on telepathy, which was rejected (A4, letter 11). File A4 contains a large group of letters to Barrett, including 35 from Myers, 15 from Mrs. Sidgwick, and four or more from Gurney, Lang, Richet, Romanes (see footnote 14 below) and Wallace. {These Archives are now held in Cambridge University Library.}

many others (for some it took the form of membership in Spiritualist churches); there was the general concern with mental processes, and the prominence of psychology, widely conceived, in the philosophical and scientific thought of the time.

2.2 Reforms at Cambridge. An important stimulus at Cambridge for this concern had been the reform of the educational system from the 1820s, led by such men as Adam Sedgwick and William Whewell [Garland 1980]. Their position was informed by a philosophy called 'intuitionism' (sometimes 'nativism'), according to which man had the capacity to form fundamental ideas transcending experience. It was the task of education to 'train the mind' by exercising this capacity. Three subjects came to receive special emphasis: logic (for the forms of thought), mathematics (for the matter of thought), and classics (for contact with the wisdom of the past). The movement was strongly anti-utilitarian, and indeed marked a revolt against Benthamism. It claimed a unity to all knowledge. It showed a strong religious colour and attempted a marriage of the truths of science, which were to be attained by inductions from facts, with the revelation of God; for example, 'natural theology' was the study of the display of Divine will in the animal kingdom [Brooke 1977, Yeo 1979]. However, religion itself was conceived of broadly, indeed as the 'Broad church', with the universities being disestablished from the Church of England.

The system of education was called 'liberal' by its proponents, although liberalism did not extend to the questioning of its own principles. Nevertheless, substantial reforms occurred at Cambridge between 1860 and 1882; Henry Sidgwick was a prominent figure in the new system [Rothblatt 1968]. The new moves included the abolishment (in 1871) of the need to affirm Christian belief to obtain university membership; a strengthening of the standards of teaching, still very low for many undergraduates; the abolition of the requirement of celibacy for Fellows; the furthering of education of women (moves that originally helped to bring Henry and Eleanor Sidgwick together); some reduction in the autonomy of the colleges; increases in the importance and prestige of the University of professorships; and the questioning of the unity of knowledge by emphasis on the variety of disciplines, including the introduction of new Triposes. The plurality of disciplines was quite a notable feature of science worldwide, and one manifestation is the creation at that time of many societies, usually national ones, for individual disciplines; thus the creation of the SPR in 1882 is typical in that respect.

Although these changes were quite substantial, several standpoints of the past were maintained, especially the prominent position of psychology and logic. The new Moral Sciences Tripos, for which Sidgwick was sometimes an examiner, included a paper on 'Psychology and metaphysics'.[3] The journal *Mind* was created in 1876 as a 'quarterly review of

---

[3] The examiners for the 1882 paper quoted at the head of this chapter were John Venn, A. T. Lyttleton, J. B. Mayor and H. S. Foxwell [Cambridge University 1883, 231]. In other words, Sidgwick happened not to be examining that year. It is

psychology and philosophy', and included notices of the SPR's publications fairly regularly.[4] And many intellectuals still affirmed Christian belief in some form or another, including Spiritualism, as we shall shortly see.

2.3 Some prominent figures. Thus psychical research, defined by the SPR in its 'Objects' as a study of 'that large group of debatable phenomena designated by such terms as mesmeric, psychical, and Spiritualistic', fits into the general ambience of English intellectual thought. For example, logic was often held to be a normative discipline concerned with how we ought to think, while psychology was a descriptive study of how in fact we do think. Thus, given the further link between psychology and psychical research, it is not surprising to find that several major logicians were sympathetic to psychical research: George Boole, with his mathematical psychology of the 'laws of thought' [Grattan-Guinness 1982] and whose wife both propagated his ideas after his death and became a founding Council member of the SPR;[5] Augustus De Morgan, who contributed a long preface to his wife's book *From matter to spirit*, one of the first extended studies of the test of mediums [De Morgan 1863]; and Lewis Carroll (C. L. Dodgson) and John Venn, both SPR members. Similarly, the interest of physicists, such as Barrett, Crookes, Oliver Lodge, Rayleigh, Balfour Stewart and J. J. Thomson [Wynne 1979], is not surprising in a climate of thought that granted theological significance to the aether(s) permeating Victorian physics [Cantor 1981]. Further, it is worth noting that a prime research area of these physicists (including much of Rayleigh's joint work with his sister-in-law, Mrs. Sidgwick), was electricity and magnetism.

Two figures of this period are notable for their attempt to weld psychical research into the fabric of their other scientific interests. One was the physicist Sir Oliver Lodge, who throughout his long career both studied the property of the (alleged) aether and saw it as the means of linking the physical and the psychical worlds, including psychical phenomena such as ESP and survival [Wilson 1971]. He claimed that 'It is in the interaction of aether and matter that the problems of psychology must find their solution.

---

amusing to wonder if the examiners set the quoted question mindful of Sidgwick's recent activities.

In an interesting detail of history, a first class was awarded for that paper to W. R. Sorley, who was to be Sidgwick's successor in the professorship of moral philosophy at Cambridge in 1900.

[4] See against the SPR in the *Mind* indexes in volumes 16 (1891) and new series 12 (1903). In addition, Gurney and Myers wrote in the journal on specific topics in psychical research.

An interesting example of the place of 'psychology' in the proliferating disciplines is F. Y. Edgeworth's study of mathematical economics. Firstly, his book on the subject was entitled *Mathematical psychics* (1881). Secondly, it was partly based on Sidgwick's criticism [1879] of Mill's wages-fund theory and also on his theory of ethics. Later he helped the authors of *Phantasms of the living* with the discussion of chance [Gurney, Myers and Podmore 1886, ch. 13].

[5] Mrs. Boole resigned at once, as she was the only woman on the Council; the next woman to join the Council, in 1901, was Mrs. Sidgwlck [Nicol 1972, 344-345].

Matter only serves as an index or pointer, demonstrating the unseen activity all around; and chief among the unseen activities are the aetheric agencies guided and controlled by Life and Mind and Spirit' [Lodge 1933, 260-261]. His concern with the means of paranormal communication, over and above any (presumed) powers of the mind and brain, has left him an overly neglected theorist of psychical phenomena.[6]

The other figure was Alfred Russel Wallace. He published his ideas on evolution in the 1850s, and held a view broadly similar to Darwin's on its processes. However, in the late 1860s two connected changes took place: he felt that natural selection was not able to explain many features of human beings (for example, the hairless skin and the power of speech); and he became converted to Spiritualism through witness of psychical phenomena [Kottler 1974, Smith 1972]. Interested in phrenology from the 1840s, he now posited the existence of a higher consciousness that served as the dynamic cause of both psychical phenomena and of those faculties of humans beyond the ken of natural selection. His enthusiasm for séances contrasts markedly with Darwin's extreme scepticism (recorded in [Pearson 1924, 62-67]).

3. Warfare between science and religion?
It is possible to combine a practically complete trust in the procedure and results of empirical science, with a profound distrust in the procedure and conclusions [...] of Empirical Philosophy.
Henry Sidgwick [1882, 543]

3.1 *The place of naturalism.* Several of the developments just described involve the relationship between science and religion, especially Christianity. The warfare between these two great regiments of thoughts is one of the best-known features of the intellectual life of the late Victorian period.

Fortunately for our understanding, recent historical research has revealed that the battle reports are largely propaganda, and that a much more complicated situation obtained. James Moore's *The Post-Darwinian controversies* [1979] is a particularly valuable source since it includes a history of the propagandising itself as well as an exceptional bibliography; and Frank M. Turner's *Between science and religion* [1974] is especially useful for the history of psychical research, as he studies in detail the positions of F. W. H. Myers, Sidgwick and Wallace among others.

---

[6] On the somewhat similar case of Crookes, see [Palfreman 1976]. For quotations of differing views on psychical research by various physicists see [Wilson 1971].

Positivists like to assert that Einstein proved that there was no aether. Einstein himself made no such claim. His special theory is an hypothesis about space-time in which the aether is not needed: his general theory is a hypothesis of a broader kind in which the status of the aether is ambiguous, as the work of Lodge and others on the (albeit controversial) topic of aether relativity has shown [Lodge 1926]. For an excellent survey of the place of the aether(s) in many areas of science, medicine and theology from 1740 to 1900, see [Cantor and Hodge 1981].

Those men, while not necessarily accepting Christianity, all reacted, in some way or another, to a range of views called 'naturalism', which were advocated in the second half of the 19th century as a philosophical underpinning for the advancement of science, Briefly, naturalism extolled Nature over God, facts over belief, reason over faith, and matter over spirit. It allowed for the ordering of ideas by the mind and their verification (if correct) by the facts. Thus its underlying psychology was associationist, usually a psychophysical parallelism.

The word 'mechanistic' was often used in naturalism: the importance of mechanical principles among the laws of nature (the law of the conservation of energy, for example), natural selection as the mechanism of evolution, the need for a mechanistic theory of the mind, and so on. Since some advocates of naturalism knew little of the science of mechanics, their use of 'mechanistic' was not always clear.

Reductionism played some role in the naturalistic framework, on at least two fronts: the desire to reduce the role of metaphysical questions, and the tendency away from Christian dogma. However, there were important differences between naturalists over the extent to which reductionism was to be taken; for example, some advocated positivism while others criticized it, and both atheists and agnostics fell within its scope. It is best understood as a *spectrum* of positions, with agreement over the basic standpoints described above [Turner 1974, ch. 2]. A brief but illuminating survey of its principles is provided in an article by Sidgwick's former student James Ward who, like his teacher, rejected naturalism as an adequate philosophy of science [Ward 1911].

Prominent figures in naturalism included Alexander Bain (the founder of *Mind*), T. H. Huxley, Herbert Spencer and John Tyndall. A significant background figure was John Stuart Mill, for his efforts to reduce all knowledge to empirically obtained data and inductions built upon them.[7] Mill held such a view even of logic, in contrast to the logicians mentioned in ¶2.3 above [Richards 1980]. But to mention Mill's philosophy recalls another important area of controversy of the time, namely, whether science actually did proceed by induction from facts or by the use of hypotheses. Some of the disputes over Darwinism were at this level, which of course is a philosophical matter, not a specifically religious one.

3.2 *The place of religion.* Indeed, the relationship between science and religion was far more complicated during this period than the warfare thesis allows. For example, many theologians accepted Darwinism as an explanation of the processes studied in natural theology [Moore (J.) 1979, part 3], while some were willing to see scripture studied by 'scientific' principles (the collection of *Essays and reviews* in 1859 was an important case). Another significant religious initiative was the encyclical 'Aeterni

---

[7] This is a good example of the complications of the period. Mill and Whewell both believed in grounding science in induction, and thought that truths would be so obtainable. However, they differed over the existence of mental states and the role of cognition [Butts 1968].

Patris', issued by Pope Leo XIII in 1879, which led to a development of relations between science and Thomism.[8]

Conversely, scientists also took initiatives to reconcile their profession with Christian faith. For example, there was the 'Scientists' declaration' of 1864-1865, offered as a complement to *Essays and reviews* [Brock and Macleod 1976]. Again, while the BAAS barely stomached Barrett's paper on psychical phenomena in 1876, it had been horrified two years earlier when Tyndall, its president for the year (and previously Barrett's chief at the Royal Institution), delivered a long address on religion and science in which he supported the latter much more than the former [Tyndall 1875]. Though he claimed in the ensuing uproar that he had had Catholicism more in mind than religion in general (he was the son of an Orange man, and the BAAS met in Belfast that year), the anger of the scientists gathered there was barely settled. In particular, in the following year two Scottish physicists, Balfour Stewart and P. G. Tait, published *The unseen universe: or physical speculations on a future state*, a work that rapidly went through several editions. They sought to refute Tyndall's materialism by claiming a role for theology in (for example) accounting for the origin of the universe [Heimann 1972]. Stewart was cautiously intrigued by psychical phenomena when approached by Barrett about the conference of January 1882;[9] he joined the SPR and indeed became its president for 1885-1887, between the two periods of Sidgwick's presidency.

Thus, while there was a marked decline in the proportion of clergymen in the "professional" science class from the 1860s onwards [Turner 1978], elements of reconciliation, or at least coexistence, between science and Christianity were considerable. Journals such as *Nature*, launched in 1869, originally made great efforts to appeal to professional and lay scientists [Roos 1981]. Organizations included the Metaphysical Society, a discussion club that, during the course of its history from 1869 to 1880, included atheists, agnostics, freethinkers, Deists, Catholics, Anglicans, positivists and naturalists, and so had Huxley, Tyndall and Sidgwick among its members [Brown 1947]. Miracles and the supernatural were prominent topics for papers; the quotation from Sidgwick at the head of this section comes from one of his contributions; and Huxley invented the word 'agnosticism' with the help of discussions there. The Society's chief

---

[8] On this encyclical see [Dauben 1979, 140-142] and [Paul 1979, 179-182]. It would be interesting to see what influence, if any, this movement had on psychical research in Europe, where Catholicism was a prominent faith. Another movement of that time, which carried most weight in the USA, was Christian Science.

[9] Barrett (see footnote 2 above), file A2, letter 19. *The unseen universe* was published anonymously; Stewart and Tait admitted authorship in 1876.

It seems slightly surprising that Scottish physicists, and indeed Scots in general, played little role in the early decades of the SPR, since the 'commonsense' philosophy that many of them followed gave some prominence both to mental processes and to theorizing by analogy [Olson 1975]. Perhaps Barrett was unlucky in reading his 1876 paper at the BAAS meeting in Glasgow ... .

founder was James Knowles, who later launched the journal *The nineteenth century*, which published some extracts of SPR reports.[10]

3.3 *The place of Spiritualism*. Seen against the alleged background of warfare between science and religion, the disagreements between the SPR and the Spiritualists appear to be a typical battle. However, with the more complex and subtle historical picture now emerging, other possibilities are available [Nicol 1972]. One difference, of course, concerned the fact that the SPR had no corporate opinion while the Spiritualists adopted a definite view on survival;[11] another concerned 'the great scandal', as Mrs. Sidgwick put it, of 'the encouragement [that Spiritualism] gives to the immoral trade of fraudulent mediumship' [1911, 708]. Thus one criticism is methodological, the other ethical (a professional concern of the Sidgwicks); religious questions are not specifically involved. Indeed, Mrs. Sidgwick, like the Spiritualists, believed in survival [Salter 1945, 4-5].

In fact, as is well known, many of the SPR's prominent scientific members were Spiritualists — hardly a proper drawing of lines for a battle with science. Indeed, there were probably more Spiritualists among its scientific members than among its humanists (that is, those with a primarily literary training). And this point leads to the remark that the main offices of the SPR were filled during its early decades by humanists: the Sidgwicks, Edmund Gurney, Myers, and Frank Podmore (who *was* a Spiritualist). In other words, the scientific representation in the SPR was somewhat limited (although, of course, greatly distinguished). This imbalance was a structural weakness of the SPR, given the nature of the subject to which it was addressed; and one consequence that emerged in the early decades of this century is the surely inordinate amount of space devoted to cross-correspondence analysis — over 2,000 pages in the *Proceedings*, sometimes reading more like extensions of the Classics Tripos than psychical research. 'We believe unreservedly in the methods of modern science', Sidgwick had written in one of his presidential addresses [1888, 272]; however, in view of the complications described above and others about to follow, I imagine that of his circle only his wife was equipped to know what these methods might be.

4. Roads of reductionism

> Our duty is not the founding of a new sect, nor even the establishment of a new science, but is rather the expansion of Science herself until she can satisfy those questions which the

---

[10] Gauld [1968, 147] lists some early extracts of SPR reports in *The nisneteenth century* and *The national review*.

[11] Another interesting comparison to explore is the question of survival in the context of the late-19th-century discussion of 'heat death', where according to the current knowledge of geology and thermodynamics the whole universe would eventually slide into one uniform temperature (a situation that the Germans more accurately named 'cold death'). This problem sunk deeply into Victorian consciousness, both in science and the arts [Brush 1979]. It is not considered in Turner's generally excellent survey of Myers [Turner 1974, ch. 5].

human heart will rightly ask, but to which Religion alone has thus far attempted to answer.

F. W. H. Myers on the aim of the SPR [1903, 305]

So far I have sketched out aspects of a picture that shows why psychical research could flourish in late Victorian science. However, there were other philosophical fragrances in the air at the time that found inimical not only psychical research but indeed the whole ambit of 'psychology and metaphysics', and wished instead to banish both by reducing them to "acceptable" categories of knowledge or regarding as illegitimate the questions there raised. For convenience I shall call them 'positivists', although they include also materialists (who wish to reduce mind to matter) and behaviourists (who reduce alleged psychological notions to patterns or states of behaviour). These positions lay at the end of the range of views called 'naturalism' in ¶3.1, which were most enthusiastic for the reductionism, and thus at the opposite end from the critics of naturalism such as Myers, Sidgwick, Wallace and Ward. These positions gained much ground at the century's turn; I shall briefly outline four movements related, respectively, to popularization, epistemology, psychology, and physiology, for they provide important background to the emergence of parapsychology, which is described in ¶5.1.

4.1 *Science and religion.* Tyndall's 1874 address to the BAAS shocked the theologically orthodox, as mentioned earlier. Both there and in other of his more popular writings on science, he, Huxley and others promulgated doctrines that, although not necessarily atheistic, urged a sovereignty of science over religion, or at least a capture of religious ground by scientific principles. My militaristic metaphor here is deliberate; the 'warfare' caricature of the relations between science and religion, mentioned in the last section, was advocated from the 1860s on, especially in some of the 'International Scientific Series' of books launched by the American E. L. Youmans, with the help of Huxley, Spencer, Tyndall, and others [Moore (J.) 1979, part 1]. *The history of the conflict between religion and science* (1874) by the positivist scientist and historian J. W. Draper was an especially influential volume in the series for prosecuting the warfare theory already described. In addition, the series helped to popularize in the USA the 'Social Darwinism' of Spencer [Sharlin and others 1976], his attempt to impose on sociology his materialist view of evolution (following the Lamarckian adaptationist tradition), and associationist psychology [Young 1970, ch. 5].

4.2 *Professional philosophy* in late-19th-century England was dominated by various kinds of idealism; it was the pasta of philosophy, psychology, and logic described in ¶2.2 cooked in the warm vapours of Immanuel Kant and (especially) G. F. Hegel. The young Bertrand Russell drank its exotic juices; but 'It was towards the end of 1898 that Moore and I rebelled against both Kant and Hegel' [Russell 1959, 54]. From then on Russell adopted an opposite extreme. In 1903 (the year of publication of Myers's

Human personality and its survival of bodily death) Russell brought out his *The principles of mathematics*, in which he tried to reduce mathematics to logic (a subject now cast in a very different form from the Boolean and other traditions mentioned earlier). He and A. N. Whitehead brought his conception to some sort of completion in their *Principia mathematica* (1910-1913) [Grattan-Guinness 1977]. The elimination of abstract objects (allegedly) achieved in that system would have earned the approval of Mill (and also of W. K. Clifford, attacker of *The unseen universe* and delight of the teenaged Russell).

Russell then started to apply his reductionist techniques to broader areas of philosophy. By 1918 he came to a philosophy called 'logical atomism', a positivist epistemology based on the idea that propositions are pictures of reality and can be put together like bricks [Russell 1959, chs. 10-12]. His position, and some related thoughts by Ludwig Wittgenstein, were taken up and generalized in the 1920s by some members of the so-called 'Vienna Circle' of philosophers. One of them, Rudolf Carnap, published as [1928] an outline of 'the logical structure of the world' (§8.4).

Reductionism was a prime epistemological aim of several members of the Circle, with the logic of relations presented in *Principia mathematica* providing much of the machinery [Grattan-Guinness 1979]. Other sciences could be reduced to physics[12] (although Carnap was curiously silent on the details of that subject; for example, he said nothing about radioactivity, quantum mechanics or relativity, major research areas of the time). Further, each science must come down to its basic principles, to be furnished from 'observation statements', that is, statements verified by direct experience. Metaphysical statements, which were not so verifiable, were thereby rendered 'meaningless'. Great emphasis was laid on language; indeed, the Circle was a source of modern language-analysis philosophy (and stood in marked contrast to late-19th-century doubts about the power of language to express the full range of possible thoughts). It led to a movement called 'logical positivism', and by the mid 1930s trendy young philosophers were diffusing its ideas elsewhere in Europe (compare §8.4). By then, however, because of Hitler, several Circle members had immigrated into the USA.

4.3 *The professionalization of psychology* owes much to Wilhelm Wundt, who instituted a research laboratory at Leipzig University in the 1870s and drew large numbers of research students, especially from the USA [Dolby 1977].[13] Acknowledging some influence from Mill and

---

[12] Carnap gave scant attention to biology, starting out from the alleged definition: 'The organisms with their essential properties and relations and events which are peculiar to organisms are called biological objects' [1928, art. 137]. Psychology was treated in more detail [arts. 19-24, 55-64], and in a manner broadly sympathetic to behaviouristic reductions, although he cast doubt on the epistemological claims of behaviourism [art. 59].

[13] Study by American students at German universities was a common feature in many disciplines in the period 1880-1914. Doctoral programmes were available (unlike the USA) for several disciplines; the quality of training was high; the fees

Spencer, he taught an associationist brand of psychology (with a Millian inductive logic alongside) with the emphasis laid on experimental studies and the physiological aspects of psychology rather than (for example) the older nativist traditions. However, he also relied on introspection theory rather than comparative techniques.

Wundt wrote a rather sceptical essay [1892] on hypnosis, and his approach to psychology was in general unsympathetic to psychical research; no 'subliminal self' would have been seen haunting his laboratory. He was not without critics — William James for one, was quite contemptuous [Perry 1935, 68] — but his influence was enormous, especially in the USA, whither his students returned to continue in his style, especially in colleges and universities.

4.4 *Incertitude*. Sidgwick is a good example of a Victorian struggling with uncertainties, in his case over religion. Another example, this time concerning psychical phenomena, is the biologist G. J. Romanes. He was very intrigued by psychical phenomena, and attended Barrett's founding conference of January 1882.[14] Later that year he published an important book on animal intelligence in the International Scientific Series [Romanes 1882], in which he accepted the mind and consciousness in theorizing while treating the subject largely in the evolutionary associationist style of Spencer. He was sympathetic to an evolutionary style of physiology ('myogenesis') then being cultivated at Cambridge by his teacher, Michael Foster [Geison 1980], a colleague of Sidgwick at Trinity College.[15] He came to reject his earlier interest in psychical research, and, with the violence of the convert, attacked Wallace viciously for advocating the subject [Moore (J.) 1979, 187-190]. However, he was opposed to naturalism and remained uncertain about the position of Christianity [Turner 1974, ch. 4].

Romanes's book became well known, passing through several editions over 30 years. Its subject matter was taken up in the USA in the early 1900s by a young graduate student who wrote a doctorate on the behaviour of rats. Previously impressed by William James's approach to psychology, these studies of animals drove introspection out of his theoretical system and led him to develop behaviourism. His name was J. B. Watson, and he propounded this influential brand of American psychology especially from

---

were very reasonable; and, as their surnames reveal, the students were often of German or Central European stock.

[14] There is an interesting sequence of seven letters of 1881-1882 from Romanes to Barrett, in which he gradually becomes more sceptical about psychical phenomena (Barrett (footnote 2), file A4, letters 104-110). And on 18 July 1881 he doubted that Sidgwick 'would be any use' on a committee of 'critical' men to investigate the telepathy of the Creery sisters!

[15] Physiology was a science that eliminated aethers from its theorizing relatively early, by 1800 [French 1981]. By Romanes's and Foster's time, myogenesis was competing with neurogenesis, according to which muscle tissue needs continuous stimulation from the nervous system [Geison 1980]. Similarly, the 'mechanistic' protoplasm theory of life was then being popularized by Huxley (and others) in opposition to vitalism [Geison 1969].

the early 1910s on [Cohen (D.) 1979]. His thesis had been published as *Animal education* [Watson 1903].

We may note here also pragmatism, a philosophy that (to speak simply) emphasizes meaning, replaces truth by usefulness, and tends to place belief above facts. It is a "soft" form of reductionism, as it can allow a status for mental states and consciousness; it has some kinship with naturalism. Thus its supporters included both believers in psychical phenomena such as William James and F. C. S. Schiller, and sceptics like John Dewey (on Dewey and James, see [Russell 1946, chs. 29-30]). One would expect a line of positive influence from Dewey to Watson, who was his pupil; but apparently Watson found Dewey incomprehensible [Cohen (D.) 1979, 25].

5. On J. B. Rhine and the 1930s

The work reported in this volume is the first fruit of the policy of naturalization of 'psychical re search' within the universities.
William McDougall, foreword to [Rhine 1934, xiii]

5.1 *Activity in the USA*. One of Wundt's pupils was J. H. Hyslop, who kept the American SPR going in the 1910s; presumably his Spiritualism triumphed over the Leipzig air. During these years, and in the 1920s, the state of psychical research in the USA was not encouraging, although Clark University managed to mount a moderately vigorous symposium in 1926 [Murchison 1928]. Divisions over Spiritualism and the mediumship of Mrs. Crandon even led to the formation of the Boston SPR separate from the American SPR [Mauskopf and McVaugh 1980, ch. 1]. Given the tendencies in various related fields described in the preceding section, it is not surprising that the Rhines quickly abandoned mediumshlp and concentrated on card-guessing experiments in their development of parapsychology, which J. B. Rhine summarized in his book *Extra-sensory perception* ([Rhine 1934]: on the words 'extra-sensory perception' and 'parapsychology', which are popularized in this book, see ¶7).

The achievements of the Rhines during this period and the controversies over parapsychology have been described in detail.[16] Their work has an interesting prehistory in an attack on behaviourism launched in the early 1920s by William McDougall (see especially [Watson and McDougall 1928]), who was soon to be the Rhines' mentor. In a similar vein a leading motivation for the Rhines' research on psychical phenomena was 'to discover the whole nature of man and to find out whether something more than just his physical, sensorimotor side exists' [Rhine and Rhine 1980, 309], in marked distinction from the behaviourist norms of their time.

On the other hand, the Rhines' experimental techniques reflect in many ways prevailing views of the time: reductions in the small range of paranormal phenomena investigated, and even in experimental details like using forced-choice rather than free-response tests; and behaviourism in the attempt to detect the phenomena from statistical analysis of (large quantities

---

[16] See especially [Mauskopf and McVaugh 1980] and [Moore (R. L.) 1977, ch. 7], and the critical review [Pratt 1978] of the latter work.

of) data rather than via direct inspection [Rhine 1977a]. There are notable similarities in method between their parapsychology and various aspects of contemporary behaviourist psychology (see, for example, [Ellis 1938] on Gestalt psychology). In particular, laboratory work was valued far above spontaneous evidence, in the hope — felt both by parapsychologists and several kinds of psychologists — of aspiring to the 'hard' sciences, where, as we have seen, physics was King. McDougall called the process of aspiration 'naturalization' in the quotation at the head of this section, doubtless having in mind the naturalistic philosophies described in ¶3.1.

5.2 *The growing influence of positivism.* As described in ¶4.2, up to the early 1930s the general philosophical atmosphere was basically inimical to both the admission of psychical phenomena and favourable consideration of psychical research, as emerged later in the decade. By then the Vienna Circle members were placed in various universities in the USA, and the Philosophy of Science Association was launched in 1934. Its journal, *Philosophy of science*, began with an article by Carnap titled 'On the character of philosophical problems', an English translation of an orthodox Circle piece on the meaninglessness of metaphysics and the reduction of biology and psychology to physics. Later in the volume, Feigl [1934] gave a 'Logical analysis of the psychophysical problem' as 'a contribution to the new positivism', in terms of the meanings of 'meaning'. Not much room for psychical research there, as the publishers of the journal made clear in an advertisement opposite page 387 of volume 2 (1935), where they announced the availability of H. C. McComas's *Ghosts I have talked with* [1935]: 'He has in investigated the most successful mediums of recent years and has revealed their tricks in every instance. He has duplicated every trick he has seen produced'.

This view is hardly new: indeed, Rhine seems to have had a similar opinion of Mrs. Crandon. But it is of special note in a journal besotted with reductionism, the view that underlies the usual forms of criticisms of psychical phenomena now made.[17] Articles in similar vein appeared in the body of the journal (and of course elsewhere) from time to time: a suable, insulting piece on 'The masquerade of ESP' [Nabours 1943], possibly marks the extreme.[18]

---

[17] In the old days fraud in psychical research was seen as an ethical problem (he or she has cheated, it is a disgrace); today it is a reductionist problem (it can duplicate the phenomena by a trick, hence it was produced by a trick). In the same style, credulousness is now regarded only as a property of belief, not of disbelief. Henri Poincaré's remark that 'there are two ways of looking at things, to believe everything and to believe nothing, and each is equally stupid' has been forgotten (or made redundant) and replaced by that monument of modern social and political philosophy: double-standardizing.
[18] By a touch of (intentional?) editorial humour, Nabours's piece was followed by one entitled 'Why psychologists tend to overlook certain "obvious" facts'. On the polemics between Rhine and the psychologists, see [Mauskopf and McVaugh 1980, ch. 9].

The same dependence on reductionism, especially reduction to physics, is clear in less polemical writings of the time from elsewhere. For example, Sigmund Freud, who held a very ambiguous attitude to psychical research,[19] wrote to Immanuel Velikovsky in 1931: 'My own experiences have led me to suppose that telepathy is the real core of the alleged parapsychological phenomena and maybe the only one [...] nothing remains to us but to await clarification of this basically *physical* problem from the hopefully not too distant future' [Velikovsky 1982, 17: my emphasis]. Russell, in his *Religion and science* (1935), while allowing as 'one line of argument in favour of survival after death [...] the phenomena investigated by psychical research' [1935, 137], preferred science in general and physics in particular to everything else discussed in the book {compare Popper at the head of §16}. Russell also accepted uncritically the historical caricature of science at war with religion discussed in ¶3, and so, in Moore's witty phrase [1979, 40], made himself a 'prisoner of war'.[20]

6. Some conclusions

This chapter is no more than a collection of notes on a much larger story concerning the development of psychical research, and also of related disciplines, over the period (say) 1870-1940, with special attention to general philosophical tendencies and inclinations. It would be a difficult story to write, but a story well worth the telling; for I am confident that the larger perspective would furnish some new perspectives on the history of psychical research. {See now [Oppenheim 1985].} In particular, it would bring out broader common features where the "local" histories stress differences. For example, the disputes between the SPR and the Spiritualists are rightly held to be important, but they took place in a general atmosphere largely *sympathetic* to metaphysical issues. Again, the attempt of the Rhines to refute materialism by demonstrating the existence of ESP was doubtless a major clash between contemporary positions: but seen in a wider view, their methods, and the narrowing down of psychical research to parapsychology,

---

[19] Freud told Hereward Carrington, in a letter of 1921, that if he had his life to live over again, he might devote himself to psychical research rather than to psychoanalysis. When the news circulated, Freud denied it indignantly; but his biographer Ernest Jones found the letter and promised to print it, as evidence, in an appendix [Jones 1957, 419-420]. However, the letter is not reproduced; it should be at p. 473. It appears in [Fodor 1971, 84].

Dr. Frank Sulloway tells me that in the American edition of the Jones biography even the promise to print the letter in the appendix is omitted. He puts the situation down to Jones's chaotic filing system while preparing the biography. Freud's behaviour may be regarded as a Freudian slip in at least two senses.

[20] This prisonership is more interesting than it appears at first sight. Positivists in science and philosophy tend to be positivist about history also. In this view, history is 'the royal road to me', a form of verificationism in which things of the past that I like (such as the conflict thesis for Russell here) are praised and the rest, being mistaken, is deplored or ignored. Among extreme positivists the notion of history at all is abhorrent {compare §9, §10 and §13}. Carnap, for example, regarded himself 'as unhistorically minded a person as one could imagine' [Cohen (I. B.) 1974, 310].

seems consistent with prevailing modes of the time against metaphysics (and, to a lesser extent, religion) and for reductionism. Naturalism, described in ¶3.1, lies *in between* psychical research and parapsychology.

The difference just noted is a different one from those brought out in the "narrower" histories of psychical research, and it strikes me as one that is more relevant to the current situation. My own correspondence and discussions with people in the field today show me that that difference is still evident in modern views. {Indeed, as in the past both supporters and opponents of psychical research and parapsychology bring to bear a wide variety of philosophical standpoints, often *quite consciously*.} Some hanker for the phenomena and the general atmosphere of the old days of psychical research; others extol the reductionism, the use of probability theory and statistics, and the experimental techniques attendant upon parapsychology (compare §18.9-§18.14). While the views are not contradictory, there are, nevertheless, noticeable differences in priority, method and even object of research. From that point of view, then, and to that extent, we have two different disciplines.

But there are those who recognize this situation, and seek a place between psychical research and parapsychology. Perhaps they stand, therefore, as the new naturalists;[21] in a sense they are successors to Myers and Sidgwick, who 100 years ago searched the ground between science and religion. I send them on their way with a timeless warning from Augustus De Morgan [1863, vii- viii]:

> Those who affirm that they have seen faith-staggering occurrences are of course supposed to be impostors or dupes. To this there can be no objection; a pretty world we should live in if the arrangement did not demand moral courage from those who offer evidence of wonders.

7. Appendix: On the history of the words 'parapsychology' and 'extra-sensory perception'

It is sometimes thought that Rhine introduced the word 'parapsychology' in his *Extra-sensory perception*. However, its origins go back to a paper by Max Dessoir [1889, 342]:

> One indicates by analogy with words like *paragenesis*, [...] *paralogism, paranoia* [...] with *para* — something, which goes beyond or [stands] beside the usual, so perhaps one can call parapsychic the phenomena which emerge out of the normal run of the psyche, parapsychology the science which treats them.

He came to visualize parapsychology as the border area between psychology and psychopathology; the corresponding objective phenomena were studied in 'paraphysics'. Richet [1907] introduced the term 'métapsychi-

---

[21] Indeed, some of the newer theories of psi functioning seem to be consonant with pragmatism, whose own kinship with naturalism was noted in ¶4.4. From other emphasized points of view, such as feedback and interaction, these theories exhibit the philosophical concerns attending branches of modern science such as systems theory, ecology and operational research.

que' into French in his SPR presidential address for 1905; covering both parapsychology and paraphysics (in their German senses) and also related physiological concerns, it was more or less a synonym for 'psychical research'.

Rhine was aware of these terms, but he used Dessoir's in his *Extrasensory perception* as a synonym: 'Parapsychology, i. e., psychical research'. He favoured the word for the etymological reasons that Dessoir had also urged, introducing cognates such as 'parapsychophysical' (including levitation and apports) and 'parapsychopathological' (psychic healing) [Rhine 1934, 7]. However, these terms did not endure, and 'parapsychology' both continued in its traditional role as a synonym and also came to be restricted to the experimental and technical aspects on which the Rhlnes concentrated. This is an odd, and indeed unsatisfactory, situation: I know of no other discipline of which the name also refers to one of its proper subsets.

Rhine did claim to introduce 'extra-sensory perception' in his book, as he later recalled: 'This term was chosen after considerable thought and some discussion with colleagues, but had I realized what a wide usage the expression would receive in the course of years, I would certainly have given it further thought in the hope of arriving at a more convenient and adaptable term' [1977b, 163]. In fact, this term also has a German prehistory: 'Aussersinnliche Wahrnehmung' seems to be due to Gustav Pagenstecher [1924], who used it as the title of his book on psychometry. Presumably Rhine missed this work, although in his book [1934, 19] he cited an earlier study [Pagenstecher 1922] on this topic! This ironic touch seems to provide a suitable point on which to close.

Acknowledgements

I am grateful to the Society for Psychical Research for permission to see the Barrett manuscripts in their possession. The draft of this chapter was kindly read by John Beloff, Geoffrey Cantor, Anita Gregory, Peter Harman and Ian Stevenson.

Bibliography

Barrett, W. F. 1877. 'On some phenomena associated with abnormal conditions of mind', *Reports of the BAAS*, (1876), 164 [title only].

Barrett, W. F. 1883. [*Ibidem*], *Proceedings of the SPR*, 1, 238-244.

Barrett, W. F. 1924a. 'The early years of psychical research', *Light*, 44, 395.

Barrett, W. F. 1924b. 'Some reminiscences of fifty years' psychical research', *Proceedings of the SPR*, 34, 275-297.

Brock, W. H. and MacLeod, R. 1976. 'The "Scientists' Declaration": Reflexions on science and belief in the wake of *Essays and reviews*', *British journal of the history of science*, 9, 39-66.

Brooke, J. H. 1977. 'Natural theology and the plurality of worlds: Observations on the Brewster-Whewell debate', *Annals of science*, 34, 221-286.

Brown, A. W. 1947. *The Metaphysical Society*, New York: Columbia University Press.
Brush, S. G. 1979. *The temperature of history*, New York: Franklin. [Built around his 'Thermodynamics and history', *The graduate journal, 7* (1967), 477-565.]
Butts, R. E. 1968. 'Introduction' in (ed.), *William Whewell's theory of scientific method*, Pittsburgh: Pittsburgh University Press, 3-29.
Cambridge University 1883. *Cambridge University calendar,* (1882-1883), Cambridge: Cambridge University Press.
Cantor, G. N. 1981. 'The theological significance of ethers', in [Cantor and Hodge 1981], 135-155.
Cantor, G. N. and Hodge, M. J. S. 1981. (Eds.), *Conceptions of ether,* Cambridge: Cambridge University Press.
Carnap, R. 1928. *Der logische Aufbau der Welt*. Berlin: Welt-Kreis. [English trans. by R. George: *The logical structure of the world*, London: Routledge and Kegan Paul, 1967.]
Carnap, R. 1934. 'On the character of philosophical problems', *Philosophy of science, 1*, 5-19.
Cohen, D. 1979. *J. B. Watson, the founder of behaviourism*, London: Routledge and Kegan Paul.
Cohen, I. B. 1974. 'History and the philosopher of science', in F. Suppe (ed.), *The structure of scientific theories,* Urbana: University of Illinois Press, 308-373.
Dauben, J. W. 1979. *Georg Cantor: his mathematics and philosophy of the infinite*, Cambridge, Mass.: Harvard University Press.
De Morgan, S. 1863. *From matter to spirit*, London: Longman. [Signed 'C. D.' with a preface by 'A. B.' (Augustus De Morgan).]
Dessoir, M. 1889. 'Die Parapsychologie [...]', *Sphinx, 7*, 341-344.
Dolby, R. G. A. 1977. 'The transmission of two scientific disciplines from Europe to North America in the late nineteenth century', *Annals of science, 34*, 287-310.
Edgeworth, F. Y. 1881. *Mathematical psychics,* London: Kegan, Paul.
Ellis, W. D. 1938. (Ed.), *A source book in Gestalt psychology,* London: Routledge and Kegan Paul.
Feigl, H. 1934. 'Logical analysis of the psychophysical problem. A contribution to the new positivism', *Philosophy of science, 1*, 420-441.
Fodor, N. 1971. *Freud, Jung and occultism*, New Hyde Park, N. Y.: University Books.
French, R. K. 1981. 'Ether and physiology', in [Cantor and Hodge 1981], 111-134.
Garland, M. M. 1981. *Cambridge before Darwin: the ideal of a liberal education 1800-1860*, Cambridge: Cambridge University Press.
Gauld, A. 1968. *The founders of psychical research,* London: Routledge and Kegan Paul.
Geison, G. L. 1969. 'The protoplasmic theory of life and the vitalist-mechanist debate', *Isis, 60,* 273-292.
Geison, G. L. 1980. *Michael Foster and the Cambridge school of physiology*, Cambridge: Cambridge University Press.

Grattan-Guinness, I. 1977. *Dear Russell — dear Jourdain*, London: Duckworth.
Grattan-Guinness, I. 1979. 'On Russell's logicism and its influence, 1910-1930', in *Wittgenstein, the Vienna Circle and critical rationalism*, Vienna: Hölder-Pickler-Tempsky, 275-280.
Grattan-Guinness, I. 1982. 'Psychology in the foundations of mathematics: the cases of Boole, Cantor and Brouwer', *History and philosophy of logic, 3*, 33-53.
Gurney, E., Myers, F. W. H. and Podmore, F. 1886. *Phantasms of the living*, 2 vols., London: Rooms of the Society for Psychical Research, and Trübner and Co.
Heimann, P. M. 1972. 'The unseen universe: physics and the philosophy of nature in Victorian Britain', *British journal of the history of science, 6*, 73-79.
Jones, E. 1951. *The life of Sigmund Freud*, vol. 3, London: Hogarth.
Kottler, M. 1974. 'Alfred Russel Wallace, the origin of man and Spiritualism', *Isis, 65*, 145-192.
Lodge, Sir O. 1926. 'Ether', in *Encyclopaedia Britannica*, 13th ed., suppl. vol. 1, New York: Encyclopaedia Britannica, 1026-1029. [Repr. with additions in 14th ed., vol. 8 (1929), 751-755; and in [Lodge 1933], 186-204.]
Lodge, Sir O. 1933. *My philosophy*, London: Benn.
McComas, H. C. 1935. *Ghosts I have talked with*, Baltimore: Williams and Wilkins.
Mauskopf, S. H. and McVaugh, M. R. 1980. *The elusive science*, Baltimore: Johns Hopkins University Press.
Moore, J. R. 1979. *The post-Darwinian controversies*, Cambridge: Cambridge University Press.
Moore, R. L. 1977. *In search of white crows*, New York: Oxford University Press.
Murchison, C. 1928. (Ed.), *The case for and against psychical belief*, Worcester, Mass.: Clark University.
Myers, F. W. H. 1903. *Human personality and its survival of bodily death*, vol. 2, London: Longmans, Green. [Cited from the 1954 printing.]
Nabours, R. K. 1943. 'The masquerade of ESP', *Philosophy of science, 10*, 191-203.
Nicol, F. 1972. 'The founders of the S. P. R.', *Proceedings of the SPR, 55*, 342-369.
Olson, R. 1975. *Scottish philosophy and British physics 1750-1880*, Princeton: Princeton University Press.
Oppenheim, J. 1985. *The other world: spiritualism and psychical research in England, 1850-1914*, Cambridge: Cambridge University Press.
Pagenstecher, G. 1922. 'Past events seership: A study in psychometry', *Proceedings of the American SPR, 16*, 1-136.
Pagenstecher, G. 1924. *Aussersinnliche Wahrnehmung. Experimentelle Studie über den sogen. [sic] Trancezustand*, Halle/Salle: Marhold.
Palfreman, J. 1976. 'William Crookes: Spiritualism and science', *Ethics in science and medicine, 3*, 211-227.

Paul, H. W. 1979. *The edge of contingency: French Catholic reaction to scientific change from Darwin to Duhem*, Gainesville: University Press of Florida.
Pearson, K. 1924. *The life, letters and labours of Francis Galton*, vol. 2, Cambridge: Cambridge University Press.
Perry, R. B. 1935. (Ed.), *The thought and character of William James*, vol. 2, Boston: Little, Brown.
Pratt, J. G. 1978. Review of [Moore (R. L.) 1977], *Journal of the American SPR, 72*, 257-266.
Rhine, J. B. 1934. *Extra-sensory perception*, Boston: Boston Society for Psychic Research. [Revised ed. Boston: Branden, 1964, with an interesting, reminiscing new preface by Rhine.]
Rhine, J. B. 1977a. 'History of experimental studies', in B. B. Wolman (ed.), *Handbook of parapsychology*, New York: Van Nostrand Reinhold, 25-47.
Rhine, J. B. 1977b. 'Extrasensory perception', in *ibidem*, 163-174.
Rhine, J. B. and Rhine, L. E. 1980. 'Afterword', in [Mauskopf and McVaugh 1980], 307-310.
Richards, J. 1980. 'Boole and Mill: Differing perspectives on logical psychologism', *History and philosophy of logic, 1,* 19-36.
Richet, C. 1905. 'La métapsychique', *Proceedings of the SPR, 19,* 2-49.
Romanes, G. J. 1882. *Animal intelligence*, London: Kegan Paul, Trench.
Roos, D. A. 1981. 'The "aims and intentions" of Nature', *Annals of the New York Academy of Sciences, 380,* 159-180.
Rothblatt, S. 1968. *The revolution of the dons: Cambridge and society in Victorian England*, London: Faber and Faber.
Russell, B. A. W. 1935. *Religion and science*, London: Oxford University Press.
Russell, B. A. W. 1946. *History of Western philosophy*, London: Allen & Unwin.
Russell, B. A. W. 1959. *My philosophical development*, London: Allen & Unwin.
Salter, Mrs. W. H. 1945. *Psychical research: where do we stand?*, London: SPR. [Myers Memorial Lecture.]
Sharlin, H., Wall, J. F. and Hollinger, D. A. 1976. 'Spencer, scientism and American constitutional law', *Annals of science, 33,* 457-480.
Sidgwick, Mrs. E. 1911. 'Spiritualism', in *Encyclopaedia Britannica*, 11th ed., vol. 25, New York: Encyclopaedia Britannica, 705-708.
Sidgwick, H. 1879. 'The wages-fund theory', *Fortnightly review*, new ser., *26*, 401-413.
Sidgwick, H. 1882. 'Incoherence of empirical philosophy', *Mind, 7,* 533-543.
Sidgwick, H. 1888. 'Opening address at the twenty-eighth general meeting', *Proceedings of the SPR, 5,* 271-288.
Smith, R. 1972. 'Alfred Russel Wallace: philosopher of nature and man', *British journal of the history of science, 6,* 177-199.
Stewart, B. and Tait, P. G. 1875. *The unseen universe: or physical speculations on a future state*, London: Macmillan.

Turner, F. M. 1974. *Between science and religion: the reaction to scientific naturalism in late Victorian England,* New Haven: Yale University Press.

Turner, F. M. 1978. 'The Victorian conflict between science and religion: a professional dimension', Isis, *69*, 356-374.

Tyndall, J. 1875. 'Presidential address', *Reports of the British Association for the Advancement of Science.* (1874), lxvi-xcvii. [Repr. in G. Basalla, W. Coleman and R. H. Kargon (eds.), *Victorian science,* Garden City, N. Y.: Anchor, 1970, 436-478.]

Velikovsky, I. 1982. *Mankind in amnesia,* London: Sidgwick and Jackson.

Ward, J. 1911. 'Naturalism', in *Encyclopaedia Britannica,* 11th ed., vol. 19, New York: Encyclopaedia Britannica, 274-275.

Watson, J. B. 1903. *Animal education,* Chicago: University of Chicago Press.

Watson, J. B. and McDougall, W. 1928. *The battle of behaviourism: an exposition and an exposure,* London: Kegan Paul, Trench, Trubner.

Wilson, D. B. 1971. 'The thought of late Victorian physicists: Oliver Lodge's ethereal body', *Victorian studies, 15,* 29-48.

Wundt, W. 1892. *Hypnotismus und Suggestion,* Leipzig: Engelsmann.

Wynne, B. 1979. 'Physics and psychics: science, social control and symbolic action in late Victorian England', in B. Barnes and S. Shapin (eds.), *Natural order,* London: Sage, 167-186.

Yeo, R. 1979. 'William Whewell, natural theology and the philosophy of science in the mid-nineteenth century Britain', *Annals of science, 36,* 493-516.

Young, R. M. 1970. *Mind, brain and adaptation in the nineteenth century,* Oxford: Clarendon Press.

18 What are Coincidences, and What are they Good For?

Coincidence is normally studied in psychical research only when the appraisal of data against chance is considered. In this chapter I record various kinds of coincidences in ordinary life, mostly drawn from my own experience, and suggest that they may exemplify forms of spontaneous phenomena, which thus require explanation. The candidacies of some theories of psi are briefly discussed.

This whole question of coincidence is one that plagues psychic research.
Russell Targ and Harold Puthoff [1977, 143]

1. Is there a problem?

In the *Handbook of parapsychology* Dr. L. E. Rhine presents an excellent survey [1977] of 'Research methods with spontaneous cases'. She coven old collections of cases such as the 'Census of hallucinations' of 1894 prepared by the Society for Psychical Research (SPR), and later studies up to her own and others' surveys, which concentrate largely on extra-sensory perception (ESP) and apparitions. However, she makes only passing mention of coincidence [p. 78]. This is not because the topic is dealt with elsewhere in the book, for it arises there only when experimental data are being appraised against chance outcomes. The same can be said of most of the introductory and general books on psychical phenomena: coincidences are usually considered as explanations of other things, but they are not themselves regarded as needful of explanation. Psychical researchers normally adopt the common, and quite correct, view that coincidences are *bound* to happen, to the annoyance of Targ and Puthoff above; even the highly unlikely is not impossible. Their interest would be aroused in someone who genuinely does *not* experience coincidences, and so exhibits 'psi-missing'.

Thus discussion of 'meaningful' coincidences has been rather limited. Alice Johnson put out a long and thoughtful essay [1898] with the SPR that contained not only many interesting cases but also discussion of the borderline between chance and non-chance, while Wilhelm von Scholz's collection of cases [1924] went through enlargements until a fifth edition appeared in 1950. Arthur Koestler wrote well on coincidence: see his [1972, 1973], the latter being a collection of good cases. Alan Vaughan published a nice ensemble of cases [1979], as did Brian Inglis [1990].

This chapter is based upon a merger of parts of two articles first published in the *Journal of the Society for Psychical Research*: 'What are coincidences?', 49(1978), 949-955; and 'Coincidences as spontaneous psychical phenomena', 52(1983), 59-71. They are reprinted by kind permission of the Society.

I suggest that some coincidences deserve attention, especially when they *clearly fulfil some current and very specific need* of the experient. To count *as* a coincidence I set tough standards on the timing (when applicable) of the relevant events, preferring literal simultaneity and rarely allowing more than one or two days' separation — preferably one or two minutes. Further, I make and keep detailed notes as soon as possible, and record everything that conceivably seems relevant, both at the time and afterwards: if not done, then the status as coincidence is denied. Much of the reportage of coincidences falls below these standards.

PART ONE   SOME TYPES OF COINCIDENCES

2. Preliminaries

I assume that, far from being an undifferentiated collection, coincidences can be divided up into types, or at least that some possess certain properties; the cases described in ¶3-¶7 exemplify five types. It is also useful to distinguish between *animate coincidences*, which involve contact between more than one living being; and *inanimate coincidences*, which concern one living being and artefacts. The distinction is not sharp; for example, inanimate coincidences could involve undetected telepathy from outside agents (including, for Spiritualists, discarnate entities). But it leads to useful points, such as that animate coincidences are similar to telepathy (two or more persons in touch with the same event) whereas inanimate coincidences resemble a form of converse telepathy (two or more events in touch with one person) and may relate to (travelling) clairvoyance.

I draw largely on my own cases, partly because I know them best and partly because to my knowledge a few contain the only published examples of some properties. Most of them are inanimate, especially finding references to relevant matters of interest without looking for them; however, already during my school days I kept meeting members of school on holiday, no matter where we went.

After a consideration of chance in ¶9, I indicate in ¶10-¶14 how coincidences may be appraised and related to some of the general theories of psi mediation. Throughout the word 'coincidence' refers only to events that seem to have more than mere chance about them, unless chance cases are explicitly included.

3. Coincidences involving psychical phenomena

3.1 John Crossley, Professor of Mathematics at Monash University, Melbourne, is interested in psychic phenomena. He arranged a visiting lectureship for me at the university in 1977, and soon after I arrived we went along to a meeting of various people with a local spoon-bender, Ori Suoray. We gave Ori keys and metal. He bent one key a little after much effort, but was on the whole fairly unsuccessful. I went with him alone into another room, and we chatted for about 20 minutes while he worked on a key that I gave him out of the desk in my room, and some other things. He may have bent that key very slightly, but said that it was a small bimetal type with which he was usually unsuccessful; and he certainly did nothing to two other keys from the desk which I kept in my possession as controls.

But I have great respect for spoon-benders, and I had left my key ring and watch in the glove compartment of John's car before we crossed over to the building for the meeting. When we came back to the car I took my watch and keys out of the glove compartment. The watch was still going, but one of the keys was bent by at least 5°! Moreover, it was the fourth of the set for my desk, the one that I used. Like the other three, it was originally completely flat, and of the type that Ori said he did not like working with! I am sure that I did not tell him that I had left my keys and watch in John's car, and the car was guarded by a rather aggressive dog during our absence. Afterwards I found that I could only just get that key to turn in the lock. I still have it.

3.2 In July 1978 the SPR organised a series of experiments in a laboratory at the City University, London with the medium Matthew Manning. I was unable to participate, being away in the West Country; but this circumstance was used to set up a telepathy experiment, in which my wife and I would try, over a distance of about 250 miles, to project to Manning feelings, passages from books, and so on every two minutes from a particular midday. The experiment was a failure, for Manning was not interested in trying to receive; but within seconds of midday there was an occultation of the infra-red beam that had been set up in the laboratory for other experiments. Moreover, it produced a shape on the trace paper of the recording instrument which was quite different from those apparently caused by Manning himself on other occasions during the experimental period. Just a coincidence, or a psychical phenomenon?

3.3 Preparing papers on coincidences seems to be a good source for *coincidences involving coincidences*. A nice example occurred for me once while writing a paper on coincidences; I was also listening to a radio programme, in which the reporter mentioned several times the coincidences that had occurred to her during its preparation. The "purpose" of this type of coincidence may be to emphasize to the experient that coincidences, whatever they are, do in fact happen. Johnson reported a nice one in [1899, 202]; and I had two from Johnson in that the contexts of two of the coincidences that she reported [pp. 225-226, 264] dealt with very obscure details in the history of science on which I had recently worked. Another coincidence of this type appears in ¶6.

4. Coincidences in sequence

4.1 The type just noted involve tiers, as it were; a different kind is where a coincidence relates to earlier ones without *involving* them. Paul Kammerer drew attention to a certain kind of coincidence of this type in his book [1919], where the same kind of event keeps occurring in the life of the experient. A relatively common form is the repetition of a certain integer; Johnson gives some good examples in [1899, 208-219]. Crossley (¶3.1) keeps running across '225' — and so did I when I stayed with him for a time. In particular, one day, we drove up behind a car with registration plate 'IGG 225' — his number, my initials!

4.2 I do not experience Kammererian runs, but I have had some pairs of coincidences that exhibited more complicated links than Kammerer discussed. For example, one Sunday afternoon in August 1974, my wife and I hired a car in Philadelphia, and left our friends there (who are interested in psychical phenomena) to drive to Montclair, New Jersey to stay with the psychical researcher Dr. Berthold Schwarz. We found a place off the road to eat our packed lunch, and noticed that we were at the entrance to the Vail House Museum. We had no idea what it dealt with, but there were several people around, so we went in. This was the home of the Vail family, in which Samuel Morse developed Morse code. Further, once a year the local history society mounted a lecture on Vail and Morse — and it was about to start in ten minutes' time. The coincidence in timing of the lecture was heightened by its content, which of course dealt with the sending of signals, on a day when we were thinking very much of psychical research.

4.3 In September 1969 in a second-hand bookshop I suddenly turned through a right angle, away from one bookshelf to another one that I had not previously seen, and without conscious intent I picked up *Bonaparte, Governor of Egypt* by one F. Charles-Roux. I opened it once, twice; and there was a reproduction of a contemporary drawing of Joseph Fourier, a civilian member of the French campaign of 1798-1801. Fourier later became a prominent mathematician, the inventor of 'Fourier analysis'. I was writing a book on Fourier at the time, and thought that I had finished my work on his Egypt period, but no! Now this drawing could appear in my book [Grattan-Guinness and Ravetz 1972, 15, 503].

Book-finding happens to me fairly often, and this case had a sequel: in 1978, a few days after the Manning case described in ¶3.2, I was in a different bookshop, ready to abandon a detailed inspection, when memory of the previous occasion surfaced; so I carried on my search after all — and within seconds had found the same book again! I had never seen it in a bookshop, or even in a bookseller's catalogue, during the intervening years.

4.4 In October 1990 I enjoyed a wonderful run of ten coincidences in as many days, and also a few afterwards, to aid a new study project on the relationship between Mozart's music and his status as a Freemason. I knew much of the music well enough, but its historical circumstances and the (massive) relevant literature were quite strange to me. So all those clues piled in, and saved me months of searching; in fact, I doubt if I would ever have found several obscure but valuable details. While recounting them at an SPR meeting on coincidences on 10 November, another coincidence took place there and then; I and another participant were carrying miniature scores (not the same edition) of the same Mozart symphony!

On Saturday 24 November 1990 came an extremely valuable addition to my Mozart cluster. I took a train from Pisa airport, which stopped next at Pisa Central. A lady with two large cases entered the almost empty carriage and started to go past me; then she stopped and shuffled around uncertainly in a strange manner that I noted at the time. Finally she placed the cases across the gangway and sat down opposite me. Then she took out the day's issue of *La repubblica*. For no obvious reason the arts section carried a section of about a dozen pages about Mozart, as I was able to see easily. Nothing pertinent to my specific interests was evident in the articles that I saw; but I made sure that I bought the paper when I descended at Empoli. The piece of information I needed to heed was at the bottom of the last page of the section, which the lady had not reached while I was on board.

4.5 Here is a strangely similar pair involving funerals of female friends. The first was Anita Gregory, a distinguished figure in psychical research. Her funeral took place on 14 November 1994. Several weeks before I had booked to see on that very evening a production of the opera *Johnny spielt auf!* by Ernst Krenek, which is very rarely staged in Britain. Completed in 1927, it deals with the life of one Anita in Berlin. Anita Gregory had been born in 1928 in Berlin.

The second friend was Cecily Tanner, a mathematician and historian. Her funeral took place on 3 December 1992. Several weeks before I had booked to see on that very evening a production of the operetta 'Princess Ida' by Gilbert and Sullivan, which is not staged very often. First produced in 1884, it is a satire on higher education for women, and the programme book included photographs of life at Girton College Cambridge. Cecily's mother had been a student there in the early 1890s, and photography was one of her hobbies.

5. Coincidences involving large-scale events

Coincidences usually concern everyday events for one person or a few people, striking for them but of no greater import. However, there are a few cases involving large-scale events, and they pose special problems for theorising, taken up in ¶12. The following example is also a case of coincidences in sequence. I am indebted to my wife for pointing it out to me; she just 'happened' to read the book in which it is described [Ridge 1965, 209-212] when I was revising a paper on coincidences.

The famous rose 'Peace' was developed just before the Second World War, in France. It was propagated and proved in the USA during the War. Still only known by its seedling number, the American Rose Society (ARS) decided some time during the winter of 1944-1945 to hold a 'name-giving ceremony' in California on 29 April 1945. As the war seemed to be drawing to an end, the name 'Peace' was chosen; but it was hardly planned that the ceremony should take place on the day that Berlin fell.

The beauty of the rose, and the appropriateness of its name, led the ARS to present a rose to each delegate of the United Nations for its first meeting in the summer of 1945; and on that day a truce was declared in Europe. Then the rose received an 'All-America Award' for roses in August, on the day that Japan surrendered; and a month later the ARS awarded it gold medal on the day that the peace treaty was signed in Japan,

Each rose-oriented event had been planned well in advance; the other things happened to come along at the same times.

6. Coincidences involving prediction

This is a fairly well recognized category of coincidences; its bearing on psychical research is largely concerned with precognition. But here is a rather unusual example, involving astrology.

At Christmas 1973 my astrological friend examined my chart, and gave me a set of predictions for 1974. By far the strongest prediction was that early June would be very favourable for publications. I was surprised at this possibility, since I doubted that any current work would be in a publishable state in June. While I did note the prediction, I did not look at the list of predictions until the following October.

Early in the previous June, after some lengthy negotiations, I was invited by the publisher to join the editorial board of a journal. We had a long meeting discussing the journal's future; and it was clear that we needed to have a joint meeting with the acting editor. The publisher phoned him up then and there, in my presence — and interrupted him typing out a letter of retirement, in which he proposed me as his successor. He knew, of course, of my recent recruitment to the board, but was not aware that our meeting was taking place.

The coincidence as such is a fairly straightforward animate case, suggestive of telepathy. However, what was the role in it of the astrological information, I wondered, when looking at the list of predictions the following October ... .

7. 'One-off' coincidences

These are coincidences that do not seem to involve other psychical effects, do not belong to a sequence, and are not on a large scale.

7.1 (Animate) I spent some months in Toronto before returning to England via New York in August 1961. 'Goodbye, Edgar', I said to a friend who was leaving to take up theological studies in New York City somewhere, 'I'll see you in New York then'. My remark surprised me as much as him; but there he was on the pavement six weeks later when I was looking around Columbia University.

7.2 (Inanimate) Among many coincidences on 7 June 1976 (in the form 7.6.76, itself a pair!), the most interesting occurred in a staffroom at the University of Western Australia in Perth. I had been looking up references on Fibonacci numbers, a branch of mathematics with applications in aesthetics, architecture and biology. An old copy of *Private Eye* seemed to be a suitable accompaniment for the coffee; and in 'Pseuds corner', a collection of pompous announcements, I saw reprinted an advertisement for a lecture where the speaker would describe her new ideas on neurophysiology derived from using random Fibonacci numbers. I duly got in touch with her later.

7.3 My time at Monash University was interrupted by a lecture tour of New Zealand. A day off, 31 October 1977, was spent at Rotorua, a centre for thermal areas of mud pools, boiling lakes, and craters. Just before visiting a buried village, I thought briefly about a history of mathematical physics that I was planning. It involved the transmission of certain ideas from France to England in the 1830s, and I made a mental note to look at Charles Babbage's *Reflections on the decline of science in England* (1830) as a source of relevant information. One of the exhibits in the village, which I then visited, contains a few books that were found there. One of them was the *Reports* of the inaugural (1831) meeting of the British Association for the Advancement of Science; and it was open at a page where Babbage's book is mentioned.

7.4 (Inanimate) While at Monash University in Melbourne, I was sent on 8 December 1977 a paper to referee for a journal. The author had traced an important collection of manuscripts to their probable destruction in a fire at a mine at Vollpriehausen, Germany, in 1945. I read the paper quickly and then had to go into Melbourne to look at the University of Melbourne's holdings of old mathematical books. While in the library stacks, I passed by a bookcase filled with thousands of booklets. I took one out to see what they were, and found them to be a huge series of reports prepared by the British and the Americans about the German economy and industry at the end of the Second World War. The one that I had selected was the second of three on the adaptation of existing underground installations. It fell open on page 5 — where there was a description of the mine at Vollpriehausen.

PART TWO   EXPLANATIONS?

It appears that, if coincidences are psychical phenomena at all, they have to be treated as spontaneous cases; I certainly have no idea how one could bring them to experimental test. Thus when theorising about them, one is faced with potentially all the variables and difficulties which attend the study of spontaneous cases. Such a prospect is too daunting for this author, and too extensive for this chapter anyway. So I shall confine myself to some outline suggestions, in the hope that those properly trained in psychical research will find them useful. I also follow the practice of psychologists in using the word 'organism' to refer to an individual's sensory and thinking equipment, when I wish to distinguish these from the individual as a whole.

8. Participator effect

Although coincidences appear to lie outside the purview of strict experimentation, they are subject to an analogue of the experimenter effect; namely, that when one takes notice of coincidences, they seem to happen more often. This, of course, is a well-known effect in many areas of life; after all, if one throws a net, then (usually) one catches fish. In this case, it is very likely that one notices coincidences that happen anyway but otherwise would have been overlooked. However, there may be other factors involved. For example, the interest of the individual in coincidences could stimulate his capacity to experience them (whatever be the process involved), perhaps by encouraging his belief that they can occur.

A particular case of participator effect is the motivation of the individual to collect coincidences in the first place. For myself, and for most other people with whom I have discussed them, the interest was sparked off by their occurrence in life, rather than by an initial, deliberate, intention to study them anyway. This would suggest that an extra-personal component is involved in coincidences, or at least that the psychological and/or extra-sensory components of the organism play an important role. (Compare [Kennedy and Taddonio 1976] on experimenter effects.) It would be worth collecting information on people's initial motivations to collect coincidences.

9. Chance

Of course, coincidences may well "just" be coincidences, as was mentioned in ¶1.2. Indeed, the amount of information available to us today, literally day and night, is so vast that, while we cannot expect to note it all as it passes, some can enter sub-conscious recording systems and re-emerge later, to be misinterpreted by us as components in an alleged coincidence.

Further, in several types of context, especially those involving relatively specific event spaces, the probability of coincidence is much greater than might naively be expected. The following result, found by the British mathematician Harold Davenport in the 1930s and published by his friend Harold Coxeter in [Rouse Ball 1939, 45-46], is relevant: of 23 people there is about an even chance that two of them will have the same birthday. It has gained (not from them) the name 'birthday paradox', unfortunately; for since any day in the year will do, 365 chances are available, so that 23 is not surprisingly small. Much more interesting is the fact that, say, you and Susan are the two; how and why did the two of you find out?

Such considerations have to be borne carefully in mind, but they need not be decisive. Firstly, explanations of coincidences in terms of response to unconsciously stored data could be just as mysterious as explanations in terms of some psi faculty or overall design. Secondly, the role of probability in the quantitative assessment of coincidences is easy to exaggerate: they often involve rather vaguely definable event-spaces, for which data to calculate probabilities are incomplete or unavailable.

I would turn the chance argument around; some coincidences are less likely to occur than we realize, and that we may overlook coincidences that happen to us. On the other hand, one must avoid the danger of seeking an explanation at all costs; for one ends up in birthday-paradox country in no time.

10. Information or perception theories

As was mentioned in ¶2.1, coincidences seem to resemble clairvoyance and telepathy most closely among psychical phenomena; and the traditional approach to ESP has been to regard it as a form of perception, and thus to formulate theories in terms of processes of communication. This approach has been criticized on various grounds — that ESP does not have the character of ordinary perception, and so on. Coincidences also offer grounds for criticising these theories, especially that the organism must process huge masses of (sensory and extra-sensory) information in order to pick out the items relevant to the coincidence.

This problem is especially severe for inanimate coincidences such as picking up the "correct" printed source without knowing why and discovering in it unexpected but highly relevant information (¶4.3-¶4.4). For how does the source manage to announce its existence (if that is the way to put it)? With animate coincidences appeal can at least be made to information passing between the organisms involved.

Among information and perception theories, perhaps Henri Bergson's filter hypothesis [1914] is the most promising form for development. According to this view, the organism is constantly blocking information obtained by extra-sensory means, and only items of an especially striking form get through; but even here a testable theory of the processes involved is a formidable undertaking. Perhaps it will gain strength when harnessed to modern brain research [Ehrenwald 1982].

It is appropriate to note here another source of perception called 'cryptomnesia', which is not given the attention it deserves. It refers to things seen and understood, but *not* noted consciously; for example, "reading" a newspaper out of the corner of your eye as you walk past it. Such information may lead you to actions that allow the coincidence to occur, but nothing paranormal is involved [Stevenson 1983].

11. Synchronicity theories

11.1 The best-known explanation of coincidences is Carl Jung's theory of synchronicity, advocated in his book *Synchronicity: an acausal connecting principle* [1955]. Impressed by the weakening of the grip of causality in theories in physics — his book originally appeared in German in 1952 together with an essay by Wolfgang Pauli — he saw acausality as a means of explaining coincidences. His rejection of causality was strong; while impressed by Kammerer's emphasis on coincidences in sequence, he rejected Kammerer's law of seriality as an explanation because of its causal taint [Jung 1955, 13].[1]

Jung described both inanimate and animate coincidences [pp. 36, 40], although without stressing the distinction between them. In fact, most of his discussion seems to apply only to the animate type which he saw as the 'simultaneous occurrence of two different psychical states', one the normal state which can be causally explained, and the other the 'critical experience', which is not so explicable and 'cannot be derived from the

---

[1] Personally I am not sure what Kammerer's law of seriality is. After giving a complicated but unilluminating classification of series [1919, 37-91] he appealed to hypotheses of imitation and of attraction, as used in the sciences, as explanations [pp. 139-167], and pointed to the occurrence of periodicity in main areas of knowledge [pp. 167-280, 371-429]. So what?

first' [p. 40]. His view is like Kammerer's except for the alleged removal of causality [Koestler 1972, 95].

The theory is limited in content in several ways. For example, it is very ad hoc, for Jung appears to define the synchronistic mechanism as the acausal source of the link between the events; but only the events specify the mechanism in the first place. Again, since the coincidences have, for Jung, to be 'meaningful' to be worth considering, the theory comes perilously close to tautology. As Anthony Flew puts it, 'Synchronicity is not a (new) species of coincidence: it is coincidence' [1953, 199]. Hans Bender [1977] defends Jung to some extent in terms of Jung's theory of the collective unconscious; this paper, incidentally, contains a few excellent coincidences in sequence.

It appears, but dimly, from Jung's book that he thought that his archetypes would provide the means of avoiding ad hocness and tautology. For example (almost the only one), he commented on a coincidence involving a flock of birds that the flock 'has, as such a traditional mantic significance' [1955, 37], But without an extensively developed, testable theory of archetypes, the prospects are bleak that synchronicity will avoid the realm of the off-the-cuff.

11.2 A related doctrine on which Jung draws to attempt to articulate his acausal theory is 'astrology, which, at least in its modern form, claims to give more or less a total picture of the individual's character' [1955, 53]. A large part of the book [pp. 60-94] is taken up with arguing for the legitimacy of astrology.

However, Jung's knowledge of astrology seems not to have penetrated beyond the surface of the horoscope. The doctrines underlying astrology — astrosophy, anthroposophy, or whatever — often assert, in some form or another, that there are 'forces' or 'spirits' in the world which guide the individual's lifeline. But these doctrines are not totally deterministic; they allow him some free will in choice of thought or action. Thus while astrology 'cannot even be readily brought within the conception, fundamental in physics, of causation' [Carter 1963, 184], it is not an obviously secure source of acausality. Indeed, one of the principal issues in the more theoretical aspects of astrology is whether the associations it detects are to be taken as causal links, triggers, synchronicities or symbolisms (see, for example, [Gauquelin 1969, part III]). This issue is well exemplified by the coincidence reported in ¶6.

The status of acausality in the theory is dubious on at least two counts. Firstly, the detection of 'meaningful' coincidences necessarily involves a cognitive element; but then the status of acausality becomes doubtful. Is not the coincidence being causally influenced by its experient(s)? Issues such as this have been discussed in, for example, [Beloff 1977]; an extended discussion is provided in [Braude 1979, 217-241]. Secondly, Jung's distinction between psychical states, just quoted, is naive, since it assumes that it is clear which state is normal and which one is critical. Suppose person X thinks of person Y and then meets him; Jung would take the thought (and its associate purposes) as normal, and the meeting as critical. This position excludes, and by silence, the interpretation that the proximity of X and Y linked up with X's thought.

Jung would not have felt happy with this last possibility because he sought to avoid psychical explanations of coincidences. This is rather surprising, since in general he was not hostile to psychical phenomena, and his co-writer Pauli was a notorious experient of spontaneous psychokinesis [Koestler 1973, 177-180].

12. Propensity theories

In recent years information and synchronicity theories have met with little favour, and alternative forms of theorising have been offered for psi events in general. I coin the term 'propensity' to cover one group of theories, since in some form or another they assume that 'nature' — including both the individual and the physical environment in which he acts — has a propensity to fall into certain states of affairs which are favourable (or unfavourable) to the occurrence of psi events. This use of 'propensity' relates to Karl Popper's propensity interpretation [1959] of probability, to which I shall allude in ¶13.

12.1 An important pioneering step in this direction is contained in G. W. M. Tyrrell's brilliant essay [1947] on 'The "modus operandi" of paranormal cognition'. Surveying a variety of spontaneous cases (though not coincidences as such), he postulated that the percipient constructed by sub-conscious paranormal means a product called a 'mediating vehicle' which is not itself 'a paranormal phenomenon but is the product of psychological machinery which all possess. It may take the form of a sensory hallucination or of an impulse or of automatic verbalization or of a dream' [p. 117]. Among other considerations, he emphasised active-agent telepathy [p. 117] and criticised with great wisdom the current tendency to restrict psychical research to experimental work [pp. 66-67].

12.2 A more recent theory of this type is Rex Stanford's 'psi-mediated instrumental response', or PMIR [Stanford 1974], in which he describes a number of ways in which the organism may scan the environment by both sensory and extra-sensory means and so cause the occurrence of psi events, including coincidences. His theory of instrumental response corresponds roughly to Tyrrell's mediating vehicles, but is more detailed concerning the modus operandi involved, both the scope and the limits of its action. He covers both ESP and psychokinesis (PK), questioning the traditional distinction between them that information theories usually prescribe.

Stanford's theory belongs to the 'propensity' group since it asserts the existence of 'dispositions in the world for the occurrence of psi events' — favourable ones (high propensity) or unfavourable ones (low propensity). He makes this especially clear in a later, related, theory called the 'conformance model' [Stanford 1978], which attends in more detail to the psychological aspects. The brain is taken to be a random event generator, whose performance depends on the dispositions of the situation in which it finds itself; it seeks the most advantageous of the available outcomes, which are appraised according to the needs of the individual. ESP and PK are to be understood as oriented towards these needs [Edge 1978] rather than based on the processing of information. This approach to psychological factors is broadly behaviouristic.

12.3 The prospects for using propensity theories as explanations of coincidences seem relatively bright. Johnson had already seen a class of coincidences 'suggestive of "Design"', where 'an intelligent will is manipulating circumstances with some purpose — some end in view' [1898, 166-167]; but her interpretation is rather narrow in teleology, for in the spirit (as it were) of late Victorian metaphysics (§17.2-§17.4) she saw the Design as 'superhuman' and a 'possible manifestation of some universal guide' [pp. 205-207]. Tyrrell's and especially Stanford's approaches seem to allow for a greater range of interaction between the individual and his environment — and thus they also avoid the synchronist's dilemma, mentioned in ¶10, of having to decide which event is critical and which normal.

Various theses in Stanford's PMIR theory seem consonant with our experience of coincidences. His emphasis on the increase of a disposition with the strength of the related need [1975, 45] is exemplified by the fulfilment of purposeful coincidences. His claim that 'PMIR tends to be accomplished in the most economical way possible' [p. 46] — a sort of least action principle for psychical research — offers an explanation of the surprising character that coincidences can take. And until noting from his theory that psi mediation is less likely to occur when the individual is following a rigid lifestyle [pp. 49-50; compare [Schwarz 1980], 281-289], I had not realized the significance of the fact, recorded in ¶2, that I had had so many coincidences at holiday time in my schooldays.

12.4 However, there are still some important difficulties to face. Firstly, the problem of processing masses of information is now transferred to the problem, not necessarily easier to resolve, of understanding how paranormal cognition or instrumental response chooses its (optimal) route among those available. Paranormal cognition 'takes place, therefore, outside consciousness', asserts Tyrrell [1947, 68], which still leaves many possibilities available. 'It is as though nature anticipates the consequences of various outcomes and weights the probabilities accordingly', asserts Stanford [1978, 208], which may be asking much both of Mother nature and of us in detecting Her at Her work. Coincidences could be a valuable source of data here, since those cases where a clear fulfilment of purpose is achieved could be interpreted in these terms.

Secondly, there is the question of the range of objects susceptible to psi mediation. In tune with his largely behaviouristic style Stanford tends to restrict the possible scope of psi mediation to effects on physical matter. For example, his 'MOBIA' ('mental or behavioural influence of an agent') interpretation of active-agent telepathy as PK is grounded on the assumption that 'the agent directly influences the brain or mind of the "percipient"' [1974, 343]. I feel inclined to extend the range to abstract objects and construe as such the needs asserted in Stanford's theory, for example. This approach would be in tune with Tyrrell, who saw paranormal cognition as 'more nearly akin to awareness of a proposition' [1947, 117]. Some of the properties of coincidences described earlier may suggest the involvement of abstract objects, coincidences involving earlier coincidences (¶3), for example, or coincidences in sequence (¶4). Important arguments in favour of admitting abstract objects in theorising in general are presented in [Popper 1972], although I am unable to endorse his theory of the 'World 3' that he claims them to inhabit (§2.12).

There may be no clash with Stanford's view here, since the purported interaction with abstract objects (in any kind of psychical phenomenon) might be reducible to the action of individuals upon their own and others' brains and minds. Thus the question of reification arises again, in a most interesting form, somewhat reminiscent of Robert Thouless's proposal to reduce telepathy (traditionally conceived) to clairvoyance in the discussion [Rhine and others 1946]. If abstract objects subsist, they could be subject to scan by clairvoyance and 'direct' modification by PK; otherwise the effects would have to be construed telepathically or in terms of Stanford's MOBIA. Among the coincidence evidence, the former hypotheses might be preferred to account for Kammererian runs involving inanimate objects such as Crossley's 225s (¶4.1). But the question is horribly difficult to tackle.

12.5 Thirdly, one must consider the measure of space-time over which psi is held to operate to effect coincidences. Usually the events constituting a coincidence have to occur within a short period of time; but some can cover an extended interval and still be striking. For example, was psi at work in the months-ahead astrological prediction involved in the coincidence described in ¶6? Again, coincidences in sequence (¶4) can last over substantial periods of time; and those involved in the 'Peace' rose (¶5) also drew on events over a large spatial scale. 'What right have we to ignore the possibility of pre-established rapport?' asked Tyrrell (in a manner reminiscent of Johnson) while honestly reporting his perplexity over paranormal cognition [1947, 116].

Very roughly speaking, the larger the piece of space-time and the greater the number of people involved in a coincidence, the more complex would be an explanation of its content. However, the concept of complexity in ordinary life situations is itself exceedingly complex, and it would be hard to extend to spontaneous cases the experimental results discovered concerning the relationship between psi mediation and the complexity of the psi tasks set, whether complexity is specified in terms of information or goals. It is, of course, Stanford's aim that such an extension be achieved.

13. Observational theories

13.1 In recent years a group of theories of psi mediation known as 'observational' theories have been introduced as another alternative to the traditional information or perception theories. They are concerned with the working of psi at the physical level, and relate to recent developments in physics, especially quantum mechanics. They stand in complement to propensity theories, which emphasise psychological factors.

There are various differences between the forms of observational theory; I confine myself to the common factors, which are well indicated (together with particular differences) in [Millar 1978]. The key to these theories is retroactive causation; when the percipient examines the target which he has attempted to guess in an ESP or PK experiment, his brain (or that of anyone else who knows the results of his guesses), acting like a random event generator (REG), retroactively influences the target objects on which he was tested. Thus while the traditional chronology reads:

Preparation of   ⇒   Test of the   ⇒   Discovery of
  target objects   percipient   his results

the observational theories advocate the order

Test   ⇒   Discovery   ⇒   Preparation

The percipient's cognizance of the results of his guesses constitutes the 'observation' of the target objects from which those results have been obtained; this event is essential for the psi event to occur. According to these theories, ESP reduces to PK, and experimenter effects can be explained.

13.2 At this point observational theories and Stanford's conformance model differ, since for Stanford psi mediation does not have to occur [Millar 1978, 305]. However, propensity and observational theories could be formulated on more broadly different bases; for the observational theories subscribe to the psychophysical consciousness-with-physics interpretation of quantum mechanics, while propensity theories could correlate with objectivist consciousness-separate-from-physics interpretation of quantum mechanics — indeed, according to [Popper 1982], they must be so correlated.[2]

In order to limit the scope of occurrence of retroactive PK, 'only pure-chance events can be affected' by it [Millar 1978, 300]. However, since individuals' brains are assumed to act like REGs, then all events involving individuals are liable to involve 'pure-chance events', and the restriction of scope of retroactive PK seems to be rather slight. For now many spontaneous psi events in general and coincidences in particular, could fall within its realm.

Presumably an observational interpretation of coincidences would claim that when the percipient notices that the coincidence has occurred, he affects something(s) in the past that caused the pertaining events to occur together. But which event might it be? For example, when I saw the car number-plate 'IGG 225' before me (¶4.1), did I retroactively cause its driver to choose the time and route that he took? If so, at which stage in his journey did the effect take place? If I had happened not to notice that the car was in front, would it have got there for reasons now independent of my

---

[2] A critique of the observational theories along the lines of Popper's opposition to consciousness-with-physics would be an interesting exercise. In an amusing coincidence of notation, state vectors in quantum mechanics have long been denoted by 'Ψ'!

presence behind it? There seem to be some deep difficulties in applying this theory to spontaneous situations such as these.

Further, there is the apparently unlimited efficacy, which retroactive PK allows to memory and recollection. Observational theories do not claim that past events can be changed, but they allow an indefinitely large range of interpretations of them. In order words, they do not make a riot of the past, but they make a riot of history; for any event of the past might have been (partially) caused by something that some one will do in the future. '[C]an PK simply strike anywhere: can I now interfere with D. D. Home's levitation just by thinking about it', asks Millar [1978, 317], and well he might.

14. Coincidences as aids to problem-solving

14.1 A depth-psychological approach has been adopted by Jule Eisenbud [1970, 1982, 1983]. He draws with great virtuosity on Freudian principles to explain several kinds of psi event; but his reliance only on Freud may be a weakness as well as a strength [Grattan-Guinness 1982].[3]

Eisenbud urged me to consider this approach as an alternative to synchronicity. It asserts that the experiment utilizes his postulated psi faculty to choose from the current or nascent events and states of affairs in his space-time neighbourhood that in conjunction appear to be a coincidence but in fact are fulfilling a pressing aim or relieving (or at least tempering) a current stress. (Normally, no claim is made that the experient has induced the existence of the relevant events or states of affairs.) This type of explanation contrasts with synchronicity in accepting causality: it is the problems as such, or ours aims or stresses, that cause the psi faculty to spring into action.

Most, if not all, of my cases described above could be fitted into the psi hypothesis; but there may be as much strength as weakness in its explanatory power, for Popper's caveat about unfalsifiable theories (§9.5) applies at least twice. Firstly, the psi hypothesis asserts the existence of some aim to be fulfilled or stress to be relieved; but this can lead the analyst to indefinitely deep quarrying in the experient's psyche for motivations. Secondly, it postulates the existence of an ontology of events and states of affairs, maybe forthcoming ones — which smacks of determinism.

14.2 But let us go long with it. In the Mozart 1990 train case (¶4.4), when the lady arrived in my vicinity, psi-me picked up (how?) the fact that normal-me needed some of the information in her newspaper. So psi-me asked psi-her to sit down opposite normal-me; hence her shuffling, since normal-her had obviously intended to go further down the carriage.

---

[3] I had a pleasant animate coincidence involving Eisenbud and Velikovsky. I wrote to Eisenbud telling him of a recent meeting with Velikovsky (§16.8), and in reply Eisenbud recalled that in 1946 he had been encouraged to develop his ideas on telepathy by Velikovsky. I reported Eisenbud's recollection to Velikovsky a few days later: he had forgotten the incident himself, and had not heard Eisenbud's name for years — until someone had mentioned Eisenbud to him on the telephone earlier that day.

But psi wins as ever; psi-her agreed with psi-me's request, and told normal-her to sit down opposite me. (Luckily there was no ordinary disincentive for her, such as broken springs in the seat.) The remainder of the story is entirely normal; psi-me had given normal-me the opportunity to notice her paper, and thereafter it was up to normal-me to profit — or not.

As regards other coincidences, the case of meeting Edgar in New York (¶7.1) is my only one of *conscious* anticipation — made, I might add, with a strong inner certainty that I have never forgotten. On the Fourier case (¶4.3), I knew enough already about his work in Egypt from primary sources to make it most unlikely to seek out a book like Charles-Roux's (indeed, the drawing was its only new relevant item for me); but had I not seen the book, I would certainly not have found the drawing. As for the information in *Private Eye* (¶7.2), I rarely read it and I found no such items in the various other issues which I then sought out.

The key-ring case (¶3.1) differs from the others in that an effect was consciously being attempted; but the mode of causality was so unorthodox that the event can as "reasonably" be explained in terms of coincidence as by some (neo-causative) theory of telekinesis. The Rotorua example (¶7.3) has a precognitive flavour to it, especially as Babbage's book was not likely to be useful for the purpose that I had in mind (hence some other reason for its consideration needs to be sought), and the 1831 British Association *Reports* is a pretty rare work anywhere, never mind in a small village that was largely inhabited by Maoris.

With regard to the Vollpriehausen mine (¶7.4), I do have some knowledge of German document storage in the last war: this was why I was sent the paper to referee in the first place. However, I had never heard of the vast collection of pamphlets, quite apart from selecting probably the only relevant one.

15. Concluding remarks

It is a pity that so few extended studies of coincidences have been made; a 'census of coincidences', like the 'Census of hallucinations' of the 1890s, would be worth compiling, to see which properties are found frequently enough to warrant special emphasis. One possible source for such a census is those spontaneous cases described in the literature, for some could also be interpreted as coincidences. But even with the limited supply of information available, it seems clear that coincidences deserve much more attention from psychical researchers than has hitherto been the case.

With regards to theorising, of the various candidates discussed above synchronicity is the best known but probably the most feeble, observational theories seem to be rather unpromising, information theories are perhaps best pursued in Bergson's filtering form, propensity theories offer some chances for progress, as does viewing coincidences as aids to problem-solving. But testing any theory of coincidences is a daunting task, since it seems unlikely that coincidences will bow to the control and restrictions of the experiment. But the effort to test them will be worthwhile; for, with their remarkable variety of form, coincidences can notably enrich ordinary life, and their study similarly enhance psychical research.

Acknowledgements

This chapter draws on material prepared for lectures delivered at an SPR Study Day in November 1981, and in the SPR monthly lecture series in June 1982. I am grateful to the sympathetic response afforded by members of the audiences, who posed difficult and thought-provoking questions. I am also indebted, for discussions of this topic over the years, to Jule Eisenbud, Anita Gregory and Berthold Schwarz. In addition, I owe thanks to an anonymous referee, whose reaction to a draft of one paper was so completely opposed to its thrusts that I have had to recast its form, with benefit to the content.

Bibliography

Beloff, J. 1977. 'Psi phenomena: causal versus acausal interpretation', *Journal of the SPR, 49*, 573-582.

Bender, H. 1977. 'Meaningful coincidences in the light of the Jung-Pauli theory of synchronicity and parapsychology', in B. Shapin and L. Coly (eds.), *The philosophy of parapsychology,* New York: Parapsychology Foundation, 66-81.

Bergson, H. 1914. 'Presidential address', *Proceedings of the SPR, 27*,157-175.

Braude, S. 1978. *ESP and psychokinesis: a philosophical examination,* Philadelphia: Temple University Press.

Carter, C. E. O. 1963. *The principles of astrology,* 5th ed., London: Theosophical Publishing House.

Edge, H. L. 1978. 'A philosophical justification for the conformist behavior model', *Journal of the American SPR, 72*, 215-232.

Ehrenwald, J. 1982. 'Psychical phenomena and brain research', in I. Grattan-Guinness (ed.), *Psychical research,* Wellingborough: Aquarian Press, 325-334.

Eisenbud, J. 1970. *Psi and psychoanalysis,* New York and London: Grune and Stratton.

Eisenbud, J. 1982. *Paranormal foreknowledge*, New York: Human Sciences Press.

Eisenbud, J. 1993. *Parapsychology and the unconscious*, rev. ed., Berkeley, California: North Atlantic Books. [1st ed. 1983.]

Flew, A. 1953. 'Coincidence and synchronicity', *Journal of the SPR, 37*, 198-201.

Gauquelin, M. 1969. *Science and astrology,* London: Peter Davies.

Grattan-Guinness, I. Review of [Eisenbud 1982], *Journal of psychophysical systems, 5*, 89-91.

Grattan-Guinness, I. and Ravetz, J. R. 1972. *Joseph Fourier 1768-1830.* Cambridge, Mass.: The MIT Press.

Gregory, A. 1982. (Ed.) 'London experiments with Matthew Manning', *Proceedings of the SPR, 56, 283-366.*

Hardy, A., Harvie, R. and Koestler, A. 1973. *The challenge of chance: experiments and speculations*, London: Hutchinson.

Inglis, B. 1990. *Coincidences*, London: Hutchinson.

Johnson, A. 1899. 'Coincidences', *Proceedings of the SPR, 14,* 158-330.

Jung, C. and Pauli, W. 1952. *Naturerklärung und Psyche*, Zurich: Rascher Verlag. [English trans.: *The interpretation of nature and the psyche*, London: Routledge and Kegan Paul, 1955.]

Jung, C. G. 1955. *Synchronicity. An acausal connecting principle*, London: Routledge and Kegan Paul.

Kammerer, P. 1919. *Das Gesetz der Serie,* Stuttgart and Berlin: Deutches Verlags-Anstalt.

Kennedy, J. E. and Taddonio, J. L.1976. 'Experimenter effects in parapsychological research', *Journal of parapsychology, 40*, 1-34.

Koestler, A. 1972. *The roots of coincidence*, London: Hutchinson.

Koestler, A. 1973. 'Anecdotal cases', in [Hardy and others 1973], 157-204.

Millar, B. 1978. 'The observational theories: a primer', *European journal of parapsychology, 2*, 304-332.

Popper, K. R. 1951. 'The propensity interpretation of probability', *British journal for the philosophy of science, 10*, 25-42.

Popper, K. R. 1972. *Objective knowledge,* Oxford: at the Clarendon Press.

Popper, K. R. 1982. *Quantum theory and the schism in physics* (ed. W. W. Bartley, III), London: Hutchinson.

Rhine, J. B. and others. 1946. 'Telepathy and clairvoyance reconsidered', *Proceedings of the SPR, 48,* 1-31.

Rhine, L. E. 1977. 'Research methods with spontaneous cases', in B. B. Wolman (ed.), *Handbook of parapsychology*, New York: Van Nostrand, 59-80.

Ridge, A. 1956. *For love of a rose,* London: Faber. [Cited from paperback ed. (1972).]

Rouse Ball, W. W. R. 1939. *Mathematical recreations and essays,* 11th ed. (rev. H. S. M. Coxeter), London: Macmillan.

Schwarz, B. E. 1980. *Psychic-nexus. Psychic phenomena in psychiatry and everyday life*, New York: van Nostrand.

Spencer Brown, G. 1957. *Probability and scientific inference,* London: Longmans, Green.

Stanford, R. G. 1974. 'An experimentally testable model for spontaneous psi', *Transactions of the American SPR, 68,* 34-57, 321-356.

Stanford, R. G. 1978. 'Towards reinterpreting psi events', *Journal of the American SPR*, 1978, 72, 197-214.

Stevenson, I. 1983. 'Cryptomnesia and parapsychology', *Journal of the Society for Psychical Research, 52,* 1-30.

Targ, R. and Puthoff, H. 1977. *Mind-reach. Scientists look at psychic ability*, London: Jonathan Cape.

Tyrrell, G. N. M. 1947. 'The "modus operandi" of paranormal cognition', *Proceedings of the SPR, 48, 65-120.*

Vaughan, A. 1979. *Incredible coincidences*, New York: Lippincott.

von Scholz, W. 1924. *Der Zufall,* 1st ed., Munich: List. [Later eds. under the title *Der Zufall und das Schicksal.*]

19 Is Psi Intrinsically Non-linguistic? Some Preliminary Considerations

The hypothesis is proposed that psi action is intrinsically non-linguistic: that is, that it involves parts of the mind that are distinct, in some way, from those parts concerned with language. After presenting a background for the hypothesis and outlining its principal features, several of its consequences for psychical research are explored.

There is, however, no *a priori* ground for supposing that language will have the power to express all the thoughts and emotions of man [...] The needs of science and of commerce have become dominant [...] and our vocabulary, based as it is on concrete objects and direct sensations, is refined for the expression of philosophic thought [...] nor can we wonder if our supraliminal manipulation leaves us with an instrument less and less capable of expressing the growing complexity of our whole psychical being.
F. W. H. Myers [1903, 99-100]

1. Hypothesis

One of the crucial factors in the performance of human beings is their deployment of language. This is carried out in various ways; for convenience I appeal to Karl Popper's classification into four modes of the principal uses made of languages [1972, 235-241]:

expressive (of a state of the organism);

signalling (to another organism, which may then respond);

descriptive (of states of affairs, both within the organism and in situations that arouse its interest); and

argumentative (as a means of critical discussion, possibly with other organisms, of problems and theories).

The word 'organism' covers non-human living beings also; a corollary to the classification is that humans are distinguished from the others by the scale of their capacity to develop the descriptive and argumentative modes.

The hypothesis outlined in this chapter is that *psi is intrinsically non-linguistic* (or languageless). In the next section I expound some of its principal and general features, and also distinguish it from other views with which it might be confused; and in the following ones I consider some of its more specific applications and aspects. {Some version of the points made here can be upheld without assuming that psi exists.}

This chapter was first published in the *Journal of the Society for Psychical Research*, 53(1985), 8-17; it is reprinted by kind permission of the Society.

2. Exposition

2.1 It is important to stress that the hypothesis is concerned with psi as and when it acts, and not with discussions and recollections of that action on other occasions; these will indeed be conducted in language in the usual way. Again, it does not rest upon the strangeness of many paranormal phenomena, nor on the inaccessibility of the objects and processes involved; these are found in various orthodox contexts (for example, pain, which is also notoriously difficult to describe). Further, the hypothesis does not raise issue with the fact that our orthodox cognitive experience, in all its range (thinking, feeling, controlling ourselves, and so on) often proceeds without any particular appeal to language [Price 1953]. However, in these contexts we often quickly reach a state in which language is prominent: not only words and statements made to others (both well-formed sentences and fragments such as indexicals) but even private compositions of linguistic objects (for contemplation, perhaps, or writing down or self-recitation).

2.2 The hypothesis proposes a deeper distinction than those involved in the situations just described. For whatever else happens when a paranormal event occurs, it seems clear that some mental action on the part of at least one living being is involved in it. The hypothesis then states that those parts of his mind and brain participating in the psi action are, *in some fundamental way*, distinct from the parts that are involved in language-oriented activity. As a corollary, *a propos* of the distinction between psi action and later linguistic discussion and recollection, it is proposed that the dichotomy between psi and language interferes with, perhaps even blocks, that linguistic expression. In other words, the psi data are in the wrong part of the head, as it were, for efficacious linguistic expression.

The scope of reference of the word 'language' needs specifying; and this is not easy. I intend it to cover all natural human languages capable of effecting all the four modes in ¶1, and also all related metalanguages; (semi-) formalised and/or symbolic languages as used in logic, mathematics, computing and some other sciences such as chemistry; and neo-languages as employed in some creative arts, especially music. Further, by 'language' I intend more than just its words, so that 'non-linguistic' is not synonymous with 'non-verbal'. However, in some sense, which is hard to specify, I wish to exclude some forms of communication from the phrase 'language-oriented activity', partly for the cases mentioned in ¶1 but mainly to grant psi its place.

2.3 In order to emphasise the distinction between language and psi-oriented thought I shall speak of the 'linguistic mind' and the 'psi mind'. The linguistic mind is those parts of the mind's and brain's activity that are oriented towards, and guided by, expression in language. The psi mind is those parts involved in activities deemed to be paranormal. The development of the hypothesis requires theorising on the distinctions between the linguistic and the psi minds, and on the ways in which they work together. Here are some initial thoughts.

3. Characteristics

3.1 Presumably the distinction would correlate with some distinction(s) between modes of working of the brain; but I propose no such simplistic demarcation as that between left- and right-brain activity, in which some important findings concerning severe epilepsy have been generalised and extended, in surely excessive optimism, into pretended characterisations of thinking in general. Nor do I claim that the mind is composed of only these two parts, or that the boundary between them is sharp or fixed: the processes of ideation possibly switch from one to the other, for example.

Further, I do not associate the linguistic mind only with ordinary consciousness, or the psi mind only with "the rest". For one reason, I do not regard the boundary line between the normal and the paranormal as sharp: indeed, examples that straddle the line arise in ¶4.4-¶4.6. Put another way, since psi is defined negatively, as the *para*normal, one should not expect to find a positive defining property for it, demarcating it cleanly from the normal (§16.5). Again, there are non-linguistic normal activities (some were mentioned in ¶1); and in the contrary case there may be linguistic paranormal activities (¶3.2).

Thus I do *not* assert that the hypothesis applies to all paranormal phenomena; or, at least, if it is a necessary condition for them to occur, then it is not also a sufficient condition. Instead, I propose it in a modal form, claiming that it applies to at least many areas of psychical research and proposing that the range of its applicability is itself an object of study.

Among areas where its credentials seem respectable, in telepathy it says that different psi minds interact with each other, and not necessarily each with its linguistic companion. In clairvoyance (and perhaps out-of-the-body experiences and precognition also) the psi mind acquires information about the physical world in some paranormal way and may then transfer it to the linguistic mind, especially if the information itself is in a language. For psychokinesis (in all its forms, including healing) the psi mind is involved, very likely without linguistic aid, in affecting the form or even the nature of physical objects themselves; an apparition is the product of the psi minds of both the person seen (or his psi residue, if dead: compare ¶4.5 below) and of the witness (which is why he can see the apparition while his linguistically dominated neighbour cannot); and poltergeist effects have a similar root in the psi mind(s) of the focus person(s) and maybe sometimes also in those of the external sources. I give these examples schematically, without taking any position on the reducibility of one class of paranormal phenomena to another, or on the role of the experimenter effect (which is discussed in ¶4.7, by the way).

3.2 In various kinds of paranormal phenomena, such as mediumistic communication or automatic writing, the psi action delivers a linguistic product; and the role of this linguistic information may be worth reconsidering. The normal view is to assign prominence to the content of this information and to try to interpret it (as evidence of survival, say, or of possession). However, there are other possibilities. The linguistic objects may be significant only as speech acts, or as phenomena as such; they may be mutants of, or even interferences in, some basically non-linguistic psi action. In other words, *there is a mental mis-match between the linguistic and the psi minds*; language unavoidably gets in the way, in a kind of 'language effect' akin to the experimenter effect. Under this view, disputes over the interpretation of linguistic information (for example, whether certain mediumistic communications are evidence for or against survival) may beg questions about its character.

Very often psi experiments are devised around some linguistic target (such as trying to read a message clairvoyantly, or think of the text that an agent is reading); but under the hypothesis entertained here, such a psi task may be maximally difficult to accomplish. As an alternative, targets should be devised of a non-linguistic character, if possible: states of mind as such, say, or emotions. (Note that women often form the majority of good subjects; and they are usually held to be the more emotional gender in general.) Under these circumstances the psi mind of the subject may best be able to engage with the targets before translating the results back to the experimenter in linguistically expressed information. However, this latter process may constitute a penalty to pay, in that linguistic accounts of interaction with such targets will be difficult to express.

Personal experience can exemplify the point raised in this clause. I do not have any special psychical ability, but I experience a number of coincidences, of which several involve unexpectedly finding books in libraries and bookshops (§18). Now books are of course containers of language; further, I am almost entirely dependent on language as a means of thinking (I "hear" everything in the head, like a radio, and cannot think "in" pictures or colours). Nevertheless, without exception all the coincidences involving books have been seemingly quite spontaneous (although very specific: I usually find not only the book but also the page required). By contrast, the various occasions when I have tried *deliberately* to find particular books or articles have always ended in failure — perhaps because I was thinking, in some language, of the task at hand, thus dooming myself to failure from the start.

3.3 A similar point may be made concerning the linguistic discussions of psi action that was mentioned in ¶1. It is well known that psychics often have very great difficulty in describing the mode of their psi action, even if they are normally fluent in linguistic expression: accounts are often found to be brief and bitty or loquacious and rambling, and incoherent in both cases. (Hence the penalty noted in ¶3.2 may not be very high!) Over and above the difficulties mentioned in ¶1 above caused by coping with the strangeness of the phenomena, the language effect of ¶3.2 may be involved again: the linguistic mind is trying to recall a part of the history of its psi companion.

Under the hypothesis, *three* kinds of competence are required of a good subject: firstly, ease of access to the paranormal via the psi mind; secondly, fluency of access from the linguistic to the psi minds to help exercise will and control over psi action; and thirdly, the converse fluency in order to effect adequate linguistic account of that action. Most people have sufficient linguistic skills but little available psi gift; and of those gifted as psychics, many seem to be weak on either the control or the linguistic sides, and very likely both.

3.4 While the evidence is far from conclusive, there are some signs of negative correlation between psychic gift and intelligence (in the usual sense of that term). For example, certain children make good subjects, but their ability often declines as they grow up (so does their capacity for visual images, sometimes) and their intelligence develops. (What about studies of baby psi, by the way?) Again, some adult psychics reveal a psychotic personality and/or (in the case of Eusepia Palladino, for example) poor linguistic powers. Indeed, in his studies of mediums James Hyslop coined the term 'pictographic process' to describe their apparent conversion of pictorially obtained psi data into verbal form [Hyslop 1918, 219]. When we note that intelligence is almost always appraised in terms of linguistic response and performance, this negative correlation is not a surprise.

4. Consequences

4.1 The principal task facing the testing of the hypothesis in psychical research is to invent situations in which a subject could inhibit his linguistic mind and thus release his psi mind, and then to appraise his capacity to effect psi action. Some efforts have already been made in this direction: I think especially of the Ganzfeld technique (where a homogenous environment is created for the percipient by putting shades over his eyes and transmitting white noise to him on headphones), and regard the experimental successes achieved with its aid as a corroboration of the hypothesis. In this section I outline a medley of aspects of psychical research upon which the hypothesis has bearing; I hope that professional researchers may be able to take up and exploit. Some of these points provide settings for situations already known in the field: others suggest alternative interpretations of such situations; a few indicate possibilities for theoretical and experimental studies, which seem not yet to have been followed.

4.2 One of the greatest difficulties for psychical researchers is to set their subjects genuine problems to solve, rather than contrived and insignificant activities such as bending forks or guessing dice. However, as an ethical point, they cannot set their subjects' lives in danger, even though psi may well then be more forthcoming (as in some cases of crisis telepathy and near-death experiences). Hence researchers should study other areas of human knowledge in which the inhibition, or even absence, of language is evident; for the hypothesis asserts that the situations so described are also psi-conducive for the subject. In addition, they may suggest more interesting tasks for him to tackle, and so reduce his boredom — a state that, one notes in passing, can inspire linguistic thinking.

Where can such lore be found? One source, I suggest, is anthropology; for reasons that can easily be fitted into my speculation on the ancient uses of psi,[1] it is well known that in peoples whom clever Western man designates as 'primitive', paranormal skills are still quite evident and trance and other states are generated [Van de Castle 1977]. Another source is creative activity in general [Koestler 1964], and especially in the arts (for evidence from composers, see [Abell 1956]). Creative acts seem to have non-linguistic origins that then seek linguistic forms of expression (or neo-linguistic in music, as mentioned in ¶2.2); 'the words come last', as T. S. Eliot is said to have said (compare his [1957, 97-98]). Creative artists are those who preserve into adulthood their child-time capacities for imagery.

Analogous situations in science are not easy to find, but Myers noted one in his discussion of genius and creative acts: mathematical calculating prodigies [1903, ch. 3]. Steven Smith [1983] has published an excellent history of these calculators: a linguistician by profession, he claims that lightning calculation and language share a common root in the unconscious [p. 25]. But equally, one might take this unconscious origin as itself non-linguistic and so its study to be of interest to psychical researchers. Smith notes another property of calculators, which reminds us of ¶3.4 above [pp. 40-44]: some of them are not specially intelligent in any usual sense of the term.

Mysticism seems to be another source of information, since it is often regarded as a non-linguistic activity and involves unorthodox modes of perception; further, as with some psychical phenomena, its induction can be helped with the use of certain drugs. However, mystical states, whether allied introvertively to the mind or extrovertly to the senses, are often held to be purely subjective experiences (although objective correlates such as changes in pulse rates may be detectable), whereas many psychical phenomena have an objective side.[2] Maybe psi and mystical states share in common the absence of language; if so, the psychical researcher can learn from the mystic for methods of attaining such states (and, conversely, the mystic could be appraised by the psychical researcher of objective experiences that can then be had). Further, both might learn much from practitioners of meditation.

---

[1] In [Grattan-Guinness 1983b; compare §16.8] I proposed that early man developed psi as a main means of (trying to) avoid disaster and peril to which he was so subject, but that it declined over the generations to become the paranormal. However, a residue remains within him and can be aroused as self-conscious 'horror by association' when psi is discussed, so that pathological effects are, to this extent, integral to psi action. This view could be related to the hypothesis of this chapter as follows: the progress of humans has been much dependent on their development of language, which is positively correlated with the decline of psi.

[2] I employ the terms proposed in Stace [1960, 61]; he proposes defining characteristics for each mode [pp. 78-79, 110-111 and 131-133]. However, his demarcation of subjective from objective experiences in terms of trans-personal experientiability and order [pp. 137-143] seems rather naive. In his ch. 7 on 'Mysticism and language' his main point is to distinguish the apparently non-linguistic mystical experience from its later linguistic recollection.

4.3 Some of the known psi-conducive conditions are those in which the linguistic mind appears not to enjoy its normal status of dominance. For example, in the hypnagogic and hypnopompic states the person is coming up to, or falling away from, the normal state of consciousness, and he experiences visions and images [Leaning 1925]. Again, the conditions induced on the subject by the Ganzfeld studies reduce language-oriented activity; in fact, it is a form of sensory deprivation, when psi (and also suggestion) can be prominent. Further, the similarities stressed in [Beloff 1974] between subliminal and extra-sensory perception may rest on their common non- or pre-linguistic character.

Conversely, the decline effect in subjects' performance may be caused, among other factors, by the return to customary dominance of their linguistic minds after the circumstances of attention to the experiment have begun to wane. Again, the sheep/goat effect might relate to the willingness or reluctance of the subject to allow that linguistic dominance to be set aside for a while. Among spontaneous cases, the fading of an apparition could be due to the falling away of the psi mind either of the witness or the organism "observed" (or both, given the scenario described in ¶3.1 above).

It would be worth finding out if depressive states, and those involving nervous breakdowns, were also psi-conducive; they do involve situations where the person's normal rationality weakens or breaks down. If so, discussing psi, even linguistically, excites associations with those non-linguistic parts of the mind that are involved in depressions, breakdowns, and the unlikeable like, that some psychics exhibit, as was noted in ¶3.4.

Another possible candidate for psi-conduciveness is kinaesthetic performance, which is a seemingly non-linguistic but surely intelligent and purposive activity. Of particular relevance are the cases when the subject is in a greatly relaxed state: under hypnosis, for example, or when receiving anaesthetics.

4.4 Consideration of the transfer from the psi to the linguistic minds relates naturally to phenomena involving other states of consciousness and multiple personalities: some dream material; examples of dissociation such as hypnosis and multiple personality; and mediumship evidence, including automatic writing and cross-corresponding evidence. In all these cases the subject is somehow taken away from his normal state of consciousness to produce information from sources to which he usually does not have normal access and of which he may later have no conscious recollection; however, this information is almost always delivered in linguistic form. It would not be possible for me to explore this difficult and complicated technical area in detail — E. R. Hilgard [1977] provides an excellent survey, apart from his credulous hostility to psychical research — but at least the following may be said.

It seems that the mind of a subject exhibiting such phenomena is divided into separate parts, of which some are consciously aware of others but not vice versa. I assume that this structure lies *wholly* outside the linguistic mind. Thus, for example, if a person P is found to exhibit personalities under hypnotic regression, then the information furnished by these personalities comes out, in language (whether or not it is P's natural language), from some mental source via his psi mind by a route that P's primary personality normally bypasses. Thus, even if the content of a personality is normal in character, such as 'Yes, I shall have a cigarette', it still passes through the extra stages that P's own 'Yes, I shall have a cigarette' does not require.

I further assume that such a subject's linguistic mind is itself not fully cognisant of the structure of his psi mind (or minds!). Hence the scrappy and disjoint character of the linguistic information is not so much evidence of the chaotic state of his psi mind but (or also) of the poor means of communication that it has with its linguistic companion.

I maintain this position both for "acceptable" topics in psychology such as multiple personality and also for "nonsense" themes like mediumistic communication. Indeed, here is a fine example of my reluctance to isolate psi phenomena as some positively definable category (¶3.1); *both* sides of the allegedly rigid line separating the normal from the paranormal are accommodated by the considerations presented in this clause. There are undoubtedly differences between these topics, but they lie elsewhere.

4.5 Extrapolating "upwards" to the supposed worlds of departed beings and discarnate entities, the hypothesis suggests that after death our survival residue takes some non-linguistic form, or at least that the residue is accessible only to the psi minds of sitters and mediums. This being supposed, the question 'What makes a good sitter or medium?' can be answered in a new way: namely, someone gifted in, or good at, non-linguistic mental activities, such as imaging and meditation, and perhaps also poor at linguistic expression and verbal response (compare ¶3.4 above). These factors should be brought into play when the psychological traits of sitters and mediums are considered.

Similar considerations could be made concerning the witnesses of ghosts and apparitions (and also the living persons whose apparitions are seen), and of focus persons in poltergeist cases. People thought to manifest reincarnation might also be appraised in this way: as many are young children, the points raised in ¶3.3 and ¶3.4 should be borne in mind.

4.6 Extrapolating "downwards" these considerations of language bring us naturally to animals. While it seems that the argumentative mode of language is usually absent from them, the other three modes appear to be present in limited versions and, moreover, in an astonishing range of individual and collective forms that animal ethology is busily revealing [Griffin 1981]. It is very likely that animal psi is embodied here and there within this range; and research into it has largely modelled itself on procedures adopted for studying human psi. The considerations presented above suggest *a reversal of role*, in that psychical researchers might consider treating their human subjects as if they were linguistically primitive, like animals. In other words, as a converse to ethologists' interest in the language learning of animals, psychical researchers could attempt to study the non-language learning of humans.

It might then emerge that the difficulties unavoidably imposed upon research in animal psi give clues for human-oriented studies. Further, the extrapolation mentioned above suggests that animal psi is an important class of paranormal phenomena anyway; for if "primitive" human society and "unintelligent" psychics can draw extensively on psi, as was suggested in ¶4.2 and ¶3.4 respectively, then animals may do so also. In such a case the psychical researcher and the ethologist have some interesting problems in common, once again crossing the line of ¶2.3 above between the normal and the paranormal.

A caveat needs to be registered, however. Humans are rich possessors of language, in all the four modes outlined at the head of this paper, much more so than other animals, so it seems; hence any similarities (or even striking differences) claimed to exist between animal performance and human psi action must be treated with caution. Further, the proposed similarity or difference is itself a linguistically expressed human theory, which may be too anthropocentric to serve as useful theorising for the psychical researcher.

4.7 From the lines suggested above, alternative and perhaps more efficacious psi experiments and investigations may be pursued. The proposed distinction between the linguistic and the psi minds may help to understand the issues involved here. Figure 1 shows the simplest case, with one experimenter and one subject; the other cases, involving several of either or both and possibly agents also, can be made up by combination. The boxes indicate objects and situations; the ellipses enclose possibilities.

Figure 1
Tableau of experimenter and subject,
using the proposed distinction between linguistic and psi minds.

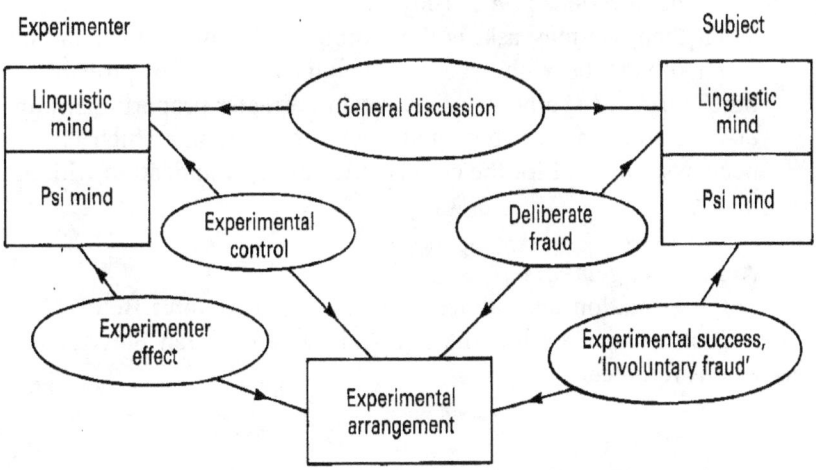

One of these possibilities is the presence or absence of the experimenter effect. According to the hypothesis, its presence in psi experimentation would be appraised statistically by comparing the results obtained when subject and experimenter are both in as psi-conducive/language-inhibitive a state as possible (so that the experimenter effect is encouraged) with those found when the subject is in a psi-conducive state while the experimenter is in a language-conducive state at the same time (when the effect is minimised). The former situation often obtains; the latter situation would be difficult to arrange. Perhaps it could be done if psychical researchers could investigate means of keeping themselves as linguistically occupied as possible during an experiment, even executing irrelevant tasks such as reading poems out loud (or to themselves); by contrast, waiting in (near) silence for the experiment to be completed is asking for the experimenter effect to occur.

However, the vulnerability still remains to psi action of so-called 'control objects' in target-control experiments (that I urged in [Grattan-Guinness 1983a] should be taken much more seriously than is the case). Even if the experimenter's psi access has been reduced by linguistic saturation, the subject's increases with the psi-conduciveness of his state, and so does the possibility of psi access to the supposedly control objects.

5. Prospects

As the title of this paper indicates, the considerations are preliminary; but I hope that I can introduce some new perspectives into those discussions of psychical research that are dominated by states of consciousness and related topics (see [Valle and Eckartsberg 1981] for a representative selection of recent views). For language is considered only with respect to the roles that it can play when *present*; the place of its *absence* does not gain attention. For example, neither in the book just cited nor in many others that I consulted does 'language, absence of' or some cognate phrase appear in the index.

Yet this topic may well be worth talking about, somehow. The quotation from Myers at the head of this chapter continues with a perfect expression of the questions [1903, 100]:

> What then, we may ask, is the attitude and habit of the subliminal self likely to be with regard to language? It is not probable that other forms of symbolism may retain a greater proportional importance among those submerged mental operations, which have not been systematised for the convenience of communication with other men?

Acknowledgements

For information and advice I am indebted to John Beloff, Emily Williams Cook, John Crossley, the late Anita Gregory, Ian Stevenson and an anonymous referee.

Bibliography

Abell, A. 1956. *Talks with great composers*, London: Spiritual Press.
Belolf, J. 1974. 'The subliminal and the extra sensory', in A. Angolf and B. Shapin (eds.), *Parapsychology and the sciences* [...], New York: Parapsychology Foundation, 103-116.
Eliot, T. S. 1957. *On poetry and poets*, London: Faber and Faber.

Grattan-Guinness, I. 1983a. 'A note on th efficacy of control objects in psi experiments', *Journal of the Society for Psychical Research*, 52, 126-128.

Grattan-Guinness, I. 1983b. 'On sources for the hostility to psychical research', *Psychoenergetics*, 5, 199-206.

Griffin, D. R. 1981. *The question of animal awareness. Evolutionary continuity of mental experience*, New York: Rockefeller University Press.

Hilgard, E. R. 1977. *Divided consciousness* [...], New York: Wiley.

Hyslop, J. H. 1918. *Life after death*, London: Kegan, Paul, Trench and Trubner.

Koestler, A. 1964. *The act of creation*, London: Hutchinson.

Leaning, F. E. 1925. 'An introductory study of hypnagogic phenomena', *Proceedings of the Society for Psychical Research*, 35, 289-411.

Myers, F. W. H. 1903. *Human personality and its survival of bodily death*, vol. 1, London: Longmans, Green. [Cited from the 1954 printing.]

Popper, K. R. 1972. *Objective knowledge*, Oxford: at the Clarendon Press.

Price, H. H. 1953. *Thinking and experience*, London: Hutchinson.

Smith, S. B. 1983. *The great mental calculators* [...], New York: Columbia University Press.

Stace, W. T. 1960. *Mysticism and philosophy*, Philadelphia: Lippincott.

Valle, R. S. and Eckartsberg, R. von. 1981. (Eds.) *The metaphors of consciousness*, New York and London, Plenum Press.

Van de Castle, R. 1977. 'Parapsychology and anthropology', in B. Wolman, (ed.), *Handbook of parapsychology*, New York: van Nostrand, 667-686.

20   Experience or Innateness? Sir Karl Popper on the Origins and Acquisition of Natural Languages

In conversations with the author in his last years, Sir Karl Popper outlined a conjecture on the origins and acquisition of a natural language. The basic standpoint is that one-word "sentences", along with body actions, gestures, and tones of voice, are the basic building-blocks not only in the learning of language but also in its formation. Further development happens (if at all, for a given person or community) in the emergence of two-, three-... word sentences (or phrases), with syntax and grammar gradually supplanting actions and tones. The conjecture contrasts with some influential philosophies of language, including Carnap's logical syntax of language, Chomsky's theory of innate structure, and Piaget's notion of autoregulation.

The question is not whether innate structure is a prerequisite for learning, but rather what it is.
    N. Chomsky, in [Piattelli-Palarmini 1980, 310]

[...] knowledge is always a modification of earlier knowledge.
    K. R. Popper, in [Popper and Eccles 1977, 425]

1. Popper's hierarchy

Popper's conjecture was based upon his claim that there are four different functions of a natural language, which may be ranged in the following hierarchy:

expression: the external manifestation of some inner state of an individual; executed in a revealing (or unrevealing) way;

registration: expression plus registration or indication of some internal and/or external states; executed in an efficient (or inefficient) way;

description: expression plus registration plus assertion of some situation; executed in the form of statements (or an equivalent) that is true or false according to the correspondence theory of truth;

argumentation: expression plus registration plus description plus assertion of some line of reasoning; executed in a valid (or invalid) way.

Further, mankind is distinguished from the other species of the animal kingdom by his capacity to execute the last three functions (including false descriptions, and invalid as well as valid argumentations). I have used the

---

This article was first published in *Languages Origins Society Forum*, no. 20 (Spring 1995), 16-25; the section titles have been added. So have ¶2.2 and ¶6.4, for which I express much thanks to Clare McLaughlin. ¶9 is a new coda for the second half of this book.

somewhat more general term 'registration' instead of his 'signalling' for the second function.

{Although Popper never mentioned him to me, he was heavily influenced by the views of the well-known psychologist Karl Bühler (1879-1963) of the Pedagogical Institute of the University of Vienna; see, for example, [Bühler 1919, esp. ch. 6]. In 1928 he wrote a dissertation on the psychology of education under Bühler's direction; it is now lost [Popper 1976, ch. 15].}

Popper presented his hierarchy in, for example, [Popper and Eccles 1977, 57-59]. He formulated it originally as a consequence of his evolutionary epistemology; it also fits in with his general methodology of conjectures and refutations and with his preference for indeterminism over determinism. He extended it to a conjecture about the origins and acquisition of natural languages by humans; he never published it, but he discussed with me on several occasions. I have elaborated it a little here, mainly in ¶5, ¶7 and ¶8. The main principles are outlined in the next section, which is followed by a string of some consequences and points of difference from more orthodox linguistic theories. For reasons of space and scope, only a few examples are taken, and always within the framework of Popper's hierarchy; placing the conjecture within linguistics in general, and within Popper's philosophy as a whole, await later occasions.

2. Building up sequences of words

2.1 Parents and child psychologists know well that the initial speech utterances by children normally take the form of single words or even noises. They are usually expressed in varied and varying tones and levels of voices, and accompanied by body signals and gestures; all these qualify as neo- or sub-languages of their own, often of a general character. According to the conjecture promoted here, these facts should be taken as the *principal* features for examining both the acquisition *and* the origins of a natural language, especially its descriptive and argumentative functions. For convenience appropriate words from modern English will be used to express utterances, but no assumptions of a phonological kind are entailed.

Initially the human actor created only single utterances (or utterance-groups) in his proto-language; for example, 'wolf' or 'kill'. Often too primitive to avoid vagueness or ambiguity, they were desimplified, after the manner of conjectures and refutations, into concatenations of utterances, such as 'beware wolf', or 'wolf kill'. This latter is still ambiguous: is the statement intended that the wolf can kill, or the request that the wolf be killed? So man came to grasp properties of word order, with 'wolf kill' distinguished from 'kill wolf', and 'I kill wolf' from 'wolf kill I'. Tones of voice, used to provide emphasis (and represented here by italics) led to further distinctions, such as '*kill* wolf' (rather than, say, '*feed* wolf') as opposed to 'kill *wolf*' (rather than, say, 'kill *sheep*'). Body language such as gestures and eye movements, and physical actions such as touching and striking, were integral parts of his natural language at early stages. They were essential for him not only to convey his own information to others but also to understand and react to their sentences.

{2.2 It is important to distinguish to distinguish two ways in which sentences (or at least declarations) can be built up. First there is the conscious adding of words to make up sentences or least to express wishes; for example, 'food' ⇒ 'food hot' ⇒ 'hot food now'. But there must also be a gradual recognition by the growing child that words that he assumed to be single are actually multiples and so require segmenting. Suppose, for example, that a father points to something and says 'big shoes'; he will presume that he has taught his child two separate words, but the child himself may think that he has learned the one word 'bigshoes' and will only register the separation of the two words after he has learnt other word-groups, such as 'green shoes' and 'big feet'. He will also have to recognise with non-examples of word separation such as 'greengage' and 'bigot'.}

2.3 From such processes of successive desimplification man and his companions came to create elements of syntax, over and above the (presumably) steady increase in vocabulary. Nouns and verbs were likely to be the most prominent components at first; the difference between them became clear, even though some words play(ed) both roles (for example, 'crack'). {The acquisition of verbs must be especially significant, for they (or their surrogates) bind together sentences; in terms of the distinction between parts and moments advocated in §12.4, verbs are moments of sentences while other words are parts.} Then adjectives and adverbs emerged as qualifiers, especially for external characteristics ('blue', 'very big') and eventually to internal ones. Then grammar began to flower, with conjunctions added (compare ¶8 below), and also prepositions (probably by convention for individual cases, given the complexity of their use in modern languages!); and types began to emerge from all these tokens. The importance of tones of voice and of body language diminished, as their roles were taken over by more refined utterances.

As a natural language developed further in a community, it took on more individual features of its own. In this way language became languages, with their many differences as well as common features undoubtedly encouraged — and sometimes maybe hindered — by contact with the natural languages of other human communities. Doubtless on occasion the language itself was the subject of discussion among members of the community, in metalanguage (as we now call it); words were invented for words ('wolf noun', and so on).

Simplifications as well as desimplifications will have ensued, especially to make the language easier to learn and to use. (The decreasing attention among young people to grammar, conjugations and declensions are also simplifications, although unwelcome ones, as they head language back to the one-word grunt.) Many so-called 'primitive' languages possess syntax and/or grammar that are far more complicated than those of the dominating natural languages today.[1]

---

[1] A similar point about simplification can be made about the standard Western musical tradition since the Renaissance as compared with its medieval forms, and some kinds of primitive and non-Western music. Indeed, the development of the

According to Popper's conjecture, *all* aspects of (a) natural language developed this way, both the "abstract" features of language as such and its external manifestations (that is, the -etic as well as the -emic aspects). The development arose chiefly in response to needs and problems: not only of survival and reproduction but also in contexts that we now recognise as religion and art, science and technology, and law and play. The category of play is of especial interest, for jokes and humour may well have been a uniquely powerful stimulus for the development of a natural language; the philosophy of wit is the most demanding (and so the least well developed) part of the philosophy of language.

3. Working with BASIL

The only innate capacity presumed by Popper's conjecture for mankind is his instinct to utter one-word sentences, and to respond to them in others. Thereafter it gives priority to experience in general; it is basically biological and social in character (though not necessarily consonant with sociobiology), perhaps more so than most theories of this kind (see [Marx 1967] on the history of biological views of natural language). For convenience I call it 'BASIL', standing for 'biological and social inspirations for language'.

While a natural language may even be said to evolve, the extent to which properties of evolution *theory* can be deployed is unclear. The processes of simplifying a language mentioned in ¶2 can be understood in evolutionary terms, as means for it to survive and develop. But the creation of new words and insights into features of syntax and decisions about grammar may have been much more sudden than is allowed even by versions of evolution that admit punctuated equilibrium. Again, there are no obvious evolutionary analogues to social policies such as codification of the grammar of Latin that was imposed because of the perceived need for uniformity across the Roman Empire. Indeed, the last three modes of execution claimed in ¶1 to be unique to mankind constitute a new level of cognitive power, and cannot be explained in evolutionary terms from the mental capacities of "lower" species. The more neutral word 'development' is used in this chapter to refer to progress (or even regress) in general.

Further, a natural language never developed or develops inevitably — that is, deterministically. Indeed, language for internal states doubtless develops with great difficulty; it still can be so hard even to express them linguistically, never mind register or describe them. Take, for example, the panoply of unclear analogies used to distinguish kinds of pain ('searing', 'gripping', and so on).

Similarly, the physiology of speaking a language must have required much experiment and (de)simplification, for it involves a most complicated interconnection between control of the larynx, the supply of breath, the use of facial muscles{, and listening to one's own speech as it

---

theory, practice and recognition of music and its languages could profitably be considered in BASILican terms from the one-note noise to the multi-note melody, and so on; compare [Popper 1976, ch. 12].

progresses [Grassmann 1890]}. Moreover, requirements for the latter vary subtly but significantly between different languages, as learners of foreign languages discover very quickly.

The consequences of Popper's conjecture are more substantial than is immediately evident. The rest of this chapter is devoted to some of them, including points of difference with three well-known figures in and around linguistics.

4. Reacting against PEG

4.1 A major influence in the philosophy of this century, and related subjects like logic and linguistics, is the advocacy from the 1900s by Bertrand Russell and others of a positivist epistemology (PEG). It started out with his 'logicist' attempt to reduce mathematics to mathematical logic, as an extension of recent efforts by Giuseppe Peano and his followers in the late 19th century to formalise some branches of mathematics in logical terms. Among its most significant influences was the philosophy of the Vienna Circle from the late 1920s onwards. Especially pertinent here is the work of Rudolf Carnap on logic, and thence on 'the logical syntax of language' in a book of 1934; see, in particular, his rules of formation and transformation in his analysis of syntax [Carnap 1937, esp. pp. 83-96].

4.2 BASIL is incommensurable with various aspects of PEG. Firstly, PEG grants a privileged place assigned to physics, and companion disciplines such as mechanics, among the sciences: 'science is physics or stamp-collecting', in Lord Rutherford's famous opinion of that time. Russell was of similar cast: he remarked in 1911, during his logicist phase, that 'anything evolutionary always rouses me to fury' [Grattan-Guinness 1977, 126]. Granted the cautions expressed in ¶2, BASIL belongs firmly to this category.

Secondly, PEG advocates a threadbare world-view, inspired by the desires to eliminate abstract objects from epistemology and to banish metaphysics from philosophy. By contrast but in line with Popper's advocacy of the World 3 that they inhabit [Popper and Eccles 1977, ch. 2], BASIL happily accommodates abstract objects, including those in a natural language and/or of a metaphysical character (compare §8.6).

Thirdly, as a rule PEG is not concerned with the origins or creation of theories of any kind: without of course disparaging such analyses, origins and acquisition are BASIL's prime concern. Finally, PEG lays a strong emphasis on formal languages, and either ignores natural languages or forces analogies with them that are usually unwarrantedly strong for BASILicans. (However, Carnap seems to have been well aware of the limitations in scope of his programme for syntax.) Some forms of PEG give priority to establishing or grasping 'meaning', but usually to a degree that smacks of essentialism and so does not belong to the BASILican view.

5. Coping with PRUs

Three further features of philosophies inspired by physics and/or obsessed with reduction (PRUs) should be mentioned. Firstly, PRUs usually grant importance to *invariants*, from homely concepts such as the volume or the mass of a body to the more sophisticated formulations in the (absurdly misnamed) 'relativity' theory, but the biologically guided BASIL does not assign invariants such a prominent position.

Secondly, as a rule PRUs seek theories formulated in terms of *cause and effect*; by contrast, BASIL favours interaction. One important advantage of this approach is that it grants appropriate status to the last two functions of a natural language in ¶1 (description, and especially argumentation) in developing competence; in linguistic theories cast in PRU terms, these functions are usually not recognised at all, or at least not sufficiently.

Finally, while BASIL accepts the importance of behaviour, it opposes PRU-like behaviour*ism;* instead, it affirms the existence of innate dispositions in mankind (for example, uttering one-word sentences), as sources for needs and the perception of problems. These may be placed under the heading of 'the soul' of a person — or, as we now prefer to say in our secular culture, his 'identity'. Indeed, religious factors are very likely to have been prominent in at least the early stages of the development of languages; for all early cultures seem to have granted high place to 'Gods' and their worship and representation, and to the death of its members. Analogy between origins and acquisition of language in most modern cultures may *not* adhere from these points of view.

6. Opposing Chomsky

BASIL agrees with some aspects of the positions advocated by Noam Chomsky, but opposes others. Certain of these objections are put forward also by other linguisticians, sometimes for similar reasons.

6.1 Accepting Descartes's advocacy of the mind-body dualism, Chomsky claims, in an anti-behaviouristic and anti-PRU spirit, the existence of mental objects. One of the principal initial offshoots of the Cartesian tradition was Port-Royal logic and grammar (1660s), in which prominence was given to the notion of the actor *judging* a state of affairs by means of an asserted statement. Following the distinction made there between the thought involved and the form of words in which it was cast in the statement, Chomsky claims that a human possesses an innate grasp of the 'deep structure' of a sentence of a natural language, which helps him to control its semantic and syntactic features; the structure is explicated by means of 'universal' grammar, syntax, semantics and phonetics [Chomsky 1967, 402-408]. This structure lies "behind" the possibly varying 'surface structures' that corresponding (statemental) utterances exhibit [Chomsky 1966, 30-36]. Evolution should have delivered deep structure to mankind, although the means involved are not yet known [Piattelli-Palarmini 1980, 35-39].

While welcoming Chomsky's opposition to behaviourism, BASIL rejects the need for deep structures even for individual sentences: on the contrary, the basic unit in BASIL is *merely but powerfully* the one-word sentence-with-actions. (Chomsky is sometimes incorrectly criticised for asserting that man possesses the deep structure of a *whole* natural language;

I did not check this point with Popper before his death, but he certainly criticised the weak form.) From this myriad of innate but modest local sources emerged structures, of varying depths (and distinct from utterances), responding to problems and needs as indicated in ¶2.

6.2 The difference of philosophical tradition noted in ¶4 arises here also; for Chomsky has been influenced by philosophies such as Descartes's, which in its general character was inspired by physics and mechanics: deep structure is offered as the (or a major) invariant for sentences in a natural language. BASIL questions the need and even the propriety of such structures. Further, the status of the *assertion* of statements has been over-asserted in these philosophies; in BASIL statements are not necessarily the primary linguistic category, especially not well-formed ones.

6.3 In addition, Chomsky's (and others') claim that intelligence causally implies the development of a natural language is replaced in BASIL by the possibility of *interaction* between intelligence and language. One example is his concept of man's *competence*, namely his ability 'to associate sounds and meanings strictly in accordance with the rules of his language' [Chomsky 1967, 398] thanks to his capacity to grasp deep structure. By contrast, in BASIL competence is held to advance in interaction with the actor's growing creation or acquisition of a natural language, and constitute part of his developing intelligence. The paper by Chomsky just cited shows the contrast with BASIL quite starkly, in that it appeared as an appendix to a book [Lenneberg 1967] on 'the biological foundations of language' with which it seems to have no contact.

{6.4 Some researches into children with cognitive disabilities may support theories of innate structure, in which the brain is assumed to be "organised" to attend to specific linguistic categories and not just word boundaries. For example, sufferers from Down's syndrome who are taught signing with sentence structures before the age of five years then go on to use spoken sentences once their articulation is sufficiently developed; by contrast, those who merely focused on speech production and missed the neuro-developmental opportunity for grammatical development by the time they began to speak after the age of five years thereafter used only one-word utterances or short telegrammatic phrases for specific situations rather than develop the ability to generate their own combinations of words and meaningful sentences.

Again, sufferers from aphasia have their cortical areas specialised for language damaged by stroke or trauma. Specific lesions may damage aspects of language: a Broca's lesion will lead to agrammatical speech, and an inability to interpret grammatical information in sentences (such as negatives, embedded clauses or passive sentences), while a Wernike's lesion will lead to fluent grammatical speech that is empty of content, with related semantic difficulties in recognising the names of objects from a group of similar items.

In addition, the processes of adding and segmenting words (¶2) must involve at some stages certain basic cognitive steps that may require innate capacities; in particular, the recognition of the special place of verbs, which convert utterances into sentences. For example, in terms of the distinction advocated in §12.3.1, verbs are moments of a sentence while other parts of speech are parts of it.}

7. On the utterance of language

As usual in linguistics, BASIL is formulated with spoken language largely in mind. Now, written language possesses different priorities and difficulties; in particular, the relative slowness of writing when compared with speaking. BASIL can help us understand the relationships between written and spoken forms of a language: Popper himself remarked to me that in ancient times written texts (especially but not only religious ones) were often *read out loud* to a company; silent reading was a later innovation, seemingly made during the early Middle Ages. BASIL can also be usefully deployed in studying, for example, the origins and development of written signs, especially hieroglyphics in ancient languages and in Eastern ones; the emergence of letters and their ordering into alphabet in Western natural languages; the development of spelling in a language; and the recognition of linguistic signs by infants.

This last example can be extended. While there is presumably no hope of setting up testable situations about the origins of language, even in its written form, the study of the acquisition of language by infants may be profitably extrapolated back, as a case study in ontogeny (not always?) recapitulating phylogeny. Again, BASIL concurs with some views on the acquisition of language but not others. For example, while Jean Piaget's objections to deep structure [Piattelli-Palarmini 1980, ch. 1] are welcome, the high status given by him to egocentricism and the autoregulation of the individual person, and his slight interest in social psychology, would both be tempered by the importance granted in BASIL to social and inter-personal interaction. Piaget regarded Darwinism as the theory of *random* mutations, and so was reserved about its applicability to linguistics [*ibidem*]; a broader conception of biological processes (allowing for accumulation, for example) could alleviate this reserve. In other words, while Chomsky seems to require too much of evolution theory (¶6.1), Piaget expects too little from it (compare §14.2.5).

More women, especially mothers, should work on these questions — if they can find the time ....

8. The role of logic

A particularly interesting aspect of the origins of a natural language concerns the growth of logical knowledge. Presumably the descriptive and argumentative functions were decisive, with their need to individuate particles such as 'and', 'aut', 'vel' (acknowledging to the Romans their superiority over English and some other languages on this detail), 'not' and especially 'therefore'. There is no deep structure for logics beyond these single words at most; however, reasons can be put forward for preferring one logic over others in a given situation (§5.4) — as with languages and theories of all kinds.

It is historically disappointing that logic/syntax and linguistics have developed largely independently: since Port-Royal days, indeed, and maybe earlier. This is also rather surprising, since from the early 19th century the tradition of regarding logic as independent of natural languages was steadily eroded. Nevertheless, they are still not usually linked closely together; maybe BASIL provides several areas of common ground.

The ways in which mathematical theories may have started and developed may be similar to those for logic — although not too much so, since its various branches outside arithmetic (mechanics, probability, topology, geometry, and so on) do not have the close links with logic that Russell thought he had established for arithmetic and set theory with his logicism (¶4).[2] BASIL seems applicable to both origins and acquisition of mathematical languages (or better, of mathematical components of language). One can easily imagine that all four functions of language were evident way back in the emergence of mathematically oriented utterances such as '*many* wolf', 'wolf far', 'not chance see wolf more', 'more here wolf', 'help wolf near', 'wol-'.

{9. Coda

The subjects examined in the last two Parts of this book have a common thread, which is well exemplified by education theory. When we teach a student, is our prime purpose to help him to organise and develop the material that we are presenting to him, or are we trying especially to bring out his innate abilities? If both aims are intended, what is the balance between the two? [Pai 1973]. For those who grant languagelessness some status and are sympathetic to at least certain psychical phenomena, innate gifts come in a greater variety and potential than is generally acknowledged. In some contrast, Popper's philosophy, including his hypothesis on the formation of language, gives prime place to the claim that theories are guesswork that need to be tested, and so stresses cognitive development more than innate gift.

The discussion in this chapter has provided evidence for both kinds. The situation in mathematics and its education is particularly important but also tricky, since among other reasons mathematics is one of the few subjects in which the young can be precocious, and where we find that most puzzling of practitioners young and old, calculating prodigies (§19.4.2).

To conclude, it seems to be particularly difficult to form refutable conjectures about the formation of natural language, including the concurrent and consequent acquisition of mathematics and logic. To invoke self-reference for the last time, the task requires of us lots of skilled ratiocination *and* innate inspiration!}

---

[2] Thus Piaget's claims about the acquisition of mathematical notions by the child draw on a none too secure understanding of Russell's programme, especially the limitations (that were inspired by both PEG and PRU). For a typical example, see [Piaget 1952, 183-184] on the equivalence of sets.

Bibliography

Bühler, K. 1919. *Abriss der geistigen Entwicklung des Kindes*, 1st ed., Leipzig: Quelle & Meyer. [English trans. of the 5th ed.: *The mental development of the child. A summary of modern psychological theory* (trans. O. Oeser), London: Kegan Paul, Trench, Trubner; New York: Harcourt, Brace. 1930.]

Carnap, R. 1937. *The logical syntax of language,* London: Routledge and Kegan Paul. [German original 1934.]

Chomsky, N. 1966. *Cartesian linguistics. A chapter in the history of rationalist thought*, New York: Harper and Row.

Chomsky, N. 1967. 'The formal nature of language', in [Lenneberg 1967], 397-442.

Grassmann, R. 1890. *Die Sprachlehre, d. h. die Lehre von den Arten von Laut-, Wort und Satzbildungen welche dem Menschen möglich sind, von ihren Formen und Gesetzen,* Stettin: Grassmann. [Repr. 1900.]

Grattan-Guinness, I. 1977. *Dear Russell — dear Jourdain. A commentary on Russell's logic, based on his correspondence with Philip Jourdain*, London: Duckworth; New York: Columbia University Press.

Lenneberg, E. H. 1967. *The biological foundations of language,* New York: Wiley.

Marx, O. 1967. 'The history of the biological basis of language', in [Lenneberg 1967], 443-469.

Pai, Y. 1973. *Teaching, learning and the mind*, Boston: Houghton Mifflin.

Piaget, J. 1952. *The child's conception of number,* London: Routledge and Kegan Paul. [French original 1941.]

Piattelli-Palarmini, M. 1980. *Language and learning. The debate between Jean Piaget and Noam Chomsky,* London: Routledge and Kegan Paul.

Popper, K. R. 1976. *Unended quest,* Glasgow: Fontana/Collins.

Popper, K. R. and Eccles, J. C. 1977. *The self and its brain,* Berlin: Springer.

PART 5

ADDITIONS

And if by chance they find themselves unable to accept
any of the existing creeds, all they can do is
to begin afresh from the beginning

Karl Popper on philosophers, preface to *Logik der Forschung* (1935)
(*The logic of scientific discovery*, London: Hutchinson, 1959, 13)

21 Types of Generality in and around Mathematics and Logics

A prized property of theories of all kinds is that of *generality*, of applicability or least relevance to a wide range of circumstances and situations. In this short survey I distinguish between two types of generality to be found in mathematical and logical theories, as a supplement to §12 on theories and notions. Various procedures, philosophies and broad theories are reconsidered and reappraised: for example, analogising, revolutions, abstraction, axiomatisation, abstract algebras, metamathematics, model theory, and the issue of monism versus pluralism. The chapter ends with a sketch of a new philosophy of numbers and arithmetic, and a claim that the concerns of this chapter may characterize mathematics and logics apart from other disciplines.

1. Omnipresent and multipresent theories

1.1 Take a mathematical theory T, and consider some wide-ranging domain D of potential reference for it, both within mathematics itself and elsewhere in applications. We distinguish two types of generality of T relative to D.

T is *omnipresent* over D when it applies *to almost all circumstances* available in D. This is the most common sense of which the word 'general' is used; 'widespread' would be an acceptable synonym. Examples relative to mathematics as a whole: the arithmetic of integers (and the arithmetic of real numbers is even more omnipresent), set theory (and multiset theory even more so in allowing for multiple membership).

T is *multipresent* over D when it finds there a wide range of applications and circumstances in which it is usable; however, *each application is quite localised*. As a synonym 'versatile' has the correct sense, but it is not normally predicated of a theory. Examples: Hermann Grassmann's calculus of extension, group theory (and indeed other abstract algebras), Felix Klein's *Erlanger Programm* for geometries, uses of π, fixed-point theorems.

The difference resembles a dale D that contains a large patch O of tulips, and also individual rose bushes M in many places, possibly some of them in the patch. That is, the domain of reference of M may be even wider than that of O, but it is more scattered.

Negation exhibits an important difference of omnipresence from multipresence. The negation of an omnipresent theory within its domain of reference may comprise only special cases or singularities. By contrast, the

---

First publication; a more extended account is in halting progress.

negation of a multipresent theory may itself be multipresent: for example, algebraic structures that are not groups.

The examples given above involve theories that are general(ish) relative to mathematics as a whole; however, the distinction could be applied to non-general theories, although there would probably be less to say.

1.2 This new distinction affects that between mathematical theories and mathematical notions made in §12.3.1. Emulating the difference between parts and moments of a multiplicity as urged by phenomenologists, I treated mathematical sub-theories as parts of a mathematical theory while notions were regarded as among its moments; wide-ranging theories were regarded as 'general' while wide-ranging moments were called 'ubiquitous'. I now modify the thesis to refer to omnipresent and multipresent theories, both to be distinguished from ubiquitous notions; and I now apply it also to pure mathematics where feasible. Several of the entries in the table in §12.4 exhibit types of generality or bear upon them. As with multipresence in ¶1.1 but unlike omnipresence, the negation of a ubiquitous notion may itself be ubiquitous: for example, the notion of non-linearity.

A distinction of category obtains here, which itself is omnipresent: between, for example, the notion of duality and any pair of dual theorems, or between the notion of correlation and any theory of correlation coefficients. Again, the theory of linear differential equations is a part of the theory of differential equations, but linearity itself is a moment. Examples: if T be the differential and integral calculus in the style founded by A.-L. Cauchy (§9.2), then the theory of limits is omnipresent, the arithmetic of real numbers is multipresent, and the notion of limits is ubiquitous. By means of counter-examples, Cauchy refuted J. L. Lagrange's foundation of the calculus, which was mistakenly based on granting omnipresence to Taylor series. If T be linear algebra, then the arithmetic of real numbers is omnipresent, determinants are multipresent, and linearity is ubiquitous. More multipresently, each means '·' of combining elements a and b in some mathematical theory is a moment of a·b while a and b are parts of it.

2. Riders

2.1 Both distinctions are intended to apply not only to some "static" state of generality that mathematics is in but also, and indeed especially, to changes of state that may accompany the introduction of new knowledge, and the further development of theories or notions already launched. For example, a theory may change status from omnipresence to multipresence when rival or alternative theories emerge and the domain of the context widens (¶5.1). Similarly, both distinctions could be useful in mathematics education, especially when guided by history-satire (§13.6).

When a theory is said to be omnipresent or multipresent, or a notion to be ubiquitous, this may be a *claim* is made about the theory or notion; critics may correctly challenge it.

2.2 Both distinctions apply also to reasonings as specified in §11.12, and especially to *logics*. That between parts and moments is omnipresent, since logical connectives and quantification are moments of well-formed formulae, linking up or binding some or all of its parts. For example, propositions P and Q are both parts of the compound proposition 'P implies Q', but the implication itself is a moment.

An important context for logic in mathematics is proof methods. Some of them are ubiquitous (for example, mathematical induction and proof by contradiction) while others are rather specific (such as infinite descent in number theory, and infinitely long proofs in some logics).

2.3 Both distinctions apply also to *algorithms*, both numerical ones and more abstract versions as found in (for example) differential operators and functional equations; efficiency of iteration is a ubiquitous moment. They apply also to *notations*: an obvious example of ubiquity is the practice of choosing symbols so that formulae are read horizontally from left to right, although there are quite a few variants such as matrices, and counterexamples such as numerals.

3. Some foundational procedures and (meta)theories

Here are some well-known aspects of the foundations or development of mathematical and logical theories, where the places of generalities and ubiquity can be substantial, and may change significantly. From now on 'topic' covers both theories and notions.

3.1 The practice of *simplifying and desimplifying* a scientific theory (§2.6, §12.7) can be particularly affected; thus it is important to know the simplified predecessors (if any) of a topic. This point applies especially to *modelling*, where the great complications of the phenomena under study involves much more interaction between the mathematical theory and the application at hand than usually obtains in traditional applied mathematics. Modellists aim to select the apparently principal features of the phenomena and leave out the rest, suggest lines of further investigation, and produce testable predictions. The models are usually designed specifically to the task, but some kinds are multipresent; for example, emulating mechanics has been fashionable. Modelling is not to be confused with model theory, which is considered in ¶3.4.

Similar remarks pertain to other procedures and relationships discussed earlier in his book: analogy and emulation between a topic P and its successor P*, especially concerning changes in analogy and emulation content between them (§12.5); importation and instantiation of topics in theories, and changes in theory as revolution, innovation or convolution (§12.3; *ROL*, §8).

3.2 *Abstraction* is a prominent means of increasing the generality of topics in mathematics (for example, abstract algebras). *Generalization* is to be distinguished from generality; it is a notion, taking the form of a relation between two topics in which the second is some expanded version of the first. When a topic or technique is enlarged on a more modest scale than a generalization, then the word 'extension' is more suitable.

Reduction, a converse of these two processes, occurs when topic P* is reduced to a special case of P, so that P may be seen as an extension or a generalisation of P*. There are also reductions within a topic, when it is shown that a particular kind of object may be reduced to a special kind of itself without loss of generality.

Axiomatisation is a special kind of reduction. When a theory T is axiomatised, then from an epistemological point of view each axiom is omnipresent in its domain by definition; however, in practice some axioms may be invoked only rarely. The discovery of previously overlooked axioms or assumptions in a topic can be a major advance, for the omnipresence, multipresence and ubiquity linked to the new axiom rise in status.

3.3 The distinction between a theory T of any kind and its *metatheories* is itself omnipresent: omnipresence, multipresence and ubiquity of or in a metatheory can be discussed, and different conclusions drawn from those concerning the object theory.

When T is axiomatised, a formalised metamathematics may be developed around it. There are three traditions in this area, similar to and yet also different from each other: abstract algebras, metamathematics and model theory. They all exhibit an interest in axiom systems as such, especially a desire to show that the axioms are independent and that the system is consistent and maybe complete. Let us note each tradition in turn.

3.4 An *abstract algebra* is a system of uninterpreted elements that are connected by one or more means of combination that satisfy certain laws (for example, associativity). It is a syntactic theory, obeying laws of combination by concatenation; its theorems are usually equations though sometimes inequalities. Generality is a principal aim; 'form' or 'structure' is a principal feature. Their wealth of interpretations and applications make these algebras impressively multipresent rather than omnipresent across mathematics, and also some logics.

3.5 *Metamathematics and proof theory* recognise as basic the difference between a theory T regarded as a collection of objects and relationships between them her and any axiomatisation A(T) of T, which took the form of assuming propositions about T. One main hope was to prove, by finitary means, metapropositions about the consistency and completeness of an A(T), and about the independence of its axioms. As with abstract algebras, syntax is much to the fore; in contrast to them, logic is explicitly laid out and studied metalogically for its own (in)consistency, (in)completeness, (in)dependence, (un)decidability, and other metaproperties. This emphasis on logic has helped formal metamathematics to be omnipresent rather than multipresent across mathematics; however, many mathematical theories are not axiomatised at all, especially in applied mathematics, and to render them so would be formidable task.

3.6 *Model theory* focuses upon the connections between the syntax and the semantics of a formal language L that contains constants (including logical ones such as connectives and quantifiers), variables, operation symbols for building up terms from them, and relation symbols to express relations between terms (for example, equality). An interpretation of L is a syntactically determined 'structure' S that includes objects of the four kinds corresponding to the quartet of notions in L (for example in arithmetic, negative integers and the predecessor relation) upon which set-theoretic relations such as (im)proper inclusion are imposed. A well-formed formula F of L lacking free variables (but maybe containing quantified ones) is a 'sentence'; if it is true in a structure S, then conversely S is a 'model' of F.

Modelhood, therefore, is a metarelation, with satisfaction as its inverse. An important class of cases is when F comprises an axiom system. Both metamathematics and model theory are potentially omnipresent rather than multipresent for mathematics, especially if truth is not relativised to a model. Examples: models of the first-order predicate calculus; the Peano axioms for the positive integers, which however are satisfied by other systems such as the odd positive integers and turned out to model only progressions.

3.7 *Relationships between these three traditions.* An axiom system is usually formulated in terms of schematic letters, or some equivalent such as arbitrary lines in a geometry. An abstract algebra can be reformulated in terms of model theory by converting the schematic letters to free variables and then quantifying appropriately. For example, the existence of an identity element e in a group G,

there is an element e of G such that for all elements g of G, g·e = g,

becomes

$$(\exists e)(\forall g)[(e \; \varepsilon \; G) \; \& \; (g \; \varepsilon \; G \Rightarrow g \cdot e = g)].$$

Among other connections, group theory has been used to develop a many-valued logic while logic and model theory are used in proving theorems in group theory. Again, model theory makes use of set theory, but is also used to make models of set theory. Without the central presence of metatheories, omnipresence would over-reach itself, and paradoxes would be in the air!

4. Definitions

Definitions of various kinds can also affect the balance of omnipresence, multipresence or ubiquity among the topics involved in the theories in which they are laid. Nominal definition is the most common type, where a concept or notion N is defined in terms of a well-formed defining expression E. As a consequence the concepts and notions in E disappear from sight when N is used, so that their omnipresence, multipresence or ubiquity in the pertaining theory T may be underrated. Again, when several different definitions of some concept or notion are available, then the choice made will affect the omnipresence, multipresence or ubiquity of all the other pertaining concepts and notions, since all the other definitions become theorems.

Another context that bears directly upon axiomatisation is when an axiom system, maybe along with some or all of its models, is held to furnish (but how?) implicit definitions of its concepts and notions. Akin to this to this kind are contextual definitions. Other types of definition used in mathematics and logic include analytic and synthetic definitions, of special interest to those of a (neo-)Kantian persuasion; definition by mathematical induction, and more generally by means of recursion; creative definitions; and so-called essentialist definitions, when it is (controversially) claimed that the essence of some concept or notion is captured. Especially in this last type but also elsewhere, there arises the issue of the scope of legitimate definability within the natural or formal language involved.

5. Limitations to generalities

5.1 *Monism and pluralism.* Monism is the claim that there is only one topic pertinent to some context, and any other candidate topic is either a special case of it, of misconceived in some way, or not within that domain of reference; by contrast, pluralism admits more than one legitimate topic. As a rule monism links to omnipresence and pluralism to multipresence. Some major branches of mathematics have long exhibited pluralities; for example, the calculus and mechanics.

The legitimacy of non-Euclidean geometries innovated the replacement of monism by pluralism in a major branch of mathematics: instead of just one geometry, there could be several, so that their domains of reference had to be reassessed. Similarly, instead of having just common algebra, which was closely attached to numerical quantities, many new algebras were created: for example, the abstract ones noted in ¶3.4.

In logic, pluralism began to emerge when logics alternative to classical bivalency were proposed in the 1910s. Logical pluralists have had to fight for status, despite the fact that that we have all long been at least logical dualists without noticing! For very often we make statements about sequences of actions and decisions *in some order in time*, with 'and' meaning 'and then'; the restatement of such actions in classical logic with each action flagged by the corresponding value of the time variable is only a formal substitute (§5.4.2). The millennia-long history of ignoring temporal logic is extraordinary (compare §5.4).

Now logics abound, many driven by applications. Several logicians, and users of logics, wish to involve reality multi- and even omnipresently in the formation and epistemology of a logic. Some even query the omnipresence of the distinction between logic and its metalogics; but then reasonable effectiveness may tip into unreasonable ineffectiveness.

These manifestations of pluralism suggest that multipresence came to prevail over omnipresence in mathematics and to some extent in logics. However, each major branch of mathematics contains domains of reference sufficiently substantial to sustain sensible talk of its "local" omnipresence.

5.2 *Some lessons from Gödel's theorems.* Kurt Gödel's paper of 1931 on the incompletability of first-order arithmetic exemplifies several features of mathematics and logics explored in this paper. Firstly, he established a central — indeed, omnipresent — place for the distinction between logic and metalogic, and thus for theory and metatheory in general. Secondly, with his first theorem Gödel ushered in a different kind of pluralism from those noted so far. For both the undecidable proposition and its negation could be adopted as additional axioms, to generate a further pair of incompletable axiomatisations of first-order arithmetic; by implication (as it were), most mathematical theories would be susceptible to such bifurcations. Further, while the proof method made much clearer the theory of recursion and thereby established its omnipresence in the theory of computation, the theorem itself showed that *reduction* of computability to recursion was impossible. These features complement those exhibited in the rest of this paper in that they disclose *intrinsic limitations* to the generality of mathematical theories, at least axiomatised ones that include first-order arithmetic and are restricted to finitary proof methods.

6. A new-old foundation for numbers and arithmetic

6.1 The common belief in the certainty of mathematics and logic(s) has helped inspire a philosophical tendency to regard both theories as examples of a priori knowledge, or least to formulate them without recourse to empirical factors. Following the different aspiration for mathematical knowledge that has guided this article, I sketch a more traditional philosophy of arithmetic; linked to human experience while remaining objective in character, it is mindful of Saunders Mac Lane's warning about avoiding excessive generality (§5.3.1) in *upholding* omnipresence and ubiquity but *denying* a prioricity or certitude. Multi-set theory and classical logic are deployed, and the phenomenologists' distinction between parts and moments (¶1.2) is prominent. There are two essential initial steps.

Firstly, ontology: to recognise that the universe that mankind inhabits is composed of (usually) *distinguishable and invariable objects*. It is this omnipresent property that leads mankind to grant arithmetic prime status among the various branches of mathematics in the first place (as it were); as was noted in §2.11.3, a sentient culture living in a fiery environment would presumably take topology and thermodynamics as dominant parts of its mathematics, with arithmetic reserved for special tasks, such as counting distant visible stars.

By 'objects' I have in mind not only physical ones such as stones and fingers, but also transitory ones such as waterfalls, experientiable ones such as psychological states and musical keys, historical events of every kind, parts of scientific theories, fictional characters in novels, and so on and on. I also include everyday arithmetic, which supplies us with, among many things, the numbers of the pages upon which this chapter is printed, and the number of digits in a numeral.

Secondly, epistemology: to assume that mankind is mentally capable of appreciating distinguishability, and thereby constructing and manipulating multiplicities of these objects.

6.2 Given this background, consider a multiplicity M composed of objects of any kind, maybe including the repeated presence of some, and not laid out in any specified order. The integer 0 is associated with absence of some kind of object in M, while 1 comes from association with any individually distinguished member. The other positive integers are associated with collective association, and may be construed as multiples of this unit 1 (following Euclid's arithmetic, incidentally).

In this theory the positive integers are moments associated with multiplicities. These associations do not require abstraction, for the approach imitates the abstract algebraists (¶3.5) in ignoring all properties apart from distinguishability, and achieves omnipresence for integers by imitating their penchant for multipresence in taking M as a multi-set of distinguishable *uninterpreted* objects. Alternatively, we could emulate the model theorists by replacing algebraists' talk of schematic letters by variables in predicate logic quantifiable over suitably wide ranges (¶3.6).

Since no order was imposed upon the members of M, then the associated integers are cardinals; however, if well-order matters, then they are ordinals, with a temporal logic to back up the order if felt useful. Alternative foundations of cardinals and ordinals, such as Richard Dedekind's theory of chains, the Peano axioms, Bertrand Russell's and Gottlob Frege's definition of cardinal numbers as sets of equinumerous sets, or a grounding in some axiomatic set theory, can have their due place. Further, the metamathematics, formal or otherwise, should be unaffected or even enriched.

To move from integers to arithmetic, addition could be specified by forming the union of the corresponding (ordered) multiplicities, allowing for multiple membership when required. Existing procedures allow us to specify the other three arithmetical operations and exponentiation, and to define rational, irrational, negative and transfinite numbers.

We should also consider means of *knowing* this version of arithmetic. The distinguishability of objects should be highlighted in the study of knowledge acquisition in young children, as indeed was done by Popper's teacher Karl Bühler (§20.1). The distinctions between parts and moments and between omnipresence and multipresence, and multi-set theory, should became better known by teachers, who could teach them to pupils. Finally, in the spirit of self-reference, omnipresence, multipresence and ubiquity should be omnipresent, multipresent and ubiquitous!

7. What makes mathematical and logical knowledge special?

This focus upon types of generality has ranged widely across mathematics and logics. Does it *help* us to distinguish these kinds of knowedge from other kinds? One may point to the major roles played by topics, and also to the ways in which they may be stitched together to formulate general theories of various kinds. Maybe we are considering a difference of degree rather than kind, but these features seem to be *much* more prominent in mathematics and logics than they are elsewhere, even in the physical sciences. Indeed, we can characterise mathematics and logics by stressing these features *far* more potently than by pointing to their unique levels of *certainty*, which grounds the traditional but perhaps optimistic explanation.

22 On Receiving the Kenneth O. May Medal and Prize

In a ceremony at the International Congress for the History of Science held in Budapest on 31 July 2009, the International Commission for the History of Mathematics (ICHM) awarded me the Kenneth O. May Medal and Prize in the History of Mathematics for my contributions to the field. The peroration is followed by my reply, in which I stressed the importance to my work of Popper's philosophy; finally come a few further career details.

1. The peroration
Prepared by Karen H. Parshall, chairman ICHM; read out by Craig Fraser.

Ivor Grattan-Guinness took a B.A. in mathematics at Oxford University in 1962, an M.Sc. in the philosophy of science at the London School of Economics in 1966, and an M.A. at Oxford University in 1967 before earning first a Ph.D. and then a prestigious D.Sc. in the history of science at the University of London in 1969 and 1978 respectively. Beginning in 1964, he served on the faculty of Middlesex University, becoming Emeritus Professor of the History of Mathematics and Logic there in 2002 at the age of sixty.

Grattan-Guinness's first book, *The development of the foundations of mathematical analysis from Euler to Riemann* (1970), was based on his Ph.D. thesis and is reflective of his deep interest in and contribution to the history of the foundations of mathematics as well as of mathematical analysis. In this study, Grattan-Guinness exhibited what have become hallmarks of his research: penetrating readings of key primary sources in a wide range of languages and broad syntheses of mathematical ideas. This work, with its extensive examination of the mathematics of Cauchy and his immediate predecessors, also served to focus Grattan-Guinness's attentions on the early-nineteenth-century French scene, a vast topic that came to dominate his research for some two decades.

Early fruits of those labors appeared in 1972 with the publication, co-authored by Jerome Ravetz, of *Joseph Fourier 1768–1830*. A survey of Fourier's life and work as well as a critical edition of the 1807 monograph on heat propagation that Fourier presented to the *Institut de France*, this book highlighted not only the Parisian mathematical and broader scientific milieu of which Fourier was a part but also developments in what Grattan-

---

This file was placed upon the website of the Commission (www.unzar.es/ichm) and has been reproduced in some journals in the field. I am very grateful to Professor Parshall for agreeing to my request that her peroration may also appear in this book.

Guinness would later style 'applicable mathematics'. He developed both of these themes, as well as that of mathematics education and its ever-evolving venues, in his magisterial, three-volume *Convolutions in French mathematics, 1800–1840: from the calculus and mechanics to mathematical analysis and mathematical physics,* published by Birkhäuser Verlag in 1990.

Grattan-Guinness's ability to synthesize and illuminate vast amounts of mathematical knowledge was also in evidence in *The Norton history of the mathematical sciences: the rainbow of mathematics* (1998), in which his subject was no less than *all* of the mathematical sciences. The latter distinction is important: the *mathematical sciences* and not just *mathematics*. In the words of Victor Katz, 'Grattan-Guinness makes absolutely clear that the use of mathematics in other fields, including economics, statistics, engineering, hydraulics, ballistics, astronomy, mechanics, optics, and so on, was a very important factor in the development of the field and to write a history of mathematics without including its applications would be highly misleading' [*Mathematical reviews* 682977, 2000d:01001].

Broad synthesis also characterized *The search for mathematical roots, 1870–1940* (2000), although in this case the topic was, as the book's subtitle reflected, 'logics, set theories, and the foundation of mathematics from Cantor through Russell to Gödel'. Another historical *tour de force*, and one that resulted from the fruits of his research during a Leverhulme Fellowship from 1995 through 1997, this study traced the complex development of a multitude of systems of mathematical logic in a wide range of national contexts — from Italy to Germany to England to the United States to Poland — as well as the broader philosophical contexts in which those theories evolved.

If profound research on the mathematical sciences in diverse national contexts has characterized his scholarly output to date, so, too, does the production of high quality reference works aimed at uniting the disparate communities of historians of science, historians of mathematics, and mathematicians. A gifted and seemingly indefatigable organizer and editor, Grattan-Guinness mobilized and oversaw the contributions of some 130 scholars to the two-volume *Companion encyclopedia of the history and philosophy of the mathematical sciences* (1994) as well as of the thousand-page collection of *Landmark writings in Western mathematics 1640–1940* (2005) with extensive historical commentary. He also served the broader scholarly community as editor of the journal *Annals of science* from 1974 to 1981, and as the founding editor of *History and philosophy of logic* (1980). These and his many other contributions to the field of the history of science have already been recognized by his election as *member effective* of the *Académie Internationale d'Histoire des Sciences*. Today, we are pleased to recognize Ivor Grattan-Guinness with the highest honor in the history of mathematics, the Kenneth O. May Prize and Medal, awarded for lifetime scholarly achievement and commitment to the field.

2. Four times lucky
I. Grattan-Guinness, in response to the award.

2.1 I am very pleased to be here to receive this medal, because I am very pleased to be here! 'You were lucky', said the surgeon after performing emergency open-heart surgery on me at the beginning of September 2008. I had experienced mild symptoms or warnings already, and yet when I had my first small heart attack I did not recognise it as such. I did know that something was wrong and that I needed to see the doctor, but I went there on my bicycle... . The moral of this absurd story is stark: be more familiar with your physical body than I was of mine, for if you are not and do not have my luck, then you will not live to regret it.

2.2 I am very pleased to be here to receive the May medal. I think that he and I started our serious interest in the history of mathematics at around the same time, in the middle 1960s. He came into a field that he saw had been in the doldrums for a very long time, but that maybe was beginning to pick up. By 1970 he was wondering about reviving this Commission, and especially about using it to launch a journal in the field that would not only carry articles but also serve as a source for the growing members to consult. He asked many of the leading figures in the field for their opinion; and luckily for me he also asked some of the youngsters who had begun to emerge. The general opinion was that the idea was attractive, though its success was not obvious.

In the end, May resolved to go ahead. This decision led to the choice of title for the journal: after some learned international correspondence about the use of the ablative case in Latin, the name '*Historia Mathematica*' was agreed. The first volume appeared in 1974, and I ran a couple of departments in it for the first few years. May's hunch proved to be correct; for 35 years later the journal is still going strong, and it is a special pleasure for me that two of my former doctoral students at Middlesex University, Niccolò Guicciardini and Adrian Rice, are now involved in its administration.

2.3 In the 1960s it was extremely unusual for anybody to take an interest in research in the history of mathematics. May's motivations lay partly in mathematics education, and partly in questions involving bibliography and information retrieval. My own motivation came entirely from a negative reaction to mathematics education. I had taken a mathematics degree for three years at the University of Oxford, but I might as well have done religious studies: paraded before me was a sequence of indeed impressive mathematical theories, but why did one need to study them in the first place? Why were they being taught in such an obviously unsatisfactory manner? Further, where had they come from? Surely nobody had sat down one Thursday afternoon and invented, for example, group theory in the way it was being served up to us.

Tuesday mornings in my first year were especially baffling. The lecture course on the calculus served up wall-to-wall epsilontics and limit theory; but it was followed at once by hydrodynamics, with infinitesimal cuboids of fluid flowing under various conditions. Both courses obviously involved the calculus, but in completely different forms. Were there two sorts of calculus? Had this always been the case? Why did neither lecturer refer to the calculus used by the other one?

I realised that these questions that I had set myself were not themselves mathematical, but philosophical in some way; and I had no idea how to handle them. Then I came across an offbeat philosopher called Karl Popper, and took a Masters course in his department at the London School of Economics.

Popper's offbeatness was perfect for my purposes, as he taught one heresy after another:
- Life is always problem-solving of some kind, and one is theorising all the time;
- Even science is guesswork, and the aim there should be to test theories severely: falsify them, or at least criticise them in some way;
- The growth of knowledge is more important than any particular state that it is in;
- We dig down to foundations of knowledge, not up from them;
- The roots of philosophical problems lie outside philosophy;
- In epistemology one seeks not only the roots of knowledge but also the roots of ignorance;
- The future is open, not only subjectively but also objectively.

In addition to all these heresies, a bonus came with the logic courses. There he stressed as fundamental the distinction between logic and metalogic, and indeed between theory and metatheory in general. One should think explicitly in terms of different levels.

Popper said hardly a word about mathematics; but he had armed me richly with means to tackle my post-religious studies questions. Indeed, there were now two contexts: not only the history of mathematics (where I chose to examine the calculus, because of those Tuesday mornings) but also the history of logic. There was even a bonus meta-question, of great interest: why were mathematics and logic so different from each other?

I do not normally discuss this background, as I know that historians of all kinds are not fond of philosophy; but I owe a huge debt to Popper, and want to acknowledge it on this occasion. A book collection of my essays, called *Routes of learning*, is due to appear later this year from Johns Hopkins University Press, and I have deliberately chosen several articles in which these issues are explicitly treated. I was very lucky to get all these insights right at the start of my career.

2.4 One of Popper's more sexist pieces of advice to students was the conjunction of imperatives 'Find a good problem, and find a good wife'. Thanks to his own teaching my first search had been successfully accomplished, indeed twice over; and while a student I also obeyed the second instruction. I am very glad that Enid can be here this afternoon, for she has not only put up with me for all of these 45 years, but also helped me in many practical and secretarial ways. Saving the world from the world's worst typist (including right now) is alone worth a medal. I was lucky for the fourth time.

3. Some further career details

At Middlesex University I supervised nine doctoral and two masters students. I am also a Visiting Research Associate in the Centre for Philosophy of Natural and Social Science at the London School of Economics.

I have published twelve books and around 250 main papers, contributions to a dozen dictionaries and encyclopaedias, and hundreds of book and article reviews. I edited the book *From the calculus to set theory, 1630-1910: an introductory history* (1980, Spanish translation 1984), and three editions: of W. H. and G. C. Young, *The theory of sets of points*, second edition (with R. C. H. Tanner, 1972); of Philip E. B. Jourdain, *Selected essays on the history of set theory and logics (1906-1918)* (1991), to complement *Dear Russell — dear Jourdain. A commentary on Russell's logic, based on his correspondence with Philip Jourdain* (1977); and of George Boole, *Selected manuscripts on logic and its philosophy* (with G. Bornet, 1997). The recent book collection of more general papers is *Routes of learning. Highways, pathways and byways in the history of mathematics* (2009).

A member of the Executive Committee of the International Commission on the History of Mathematics from 1977 to 1993, I have been an associate editor of *Historia mathematica* almost continuously from its inception in 1974. I am an advisory editor to several other journals and book series, especially the editions of the writings of C. S. Peirce and Bertrand Russell since their respective inceptions in the late 1970s. I served as the Associate Editor for mathematicians, statisticians and computer scientists for the *Oxford dictionary of national biography* (2004).

I have given nearly 600 lectures in over 20 countries, including lecture tours undertaken in Australia, New Zealand, Italy, South Africa and Portugal. From 1986 to 1988 I was the President of the British Society for the History of Mathematics.

From 1980 to 1988 I was a member of the Council of the Society for Psychical Research. For its centenary in 1982 I edited *Psychical research. A guide to its history, principles and practices* (1982), from which §16 is drawn.

Index

Passing mentions of persons and topics are not normally indexed. In page references of the form 'm~n', the symbol '~ shows that the (sub-)entry in question appears on *most* pages between m and n and is relevant throughout; 'm-n' cites *all* pages as usual.

Abstraction 55 325 329
Acoustics 55 183 184 201 204
Algebra, common 178-180 206 219 227 228
Algebras, abstract 244 323-327
Ampère, A. M. 149 249 250
Analogy 9 198~207 211 325
   content 202 204 211 325
Anomaly 253 254
Applied mathematics 9 10 177-191 195-211 325 326
Arithmetic, foundations of 61 238 239 323 324 327-330
Astrology 159 160 286 2 90
Astronomy 42 148~154 184 203
Axiomatisation 187 222 223 236 323 326 328

Babbage, C. 81 287 296
Bachelard, G. 31
Barrett, W. F. 262 264 267 271
Bartley, W. W., III 1 19 20 50 57 59 95 96 114 130
Bashmakova, I. G. 227 228
Behaviourism 60 136 249 250 258 270-273 316
Bergson, H. 289 296
Bernoulli, D. 183
Birthday paradox 288
Blackmore, J. T. 32
Bohr, N. 21 22 151
Boole, G. 179 185-187 264
Boole, M. E. 264
Born, M. 42 43

British Association for the Advancement of Science 262 267 269 287
Brouwer, L. E. J. 92
Bühler, K. 16 136 312 330

Calculus, differential and integral 55 148 149 177 181 182 188 189 324 *See also* Mathematical analysis
Cantor, G. F. L. P. 235 243
Carnap, R. 3 17~21 74 96 128 132 133 134 138 139 270 273 274 311 315
Cauchy, A.-L. 145 146 149 180-182 203 207 211 222 225 226 243 324
Causality 82 86 289 295 296 316
Chapman, S. 232
Chomsky, N. 11 135 136 311 316-318
Coincidences 11 281-297 302
Complex-variable analysis 180 206 207 226 227
Comte, A. 249 257
Confirmation 3 79 123
Continuity of nature 46 82 86
Convolution 199 200 325
Courant, R. 231
Cournot, A. A. 180
Criticism 91 162
Crookes, W. 262 264 265
Crossley, J. C. 282 283 293
Cryptomnesia 289

Dalton, J. 30 59
Dantzig, G. B. 184 207
Darwinism 266 269
Definitions 163 189 327 328
Descartes, R. 177 220 221 316 317
Desimplification 7 40 41 45-47 61 150 153 178 207-209 313 314 325
Dessoir, M. 275 276
Dewey, J. 272
Dienes, Z. P. 240
Dijksterhuis, E. J. 219
Dingler, H. 16
Dirac, P. 21 206 208
Dirichlet, J. P. G. 223
Discovery in science 42 149 163 232 240 256
Dowsing 256
Dugas, R. 145 189
Dunninger, J. 256

De Morgan, A. 186 264 275

Eccles, J. C. 3 20 28 59 60 136 137
Economics 209 210
Eddington, A. 232-234
Edgeworth, F. Y. 264
Effectiveness of mathematics 9 10 187 188 195-211
  ineffectiveness 203 208-211
Eisenbud, J. 295 296
Electrodynamics 149 249
Empiricism 5 26 31 51 73 114 132 137-139 161 257
Equilibrium 200 206 209
Essentialism 6 39 42 44 47 106 241 315
Euclid 218~223 225~228 230 237 238 329
Euler, L. 183 184 203 204 222
Experience, limitations of 30~33 73 105 155 156 163
Experimenter effect 287 288 294 301-303 307

Extrasensory perception 251 254~258 264 272-276 281 288 291~294 301~306

Facts 5~9 26 27 37-44 47 49 51 56 68 73 86 92 104-107 131 134 144 147 151-159 162 246 250 257 263 266 272 273
Field, H. 56
Forchheimer, P. 25
Forms in mathematics 178 187-191
Fourier analysis 55 179 182 183 201 204 205 209 223 225
Fourier, J. B. J. 55 149 150 179 182 183 190 201 204 205 209 223 225 284 296
Fourneyron, B. 28 29
Frege, F. L. G. 57 237 240
Fresnel, A. J. 73 149 205
Freud, S. 274 295
Functional analysis 225

Gellner, E. 19
Generality of theories 187 323-330
Geometries 16 179 180 323 328
  See also Euclid
Gödel, K. 57 76 92 134 135 138 206 224 328
Goethe, J. W. von 13

Haack, S. 97 99 103-105
Hammersley, J. M. 233
Hansel, C. E. M. 254 255 258
Heat theory 31 55 150 182 183 204 209 *See also* Fourier
Heath, T. L. 219 220
Hempel, C. 18
Hilbert, D. 134 188 196 205 223 223 238
Hintikka, J. 196
Hinton, J. 48
Historiography 51 146-153 164 171-175 223-227 243
History and heritage 9 51-53 173 199 217~220 225~228 231

History of mathematics 9-11 54
    144 197 208 217 223 225 228
    231 240-244
History-satire 53 76 225 244 245
Hypnagogy and hypnopompy 305
Hyslop, J. H. 272 303

Idea-lism *or* ideal-ism 33
Ignorance 27 47 50-54 59 224 225
(In)determinism 7 28 73 83 86 92
    224 312
Induction (in science) 5~10 15 16
    28 36 68 69 79~86 89 91 93 94
    99 111 114 121 131 133 138
    150 152 189 201 257 263 266
  inventory problem 8 93 94
Innovation 199~204 325
Instrumentalism 7 26
Instruments 29 148 153 209

Jevons, W. S. 185-187
Johnson, A. 281 283
Jourdain, P. E. B. 128 137
Jung, C. G. 247 289-291
Justificationism 4 5 234 235 245

Kamke, E. 18
Kammerer, P. 283 284 289 290
Kant, I. 1 16 112 116 119 127 137
    210 269
Kantian philosophy 1 16 32 47 57
    99 249 328
Kaushal, R. S. 200 202 208
Knowles, J. 268
Knowns and unknowns 35 258
Koestler, A. 232 233 281 290 291
Kuhn, T. S. 8 41 54 70 143 147-
    150 160 161 172-174 253

Lagrange, J. L. 50 149 150 187
    203 204 207 208 222 324
Lakatos, I. 8 9 20 143~164 190
Language
  acquisition *or* formation of 11
    135 311-319
  formal 103 300 312
  natural 5 7 13 59 60 156 311–319

Languagelessness 299-308 319
Laplace, P. S. 46 149 181 182
    208 222
Laudan, L. 48 153 164
Linearity 182-185 203-205 324
Linguistics 312 315~319
Locke, J. 113 116 120 121
Lodge, O. 264 265
Logic
  classical *or* bivalent 8 38 39 44
    45 83~98 106 135 158 318
    328 329
  fuzzy 95-97 106
  temporal 328 330
Logic ≠ mathematics 56 57 100
    188 190 319
Logical monism 8 45 73 89 91
    323 328
Logical pluralism 8 38 39 57
    94~98 106 189 325~330

Manning, M. 283
Marx, K. 112 115 116
Mathematical analysis 145 146
    203-206 *See also* Calculus
  convergence of infinite series
    146 148 223
  Differential equations 56 182
    200 203 204 324
Mathematical physics 5 55 149
    150 179 182 205 207 250 287
Mathematics education 217 219
    224-228 231-246 312 319 324
  New Mathematics 232-235 238
    241 245
Matrix theory 203 205 222
May, R. M. 182
Mechanics 25 26 46 181 182 201
    258 315 317 319 *See also*
    Mathematical physics
  Newtonian 56 57 68-73 148
    154 159 160 181 187 200
Medawar, P. B. 20 27 47 74 80
    207
Metalanguage 129 300 313
Metamathematics 9 56 92 95 134
    140 223 323 326 327 330

Metaphysics 29 32 33 48 61 83 159 249 274 275 292
Metatheory 9 52-54 60 91-95 99 100 210 223 224 326 329
Methodology 28 36 48 50 52 144 147~155 159~163 171~175 234 235 241 312
Mill, J. S. 251 264 266 270
Model theory 188 189 323 325–327
Modelling 31 202 325
Moments *See* Parts
Morris, C. 17 20
Multiplicity 9 324 329 330
Multipresence and omnipresence 10 323-327 330
Music 16 314
Myers, F. W. H. 29 265 268-270 299 304 308

Mac Lane, S. 91~95 329
McDougall, W. F. 272 273

Nativism 263
Naturalism 265 266 269~272 275
Nominalism 56 57
Non-scientific areas 144 160 162 175
Normal science 8 41 42 69 70 75 147 150 172 173 253
Notions 195 199~205 210 211 323 324 325 327 328
  ubiquity of 195 211 324-330

Ohm, G. S. 183 204
Ontologically correct theories 6 43-46 49 54~61 105 106 154~162 178 256
Ontology 6~9 25 26 32 42~51 54 55 74 137 144 155~162 189 239 253 256 295 329

Pagenstecher, G. 276
Paradigms 70 147-150 163
Parapsychology 10 254 261 262 269~276 281

Parts and moments 39 198 324 325 329 330
Part-whole theory 185 221
Pauli, W. 21 291
Peano, G. 315
Peirce, C. S. 11 98 137 143 144 181 186 210 212
Pendulum theory 41 74 208
Phenomenology 198 324 329
Philosophy of mathematics 59 60 100 143-146 161 177-179 185~191
Piaget, J. 233 239 240 311 318 319
Plato 112 115 116
Poisson, S. D. 149 181 208 209
Polya, G. 144 145 190 191 196-198 200 202 211
Popper, K. R., career *or* books 1 2 15~20 69 91 111 131 312
  Logik or *Logic* 1-3 42 67 95 114 132 133 157
  Postscript 122 130 131
  The open society 18 112 115 116 130 131
Popper, K. R., fallibilist philosophy 5~11 15 26~29 32 53 59 76 79 100 133 135~138 159 197 224 234 236 312
  comprehensively critical rationalism 50 95
  corroboration 4 6 9 10 17 26 33 34 42~47 58 67~75 79 80 96 133 154 155 160 175 176 199 205 303
  evolutionary epistemology 5 7 28 53 59 136 137 312
  schema for problem-solving 53 73 80 93
  three worlds theory 106 137 155-157 293 315
  verisimilitude 4 7 26 36 45 47 68 75 96 156~160 163
Positivism 8 10 16 19 31 32 49 131 139 140 183 197 204 249 256 257 265 269 270 273 274 315

Potential theory 26 187 204 205
Pragmatism 82 272 275
Probability 3 4 17 18 28 42 67 71
 80 123 133 158 190 197 212
 225 275 288 291
Problems *as* problems 92 93
Proofs 144 145 160 181 189
Proposition *or* statement 28 106
 134 135 154 156 185 312 316
 325 329
 universal 33 94
Pseudo-science 144 159 164
Psi 281 282 288 291–295 299–308
Psychic healing 252 254 259 276
Psychical research 10 30 251 258
 262 281 284–287 291 292 296
 Ganzfeld technique 303 305
 information theories 291 296
 observational theories 258 293–296
 propensity theories 291–293
 response theories 288 291 292
Psychokinesis 251 258 291 294
 295 301
Psychologism 6 239
Psychology 1 16 28 54 58–61 98
 157 163 185–187 196 210 240
 255 259 268–274 289 294 295
 316–319
Pythagoras's theorem 218 219

Quantum mechanics 18 21 22 205
 206 293
Quine, W. V. O. 31 33 39 56 93
 95 96 130 133–136 153

Rationality 60 144 150 152 162–
 164 234 235 305
 outside science 49 175–176
Realism 32 58 73
Reasonings 189–191
Reductionism 38 46 49 51 135 189
 199 247 263 273 274 315 326
 329
Refutation 37–43 69 144 147–155
 157 159–163

Reification 7 26 27 31 32 40 42
 55 60 74 75 178 181 184–187
 190 293
Relativity theory 26 44 46 69 158
 197 265 270
Reliability science 67 71
Research programmes 147 153
 154
Revolutions in science 8 41 47
 150 163 173 175 199 255 258
 259
Rhine, J. B. 261 272 274
Rhine, L. E. 261 272 274 281 293
Richet, C. 262 275
Robinson, A. 145 151
Romanes, G. J. 262 271
Rostand, F. 145 189 190
Russell, B. A. W. 8 11 17 19 21
 27 87 111–124 127–139 143
 161 237–240 269–274 281
 315 319 330
 History of Western philosophy
 8 111 115 116 130 139

Satisficing 95
Schlick, M. 16 18 21 22 132 138
 173 175
Schröder, F. W. K. E. 181
Schrödinger, E. 205
Science education 53 54 75 176
Sciences, established 251–257
Seidel, L. 146
Self-reference 54 60 162 163 224
Set theory 56 137 221 222 231
 235–239 242–245 260 319
 323 327 329
Shearmur, J. 58
Sheffer, H. M. 129 134
Shewhart, W. A. 71
Sidgwick, E. 262 264 268
Sidgwick, H. 263–269 271 275
Small effects 67 72 74 207 209
Society for Psychical Research
 10 255–258 261–263 267
 281–284 299
Sorley, W. R. 264
Spencer, H. 266 269 271
Spiritualism 262–265 268 272

Stanford, R. G. 291-294
Statistics 56 71 72 92 189 190 206 212 272 273
Structure-similarity 9 54-56 177-190 200
Symmetry 203
Synchronicity 289-291 295 296

Tarski, A. 4 7 37 38 58 72~76 92 95 96 99 103-107 134 139 140 144 156 224
Technology 17 70-72 98 210 211
Testing of theories 3 36-40 47 82-85 92~100 153
Theory-building 88 190 191 195~198 200 202 207 210 211
Thirring, H. 22
Thomson, W. (Lord Kelvin) 183 190 204 250
Toeplitz, O. 225
Truesdell, C., III 50
Truth 4~7 26-28 36-39 42-45 54~61 80 92 95-97 103-107 119 123 135 139 140 144 154~163 175 178 189 197 200 232 235 237 247 249 257 272 311 327
  correspondence theory of 26 38 39 73 103~106 134 157
  semantic theory of 7 68 72 76 92 95 104 107 134 140 163 316 317
  tri-distinction involving possession of 4 39 45 106 159 256
Turing, A. M. 40
Turner, F. M. 262
Tyndall, J. 266 267 269
Tyrrell, G. W. M. 291-293

Unfalsifiability 4 39 83 85 152-154 159~163 175 295
Unguru, S. 219 228
Unveilism 51 52 173-175

Vaihinger, H. 74
Velikovsky, I. 258 259 274 295
Velupillai, V. 209

Verification 5 6 15 31 51 68 75 133 152 266
Vienna Circle of philosophers 1 17 275 315
Viète, F. 179 220
Vision 30 31

von Neumann, J. 205 207

Wallace, A. R. 262 265 269 271
Ward, J. 266 269
Watson, J. B. 271 272
Weierstrass K. T. W. 146 203
Whitehead, A. N. 29 114 124 132 251 254 257 270
  Whitehead and Russell, *Principia mathematica* 131 132 270
Wigner, E. P. 188 195~203 206~212
Wittgenstein, L. 8 17 19 99 112 113 120 121 128 129 134 270
Wundt, W. 257 258 270-272

Yeats, W. B. 169

www.ingramcontent.com/pod-product-compliance
Lightning Source LLC
Chambersburg PA
CBHW080238170426
43192CB00014BA/2484